Microcomputer Control of Thermal and Mechanical Systems

Microcomputer Control of Thermal and Mechanical Systems

W. F. Stoecker
University of Illinois at Urbana-Champaign

P. A. Stoecker
Hewlett-Packard Company

Copyright © 1989 by Van Nostrand Reinhold

Library of Congress Catalog Card Number 88-20869

ISBN 0-442-20648-8

All rights reserved. No part of this work covered by the copyright hereon may be reproduced or used in any form or by any means—graphic, electronic, or mechanical, including photocopying, recording, taping, or information storage and retrieval systems—without written permission of the publisher.

Printed in the United States of America

Van Nostrand Reinhold
115 Fifth Avenue
New York, New York 10003

Van Nostrand Reinhold International Company Limited
11 New Fetter Lane
London EC4P 4EE, England

Van Nostrand Reinhold
480 La Trobe Street
Melbourne, Victoria 3000, Australia

Macmillan of Canada
Division of Canada Publishing Corporation
164 Commander Boulevard
Agincourt, Ontario M1S 3C7, Canada

16 15 14 13 12 11 10 9 8 7 6 5 4 3 2 1

Library of Congress Cataloging-in-Publication Data

Stoecker, W. F. (Wilbert F.), 1925-
   Microcomputer control of thermal and mechanical systems / W. F. Stoecker, P. A. Stoecker.
     p. cm.
   Bibliography: p.
   Includes index.
   ISBN 0-442-20648-8:
     1. Automatic control. 2. Microcomputers. 3. Heat engineering—Data processing. 4. Mechanical engineering—Data processing.
I. Stoecker, P. A. (Paul A.)   II. Title.
TJ223.M53S76 1988
629.8'95—dc 19   88-20869

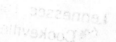

Preface

Microcomputers are having, and will have in the future, a significant impact on the technology of all fields of engineering. The applications of microcomputers of various types that are now integrated into engineering include computers and programs for calculations, word processing, and graphics. The focus of this book is on still another objective—that of control. The forms of microcomputers used in control range from small boards dedicated to control a single device to microcomputers that oversee the operation of numerous smaller computers in a building complex or an industrial plant. The most-dramatic growth in control applications recently has been in the microcomputers dedicated to control functions in automobiles, appliances, production machines, farm machines, and almost all devices where intelligent decisions are profitable.

Both engineering schools and individual practicing engineers have responded in the past several years to the dramatic growth in microcomputer control applications in thermal and mechanical systems. Universities have established courses in computer control in such departments of engineering as mechanical, civil, agricultural, chemical and others. Instructors and students in these courses see a clear role in the field that complements that of the computer specialist who usually has an electrical engineering or computer science background. The nonEE or nonCS person should first and foremost be competent in the mechanical or thermal system being controlled. The objectives of extending familiarity into the computer controller are (1) to learn the characteristics, limitations, and capabilities of the microcomputer as a controller, and (2) to be able to communicate intelligently with the computer specialist about the control application. Practicing engineers see microcomputer control penetrating their fields by complementing and in some cases supplanting former control concepts. These engineers can benefit from a book that starts at an elementary computer and electronic level and proceeds step-by-step until it explains the functioning of commercial control hardware.

The major sections of the book and their sequence reflect the assumed background of the reader and the goals to be accomplished. These major sections are: analog electronics, digital electronics, elementary microcomputers, higher level microcomputers typified by the personal computer, and control analysis—both continuous and digital. The background assumed is that of basic electricity and dc circuits. The reader with a more-extensive experience in electronics and/or instrumentation may move rapidly through Chapters 2 through 4 on dc circuits, operational amplifiers, and transistors. The topics in Chapters 5 and 6 on transducers and actuators are crucial to computer control, because the quality of control can be no better than the fidelity of the sensed data and the ability to regulate an actuator.

The section on digital electronics includes Chapters 7 and 8 which explore gates, digital-to-analog and analog-to-digital converters, as well as some chips frequently encountered in the circuitry external to the computer. We elect to devote a separate chapter (Chapter 9) to memories which are a different category than most other chips in that memories never exist except as an integral part of the computer. Focusing on memories detached from the computer permits identification of the characteristics of several types of memories.

The progression into the computer begins in earnest in Chapter 10 with the study of binary arithmetic, the hexadecimal system, signed and unsigned numbers, and carry/overflow operations in binary arithmetic.

Chapter 11 explores programming of microprocessors by first devising a generic microprocessor. After some simple instructions are illustrated with the generic microprocessor, the chapter moves to two commercial microprocessors—the Intel 8080/8085 and the Motorola 6800 family. Chapter 12 steps up to a programming language one level higher than machine language by considering assembly language. Elementary computers as covered in Chapter 13 are exemplified by the Intel System Design Kit and the Motorola Evaluation Kit. These are small computers with rudimentary input and output capabilities. They are programmed in machine language and facilitate a working knowledge of the internal operation of the microcomputer. The modest experience in machine-language programming is considered a good investment of effort for future work with higher level computers, as well as for any later programming of small computers dedicated to controlling a single machine.

The subjects addressed in Chapters 14 and 15 fit into the category of communication, as Chapter 14 explains parallel input/output to and from the elementary computer. Chapter 15 addresses serial data transmission, describing how the microcomputer converts data beteween parallel and serial forms, and explaining the commercially-used convention of RS-232-C. It is often convenient and even necessary to transmit serial data over telephone lines. This is facilitated by the use of modems.

Chapters 16 and 17 return to mechanical and thermal systems, now with an understanding of the computer as a controller and how to interface the computer with the system. Chapter 16 on continuous control is to serve several purposes. It is a review for those who have already studied automatic control and a preparation for proceeding to the next chapter on digital control. Two further emphases in Chapter 16 are inversion of the mathematical representation of control systems back to the time domain for greater physical insight, and secondly to develop more experience in translating the behavior of a physical facility into block diagrams and transfer functions. Also, the chapter treats several frequently-used modes of control—proportional and proportional/integral—to prepare for the execution of these modes using computer control.

Chapter 17 concentrates on digital control, introduces and provides some physical interpretation of the z-transform, and applies it to computer control. Using the same approach in Chapter 17 as in Chapter 16, the inversion of the z-transforms representing the performance of control loops provide a physical interpretation of computer control. The z-transforms will be used for stability analysis of computer-control loops and will show the influences of such variables as the sampling interval on stability.

To bring the subject of microcomputer control into focus on industrial systems, the authors asked Clay Nesler, Manager of Systems and Controls Research, Johnson Controls, Incorporated, Milwaukee, to write the final chapter in the book. Nesler is well-grounded in both the hardware and control mathematics and in addition is experienced in tuning and commissioning computer control systems in large buildings. His chapter on "Field Application of Microcomputer Controllers" completes the cycle encompassed by the book which starts with fundamentals, studies individual componenets, puts the components together in an analytical fashion, and finally puts the controller into use industrially.

The intended audience of this book is the engineer whose major interest is in applying microcomputer control to a mechanical or thermal system. The audience would thus include engineering students at the senior or first-year graduate level who are mostly in fields other than electrical or computer engineering. The chief segment of the intended audience is the practicing engineer who has a need to apply microcomputer control, but has not had formal courses in the subject. It is the plan that the text starts at an elementary level and moves step by step so that an experienced engineer can use the book in a self-taught manner. The self-paced approach will be particularly enhanced if the reader can gather some modest equipment, wire some circuits, and enjoy the experience of witnessing some of the analog, digital, and microcomputer equipment in operation.

Many students and colleagues have provided valuable input during the preparation of this book. Special acknowledgment is due Professor Roy R. Crawford, a colleague of WFS at the University of Illinois at Urbana-Champaign, who is one of the instructors in the course at that institution. He has performed valuable service in bringing in new ideas and concepts to our course, many of which are also reflected in this book.

Wilbert F. Stoecker
Paul A. Stoecker

Contents

1	**Microcomputer Control**	**1**
1-1	The Penetration of Microprocessors into Engineering Fields	1
1-2	The Path by which Microcomputer Control Has Grown . . .	2
1-3	Chemical and Process Industries	3
1-4	Environmental Control of Buildings	4
1-5	Automobiles	6
1-6	Home Appliances	6
1-7	Computer Control in Manufacturing	7
1-8	Electric Power Generation and Regulation	7
1-9	Agricultural Applications of Computer Control	8
1-10	What the Engineer Who Applies Computers Needs to Know	8
	References	9
2	**DC Circuits and Power Supplies**	**11**
2-1	Understanding Circuits	11
2-2	Kirchhoff's Laws	11
2-3	Thévenin Equivalent	12
2-4	Norton Equivalent	14
2-5	*RC* Circuits	16
2-6	Resistors	18
2-7	Diodes	20
2-8	Rectifying Circuit	22
2-9	Voltage Ripple	23
2-10	Commercial Power Supplies	24
2-11	Voltage Regulators	24
	Problems	25
3	**Operational Amplifiers**	**29**
3-1	Application of Operational Amplifiers	29
3-2	Basic Characteristic of the Op Amp	30
3-3	Comparator	30
3-4	Inverting Amplifier	30
3-5	Choice of Resistances	32
3-6	Non-inverting Op Amp	32
3-7	Buffer or Follower Amp	33
3-8	Signal Conditioning	33

	3-9	Summing and Multiplying Amplifier	34
	3-10	Generalized Circuit for an Op Amp	36
	3-11	Integrator	37
	3-12	Pin Diagram of 741 Op Amp	39
	3-13	Limitations and Ratings of the Op Amp	40
		General References	41
		Problems	41

4 Transistors — 45

	4-1	Impact of the Transistor	45
	4-2	Symbols and Terminology	45
	4-3	Current Characteristics	47
	4-4	Bipolar-Junction and Field-Effect Transistors	49
	4-5	Voltages at the Transistor Terminals	49
	4-6	Voltage Amplifier	49
	4-7	Transistor as a Switch, and Saturating the Transistor	51
	4-8	Common Emitter and Common Collector Circuits	52
	4-9	Zener Diode	53
	4-10	Constant-Current Source	54
	4-11	Designing a Constant-Current Source	55
	4-12	Operating Limits of a Transistor	56
	4-13	Transistor Packages	57
		References	58
		Problems	58

5 Transducers — 61

	5-1	Importance of Good Instrumentation	61
	5-2	Thermocouples	62
	5-3	Thermocouple Reference Junction	63
	5-4	Metal and Thermistor Resistance-Temperature Devices	63
	5-5	Series Circuit	64
	5-6	Bridge Circuits	67
	5-7	Amplification of a Bridge Output	69
	5-8	RTD Circuits Supplied with Constant Current	70
	5-9	Temperature-Dependent Integrated Circuits	71
	5-10	Application of Sensors—Liquid Temperature	71
	5-11	Application of Sensors—Temperature of Air and Other Gases	75
	5-12	An Overview of Temperature Sensors and Transducers	77
	5-13	Flow Rate and Velocity Measurement	78
	5-14	Venturi Tubes—Liquid Flow Measurement	79
	5-15	Orifice—Liquid Flow Measurement	81

5-16	Flow Measurement of a Compressible Fluid in a Venturi or Orifice	83
5-17	Pitot Tubes	85
5-18	Hot-Wire Anemometer	85
5-19	Turbine Flow Meter	86
5-20	Ultrasonic Flow Meters	86
5-21	Vortex-Shedding Flow Meters	87
5-22	Evaluation of Flow-Measuring Devices	89
5-23	Pressure Transducers	90
5-24	Evaluation of Types of Pressure Transducers	90
5-25	Force	91
5-26	Torque	91
5-27	Electric Current	91
5-28	Humidity Sensors	92
5-29	Chemical Composition	92
5-30	Liquid Level	93
5-31	Position and Motion Sensors	94
5-32	Rotative Speed	97
5-33	How to Choose Transducers	97
	References	98
	General References	99
	Problems	99

6 Actuators — 105

6-1	Actuators for Computer Control Systems	105
6-2	Two-Position DC Electric Switch	106
6-3	Silicon-Controlled Rectifier (SCR) for DC Switching	107
6-4	Triac—Alternating Current Switching	108
6-5	Optically Isolated Switch	110
6-6	Solid-State Relays	110
6-7	Electric-Motor Actuators	112
6-8	Magnetic Operator	113
6-9	Hydraulic Actuator	114
6-10	Pneumatic Valve and Damper Operators	114
6-11	Electric-to-Pneumatic Transducer	115
6-12	Stepping Motors	115
6-13	Performance of Stepping Motors	118
	References	121
	Problems	121

7 Binary Numbers and Digital Electronics — 123
- 7-1 Transition to Digital Electronics — 123
- 7-2 Binary Numbers — 124
- 7-3 Conversion between Binary and Decimal Numbers — 125
- 7-4 Addition of Binary Numbers — 126
- 7-5 Basic Logic Operations — 126
- 7-6 OR Gate — 127
- 7-7 AND Gate — 127
- 7-8 Inverter — 128
- 7-9 NOT-OR (NOR) Gate — 128
- 7-10 NOT-AND (NAND) Gate — 129
- 7-11 Exclusive-OR (XOR) Gate — 129
- 7-12 Combining and Cascading Gates — 129
- 7-13 De Morgan's Laws — 130
- 7-14 Gate Chips — 131
- 7-15 Ladder Diagrams for Conditional and Sequential Control — 131
- 7-16 Ladder Diagram Using Gates — 135
- 7-17 Sequential Logic Circuits — 137
- 7-18 Binary Addition with Gates — 137
- 7-19 Pull-Up Resistor — 138
- 7-20 Three Classes of Outputs Found on Inverters and Buffer Gates — 139
- 7-21 Debounced Switch — 141
- 7-22 Clocks and Oscillators — 142
- 7-23 Flip-Flops — 143
- 7-24 Divide-By Counters — 144
- 7-25 Schmitt Trigger — 145
- 7-26 Monostable Multivibrator — 148
- 7-27 Low-Frequency Pulses — 148
- 7-28 Latches — 149
- 7-29 Comparators — 150
- 7-30 Analog Switches—Field-Effect Transistors — 150
- 7-31 Binary-Coded Decimal (BCD) — 151
- 7-32 Seven-Segment LEDs — 151
- 7-33 Summary — 153
- General References — 153
- Problems — 154

8 Conversion Between Digital and Analog — 159
- 8-1 Elements of a Microcomputer Controller — 159
- 8-2 A Simple DAC — 160
- 8-3 DAC Using R-$2R$ Ladder Circuit — 162

8-4	The 1408 DAC	163
8-5	Applying the 1408 DAC	165
8-6	Multiplexers	167
8-7	Fidelity of Voltage Transmission Through a MUX	169
8-8	Sample-and-Hold Circuits	170
8-9	Operating Sequence with Multichannel Control	172
8-10	Where Analog-to-Digital Conversion Is Needed	173
8-11	Internal Functions of One Class of ADCs	174
8-12	More Complete Description of the Internal Functions of an ADC	175
8-13	Staircase and Successive Approximation Search Routines and Dual-Slope Integration	176
8-14	Pin Diagram of an 8-Bit ADC	177
8-15	Characteristics of the ADC 0800	178
8-16	Analog-to-Digital Conversion Using a DAC in Combination with Software	180
8-17	Choosing the ADC	180
	Problems	182

9 Memories — 185

9-1	Function and Types of Memories	185
9-2	ROMs	186
9-3	EPROMs	187
9-4	RAMs	187
9-5	The MCM6810 RAM	189
9-6	Four-Bit RAMs—the MCM2114	190
9-7	Dynamic RAMs	192
9-8	EEPROMs	192
9-9	Memories on the Microcomputer	192
	General References	193
	Problems	193

10 Binary Arithmetic — 195

10-1	The Eight-Bit Microcomputer	195
10-2	Two's Complement Arithmetic—Subtraction	195
10-3	Multiplication	196
10-4	Hexadecimal System	197
10-5	Labeling Conventions	197
10-6	Signed and Unsigned Numbers	198
10-7	Unsigned Numbers—The Carry Flag	200
10-8	Signed Numbers—Two's Complement Overflow	204
10-9	Status Registers on Microprocessors	206

References	207
Problems	208

11 Programming a Microprocessor — 209

- 11-1 A Generic Microprocessor … 209
- 11-2 Data and Address Buses in a Generic Microcomputer … 209
- 11-3 The Accumulator with its Arithmetic, Logic, and Transfer Operations … 210
- 11-4 The Fetch-Decode-Execute Sequence … 211
- 11-5 Preliminary Instruction Set … 212
- 11-6 Program Counter … 214
- 11-7 Status Register and Jumps … 214
- 11-8 Another Accumulator—Incrementing and Decrementing … 217
- 11-9 Additional Addressing Modes … 218
- 11-10 The Index Register and the Use of Register Addressing … 219
- 11-11 Subroutines and the Stack … 221
- 11-12 The Intel 8080/8085 Microprocessor … 222
- 11-13 Loading Into and Storing From the Accumulator … 228
- 11-14 Forms of Addressing on the 8080/8085 … 228
- 11-15 Flag Register … 229
- 11-16 Subroutines … 231
- 11-17 The 8080/8085 Programming Guide … 232
- 11-18 The Motorola 6800 Family … 232
- 11-19 Registers in the 6800 Microprocessor … 232
- 11-20 The Instruction Set of the 6800 … 234
- 11-21 Condition Codes … 237
- 11-22 Forms of Addressing … 238
- 11-23 Branches—Relative Addressing … 238
- 11-24 Index Register—Indexed Addressing … 240
- 11-25 Loops … 240
- 11-26 Stack Pointer … 242
- 11-27 Subroutines … 243
- 11-28 The 6800 Microprocessor Programming Guide … 243
- 11-29 Summary … 244
- References … 246
- Problems … 246

12 Assembly Language Programming — 249

- 12-1 Machine Language and Assembly Language … 249
- 12-2 An Overview of the Assembly Process … 249
- 12-3 Major Components of the Program … 250
- 12-4 Assembly Language Statements … 251

12-5	Assembler Directives	253
12-6	The Location Counter	254
12-7	Using Assembler Labels and Symbols	255
12-8	Relocating Assemblers and Loaders	257
12-9	The Operation of an Assembler	258
	References	259
	Problems	259

13 The Structure of an Elementary Microcomputer — 261

13-1	Definition of an Elementary Microcomputer	261
13-2	The Bus Structure	262
13-3	Flow of Information on the Buses During Execution of a Program	263
13-4	The Intel 8080 Microprocessor	264
13-5	Structure of the SDK-85 System Design Kit	266
13-6	Memory Map of the SDK-85	266
13-7	The Motorola 6802 Microprocessor	268
13-8	Structure of the MEK6802D5 Evaluation Kit	269
13-9	Memory Map of the D5 Evaluation Kit	270
13-10	Common Features of an Elementary Microcomputer	271
	References	272
	Problems	272

14 Parallel Input/Output and Interrupts — 273

14-1	Parallel Input/Output	273
14-2	A Generic Parallel I/O Chip	274
14-3	Processing Interrupts	275
14-4	The Motorola Peripheral Interface Adapter (PIA)	276
14-5	Registers in the PIA	277
14-6	Preparing the PIA to Send and Receive Data	278
14-7	Interrupt from a Peripheral—An Overview	279
14-8	The Control Register and the Control Lines	281
14-9	Setting the Microprocessor to Receive an Interrupt	283
14-10	Structure of an Interruptible Program	283
14-11	User I/O Socket	285
14-12	Intel 8155/8156 RAM with I/O	286
14-13	Intel 8212 I/O Chip	286
14-14	Rudimentary Control Capability Now Available	287
	Problems	288

15 Serial Input/Output and Modems 289
15-1 Serial Data Transmission 289
15-2 Mark, Space, and Baud Rate 290
15-3 Synchronous and Asynchronous Communication 291
15-4 Parity 292
15-5 Shift Register 292
15-6 A Generic Universal Asynchronous Receiver/ Transmitter (UART) 293
15-7 The MC6850 Asynchronous Receiver/Transmitter (ACIA) . 297
15-8 Registers in the ACIA 297
15-9 The Control Register 299
15-10 The Status Register 300
15-11 Transmitting and Receiving with the ACIA 301
15-12 The Intel 8251A Programmable Communication Interface . 302
15-13 The Control and Status Register on the 8251A 303
15-14 Communicating Using RS-232-C and Modems 305
15-15 RS-232-C Interface 306
15-16 Level Conversion Between RS-232-C and TTL 308
15-17 Communicating Between Two Elementary Microcomputers Using RS-232-C 310
15-18 Transmission over Telephone Lines Using Modems 310
15-19 Dial-Up Modems 313
15-20 ASCII Characters 313
15-21 One-on-One Communication 313
References 315
Problems 315

16 Dynamic Behavior of Systems 317
16-1 Returning to the Thermal and Mechanical System 317
16-2 On/Off Controls 318
16-3 Make/Break Sensor with On/Off Actuator 319
16-4 Analog Sensor with On/Off Actuator 320
16-5 Modulating Control Strategies 321
16-6 Proportional Control 322
16-7 Proportional-Integral Control 324
16-8 Proportional-Integral-Derivative (PID) Control 327
16-9 Dynamic Analysis 327
16-10 Laplace Transforms 327
16-11 Inverting a Transform 328
16-12 Transforms of Derivatives 330
16-13 Solving Differential Equations by Means of Laplace Transforms 330

Contents   xvii

    16-14 Transfer Functions 332
    16-15 Feedback Loops 333
    16-16 Stability Criteria for a Feedback Control Loop 334
    16-17 A Proportional Controller Regulating the Pressure in an
           Air-Supply System 336
    16-18 Response of a Proportional Air-Pressure Controller to a
           Disturbance in Air-Flow Rate 339
    16-19 The Integral Mode of Control 344
    16-20 The Proportional-Integral (PI) Mode of Control 350
    References 354
    Problems 354

17 The Computer and Its Sampling Processes   357
    17-1   Unique Features of Computer Control 357
    17-2   Numerical Simulation 358
    17-3   Sampled Data 362
    17-4   Responses to Sampled Values 365
    17-5   The z-Transform 370
    17-6   Response to a Series of Impulses 376
    17-7   The Zero-Order Hold (ZOH) 378
    17-8   Inverting a z-Transform 380
    17-9   Cascading z-Transforms and Transforms of a Feedback Loop   382
    17-10 How a z-Transform Can Indicate Stability of a Control Loop   386
    17-11 Proportional Control 390
    17-12 Proportional-Integral Control 397
    17-13 Forms of Actuator Signals 399
    17-14 Non-linearities—Dead Time 400
    17-15 Non-linearities—Hysteresis 401
    17-16 Summary 401
    References 402
    Problems 402

18 Field Application of Microcomputer Controllers   409
    18-1   Applying Microcomputer Controllers to Field Processes ... 409
    18-2   Practical Control Algorithms 410
    18-3   Incremental PI Control Algorithm 411
    18-4   Position PI Control Algorithm 412
    18-5   Criteria for Tuning 414
    18-6   Manual Control Test 418
    18-7   Trial-and-Error Tuning 419
    18-8   Closed-Loop Tuning 422
    18-9   Open-Loop Tuning 423

18-10 Hysteresis Compensation	426
18-11 Summary	427
References	428
Problems	429

Microcomputer Control of Thermal and Mechanical Systems

Chapter 1

Microcomputer Control of Thermal and Mechanical Systems

1-1 The Penetration of Microprocessors into Engineering Fields

The designations "microprocessor" and "microcomputer" are sometimes used interchangeably, but at the outset we shall distinguish between them. We shall understand a microprocessor to be a single chip that performs various operations when provided with a program configured in a prescribed structure. The microcomputer combines the microprocessor with other components and circuitry, such as memories and input/output (I/O) devices. While it is true that there are chips now available that contain memories and I/O capabilities as well as the microprocessor, we shall maintain the distinction between the two terms.

The technical world is indebted to electrical engineers, computer scientists, and solid-state physicists for the invention of the microprocessor and microcomputer. The application of these devices has spread rapidly to virtually all fields of engineering so that engineers in many fields now need to know about the characteristics of computer controllers. From the vast arena of computer applications in engineering, this book selects one specific field—control. The scope is limited further by concentrating only on the control of thermal and mechanical systems, although because the computer software and hardware are similar in many control fields, knowledge gained here has broader applications.

This book is written out of the conviction that the engineer who designs thermal and mechanical systems should also be able to select components, circuits, and the microcomputer to control these systems. It is not envisioned that these engineers will conceive and manufacture a new integrated circuit or a new chip. Rather, they should know what chips are available, how they should be used, and which of their characteristics are important for applications. Maintaining a rigid boundary between the computer equipment and the thermal/mechanical system complicates matching the controlling microcomputer with the thermal/mechanical system.

The types of processes associated with thermal/mechanical systems include physical motion, application of force, generation of power, and transfers of thermal and mechanical energy. Robotics and numerical control of machines are not specifically addressed, but the subjects covered in this book provide a background for further work in those fields.

1-2 The Path by which Microcomputer Control Has Grown

Computer control first appeared in activities and industries that operate on a high technical level. As Fig. 1-1 shows, one of the early appearances of computer control was in the research laboratory for data acquisition and control of experiments. About the same time the chemical and process industries began installing computer control systems, and several years later energy-management systems in large buildings became popular. Those computer

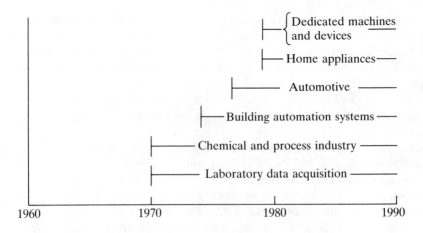

Fig. 1-1. Emergence of some of the applications of computer control.

1-3. Chemical and Process Industries

control systems were built around minicomputers, and the large installations cost hundreds of thousands and in some cases over a million dollars.

A significant shift that began in the late 1970s was driven by the dramatic improvements in computers. Lower-cost microcomputers began to rival the speed, memory capacity, and power of minicomputers. The price of the total system dropped, and one large central computer began to be complemented by a number of smaller remote computers in communication with the central one. Of equal importance was the penetration of computer control into smaller-scale applications, such as automobiles, home appliances, and special purpose machines, and into rather small and low-cost devices. This created new career opportunities for engineers in the control of individual machines and devices in addition to the control of large plants.

1-3   Chemical and Process Industries

The chemical and process industries (chemical plants, petroleum refiners, petrochemical plants, food processors, etc.) were early users of computer control. The earliest applications were directed toward sensing and regulating in order to achieve safe operation and to maintain quality control of the product. The calculation capabilities of the computer opened yet another attractive challenge—to optimize the process, system, or plant based on existing conditions. Specifically, for a certain composition of the raw materials and their available flow rates, the temperatures, pressures, and other controllable variables can be set to obtain maximum yield of product. Another typical optimization is to operate the plant in such a way that it produces at a specified rate with the minimum energy requirement.

When multiple compressors serve an industrial refrigeration system, there is always the question of how to sequence them and proportion the load among them in such a way that the combination operates efficiently. Furthermore, large electric motors should not start and stop more frequently than perhaps every half-hour or hour, so short cycling should be prevented. Also, many compressors have a minimum capacity, and excessive wear results when operating below that capacity. On the other end of the scale, requiring too great a capacity may overload the motor. A microcomputer controller can be programmed to consider all of these factors and decide which compressors should be started and stopped and how they should be loaded.

A microcomputer has been used to control the sequence of events in a batch-type vacuum process[1] used for sterilization of products. Some of the individual processes in the cycle include evacuation, preheating, filling the chamber with steam, holding for a specified period, venting, re-evacuation, drying, and final bleeding in of air. The timing of the processes in this cycle varies with the product, and a unique program can be developed for each product. The operator simply calls the applicable program and the controller executes it.

In a certain chlorine plant,[2a] a microcomputer monitors 52 chlorine cells and regulates the electric voltage and current to each. In industrial drying processes, such as those encountered in paper making, the humidity of the drying air must be controlled, and a microcomputer has been used[2b] to convert sensed values of drybulb and wetbulb temperatures into magnitudes of absolute moisture content. In a steelmaking basic oxygen plant,[3a] a microcomputer controls the oxygen flow rates, the rates of addition of additives, and the cooling-water flow, as well as monitoring emissions from the facility. In an automobile assembly plant the electric energy required for compressing air may constitute 10 to 20 percent of the total electric load. A microcomputer has been used[3e] to lower the supply pressure when possible and reduce the no-load running time of compressors.

In the food processing industry there are often multiple operations that must be controlled simultaneously which lend themselves to the power of the microcomputer. Yung[4] described a rice-processing plant where a microcomputer controls the input proportions of rice and water, the total amount charged in the cooker, the temperature of the cooker, the time in the cooker and hold time at each given temperature, and the water content of the finished product.

In oil and gas pipeline technology, computer control of remote pumping stations and communication between these individual computers and a master computer[5] fosters efficient operation.

1-4 Environmental Control of Buildings

Heating and cooling systems for buildings must respond to a variety of heating and cooling loads and ideally must meet these loads with a minimum energy requirement. Computer control of the thermal system of large buildings is now a well-established practice.[6,7] Energy management systems were first introduced about 1969 and became established in the mid-1970s. The computer control system can start and stop equipment such that comfortable conditions are maintained, without requiring extra energy; can reset temperatures of conditioned air and chilled or hot water so that the heating or cooling can be performed with minimum cost; and can also limit the peak electrical demand by interrupting services in noncritical areas. In addition to regulating the thermal system of a large building or a complex of buildings, these computers are now being used to monitor the spaces for security purposes, such as for smoke, fire, or intrusion.

1-4. Environmental Control of Buildings

The early approach to computer control in large buildings and in the process industries was to employ one central computer. All sensors fed their information to the central computer, which processed the information, made decisions according to its programmed instructions, and relayed signals to the actuators that controlled the components in the system. This concept is now being supplanted by the "hierarchical" or "distributed control" concept.[8-10] Figure 1-2 shows a schematic diagram of distributed control indicating that in addition to the central computer there are a number of satellite microcomputer controllers. The microcomputers receive information from the sensors and control their associated actuators. The central computer is the overall supervisor of the microcomputers; it can poll the status of any sensor through the appropriate microcomputer and can change the memory in any microcomputer. One strength of the distributed control concept is that if the central computer fails, the microcomputers continue to control with the latest instructions they have available. If one of the microcomputers fails, its memory can be refreshed by the central computer. Each of the microcomputers may be responsible for up to about 20 to 50 sensors and actuators. The communication lines between the microcomputer and the central computer are not heavily loaded, because they carry only exceptional information, such as an alarm from one of the microcomputers, specific requests for information, and changes in programs to be made in the microcomputers.

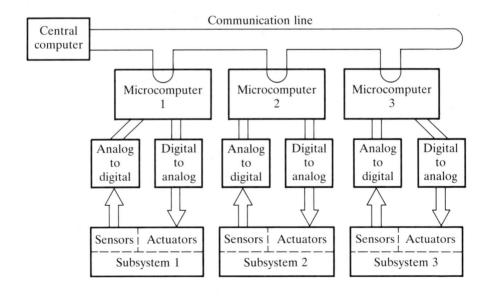

Fig. 1-2. Distributed control or hierarchical concept.

1-5 Automobiles

In the early 1980s microcomputers began to assume many control functions on automobiles.[11-13] The major categories are engine control for economy and emission regulation, dashboard monitoring, and enhancing the performance perceived by the driver. Some engine functions being taken over are (1) choke control to provide the leanest mixture possible during warm-up, since 60 to 70 percent of the car's emissions occur during startup, (2) adjusting the air-fuel mixture at the carburetor by sensing the oxygen level in the exhaust, (3) regulation of the idle speed in response to such loads as the air-conditioning compressor or rear-window defroster, and (4) optimized spark control, which adjusts the timing just short of "knocking" and replaces the vacuum system as the actuator of the spark advance and retard.

In the passenger compartment the microcomputer is taking over the supervision of the dashboard and is a key element in the electronic instrument cluster. Digital readouts and message displays are replacing the analog indicators and the on/off warning lights. Critical warnings take priority for such alarms as loss of brake or oil pressure or low alternator output. Electronic odometers measure fuel consumption rates and the road speed and then compute the instantaneous fuel consumption per unit distance.

An automatic transmission wastes fuel because of slippage in the fluid link between input and output shafts. At the proper time the computer control engages a clutch to lock the engine firmly to the drive shaft to eliminate slippage losses.

As more instrumentation appears on the engine of the automobile, the way is paved for the repair person to plug into the system to obtain more information when diagnosing a problem. If the fault lies within the microcomputer or electronic circuits themselves, an individual circuit board is replaced.

1-6 Home Appliances

One of the early applications of the microcomputer in the home was for controlling microwave ovens. The microcomputer^{2d} is capable of controlling variable-power cooking, defrost followed by cooking, and temperature-controlled cooking.

To control the heating system in an energy-conservative manner, a microcomputer^{3d} can sense the temperatures in each room and regulate the flow of warm air by means of a damper in each duct supplying a room. Space heaters using digital electronics[14] can control temperatures within $\pm 1°C$ in contrast to electromechanical controls that sometimes have a $\pm 9°C$ range of error. A microcomputer that controls a burner in a residential furnace[15] regulates both safety and performance. A microcomputer supervising a residential air conditioner[16] controls the compressor speed by means of a variable-frequency inverter to regulate capacity rather than on/off control. In residential heat pumps that extract heat from outdoor air in the winter, the outdoor air coil may become blocked with frost, so the coil must be defrosted periodically. A microcomputer[17] can be used to determine when to defrost and how long the defrost should proceed. The microcomputer can monitor a number of other functions, and when there is a malfunction it stores this information in its memory so that it can be retrieved by a repair person.

Other appliances such as clothes dryers and washing machines[3f,18] are now using microcomputers to replace the electromechanical control traditionally employed to regulate the sequence of operations. Some of the operations are time-programmable by the user. Refrigerators[18] are now available that incorporate a microprocessor to supervise such monitoring actions as alarms for the door remaining open too long and for excessively high temperatures in the refrigerator compartments.

1-7 Computer Control in Manufacturing

In manufacturing operations there is an enormous variety of machines performing complex functions. In these machines decisions are continually being made in order to react to changes in assignments or changes in the behavior of the materials being processed. Computer controllers of these machines might have the overall instructions stored in memory but revise the basic instructions on the basis of variations that occur in the process. Microprocessor-based devices, interconnected by local area networks (LANs) that may include hierarchies of digital computers, can now unite discrete and analog controls in an automated factory.[19]

Complicated stitching patterns for production-type sewing machines can be programmed on a microcomputer.[20] The stitching machine when activated moves the material to the proper position, begins the stitching, and moves the position and controls the stitch length to achieve the desired pattern.

In the transportation of materials throughout a factory, it is no longer necessary for a human to drive the individual vehicles. The trucks can be computer controlled[21] under the direction of a central computer.

1-8 Electric Power Generation and Regulation

Industrial consumers of electricity are usually assessed demand charges based on their peak use of electricity during a month. Numerous commercial controllers are available to shed certain preassigned electrical loads in a plant or building when the total demand in kilowatts begins to approach the point where a new peak would be established.

At a NASA experimental installation of a wind turbine generator,[2c] a microcomputer controls the yaw and blade pitch angle. Steam serving power plant turbines is usually superheated, and a microcomputer[3b] has been used to control both the temperature of the steam and the rate of change of this temperature—a necessary adjustment to prevent nonuniform turbine metal expansion. Aboard ships with variable pitch propellers driven by gas turbines, abrupt changes in pitch and/or speed must be programmed such that a maximum torque is not exceeded that would damage the drive train.[3c]

1-9 Agricultural Applications of Computer Control

As farmers move to remain competitive, many are turning to electronics and microcomputer control to enhance their machines and regulate processes. Examples include a feeding system for dairy cows in which each cow wears a transponder that outputs a unique radio-frequency signal.[22] This signal identifies the animal to a microcomputer reader, which dispenses the amount and formula of its feed. Farm implements such as fertilizer sprayers when equipped with appropriate microcomputer control can sense the ground speed of the sprayer and the character of the soil to make continuous decisions on the amount of fertilizer to spray.

1-10 What the Engineer Who Applies Computers Needs to Know

While there is probably no limit to the knowledge that could be useful, certain topics are particularly important. The emphasis in our study will be on the lower level of the spectrum of computer sophistication. We will concern ourselves with thermal and mechanical systems or subsystems, sensors and instrumentation, actuators, signal conditioning, and the operation of the microcomputer. Knowledge of the system is expected to be the traditional background of the mechanical, chemical, civil, or agricultural engineer. No computer control, regardless of its level of refinement, can properly control a poorly instrumented system. Furthermore, while indicating types of instruments (dial-type pressure gauges, mercury-in-glass thermometers, etc.) may have been adequate at one time, now all the information must be converted into electrical form through the use of transducers. Before this information can be sent to the computer, it must be converted into digital form. Within the microcomputer an understanding is needed of how to receive, process, and send data back out in order ultimately to control an actuator.

References

1. Vacudyne Altair, Inc., Chicago Heights, IL, 1981.

2. Proceedings of Fourth IECI Annual Conference, *Industrial Applications of Microprocessors*, Inst. of Electrical and Electronics Engineers, Inc., New York, 1978.

    a. J. L. Hilburn, P. M. Julich, R. L. Mitchell, "A Micro-computer-based Supervisory System for a Mercury-Anode Chlorine Plant," pp. 46–49.

    b. G. Totates, S. Klein, C. Rupp, and R. P. Jefferis, "Design of a Psychrometric Computer," pp. 83–86.

    c. A. J. Gnecco and G. T. Whitehead, "Microprocessor Control of a Wind Turbine Generator," pp. 143–147.

    d. T. Yoshioda, Y. Tanji, S. Watanabe, and T. Yamada, "A Microprocessor Control for Microwave Oven," pp. 159–163.

3. Proceedings of Fifth IECI Annual Conference, *Industrial and Control Applications of Microprocessors*, Inst. of Electrical and Electronics Engineers, Inc., New York, 1979.

    a. D. L. Browne, "Microprocessor Control for Industrial Steelmaking BOP Shop," pp. 29–31.

    b. F. Behringer and J. Bukowski, "Microprocessor Controls for Turbine Reheat Steam Temperature," pp. 32–36.

    c. C. J. Rubis, K-L. Li, and C. R. Westgate, "A Microprocessor Torque Computer for Gas Turbines," pp. 37–40.

    d. S. K. Kavuru, "A Microcomputer Controlled Residential Energy Conservation System," pp. 88–91.

    e. N. Iwama, Y. Inaguma, S. Kuroiwa, and M. Watanabe, "A Microprocessor-Based Compressed Air Supply System," pp. 166–171.

    f. B. Ketelaars and P. Tharma, "A Universal CMOS Controller for Washing Machines," pp. 202–207.

4. S. H. Yung, "Controlling Multi-variable Processes," *Food Engineering*, pp. 82–84, October 1980.

5. A. E. Whiteside, "Petroleum Pipeline Control"; C. C. Jennings, "Booster Station Control Systems"; and W. F. Cruess, "Data Transmission," *Mechanical Engineering*, pp. 59–72, June 1982.

6. *Intelligent Control for HVAC Systems*, Symposium, *ASHRAE Transactions*, vol. 88, part I, pp. 1229–1255, 1982.

7. T. F. Meinhold, "Computerized Energy Management Systems," *Plant Engineering*, Aug. 10, 1982, pp. 54–65.

8. "Electronic Boon for the CPI-Distributed Control," *Chemical Engineering*, vol. 86, no. 17, pp. 91–93, Aug. 13, 1979.

9. *Energy Monitoring and Control Systems*, U. S. Army Corps of Engineers, Huntsville Division, Huntsville, AL, 1981.

10. "Users Prefer Local Control of Industrial Processes," *Energy User News*, vol. 6, no. 31, pp. 1 and 12, Aug. 3, 1981.

11. G. Bassak, "Microelectronics Takes to the Road in a Big Way: A Special Report," *Electronics*, vol. 53, no. 25, pp. 113–122, Nov. 20, 1980.

12. "A CMOS Engine Controller," *Popular Electronics*, pp. 98–99, March 1982.

13. R. Valentine and J. Stewart, "Automakers Shift to Processors," *Electronics Week*, pp. 61–65, Dec. 10, 1984.

14. "Space Heaters Look to Digital Control," *Electronics*, vol. 54, no. 3, p. 45, Jan. 31, 1980.

15. "Microcomputer Burner Control Handles Safety, Performance," *Air Conditioning, Heating, & Refrigeration News*, pp. 3 and 19, Nov. 8, 1982.

16. "Inverter Controls Home Air Conditioner," *Electronics*, p. 86, Mar. 10, 1982.

17. "Heat Pumps Will Use New Microprocessor Control System," *Air Conditioning, Heating, & Refrigeration News*, pp. 1 and 7, Jul. 16, 1979.

18. *Whirlpool News*, pp. 2–3, April 10, 1981.

19. J. A. Moore, "Instrumentation and Control—Toward the Automated Factory," *Mechanical Engineering*, pp. 26–32, October 1984.

20. Union Special Corporation, Chicago, Illinois, 1981.

21. H. Kulwise, "Automatic Guided Vehicle Systems," *Plant Engineering*, pp. 50–57, Jan. 7, 1982.

22. W. R. Iverson, "Electronics Making Its Mark on the Farm," *Electronics*, pp. 110 and 112, Dec. 1, 1983.

Chapter 2

Direct-Current Circuits and Power Supplies

2-1 Understanding Circuits

This chapter concentrates on simple circuits, and in most cases this material will be a review for the reader. The understanding of direct-current (dc) circuits covers the major needs in the computer control of mechanical engineering systems, because much of the analysis is of steady-state conditions. Alternating current (ac) circuits are used particularly in the control of electric power and as the source for obtaining direct current, as well as in interfacing dc equipment. Not all control circuits are steady-state, however, because they may be required to respond to time-varying signals. Capacitors influence the behavior of circuits during changes with respect to time.

The need for a dc voltage source appears continually in the analog sections of computer control, and in most cases this dc voltage is derived from alternating current through a "power supply." It is often crucial that the dc source be pure and constant, so this chapter will include a discussion of power supplies.

2-2 Kirchhoff's Laws

One technique for analyzing a circuit (computing currents in all conductors and voltages at all points) is through the use of Kirchhoff's laws. Kirchhoff's voltage law states that in any loop the sum of the source voltages must equal the sum of the voltage drops. Another law that is combined with Kirchhoff's is Ohm's law, $V = IR$. Kirchhoff's current law states that the sum of all currents leading into a point in a circuit must equal the sum of the currents leaving the point.

Example 2-1. Determine the currents I_1, I_2, and I_3 in the circuit shown in Fig. 2-1.

Solution. Two equations result from Kirchhoff's voltage law:

$$\text{Loop } A\text{-}B\text{-}C \qquad 10 = 6I_1 + 20I_2 \qquad (2\text{-}1)$$
$$\text{Loop } A\text{-}B\text{-}D\text{-}C \qquad 10 - 2 = 6I_1 + 10I_3 \qquad (2\text{-}2)$$

Kirchhoff's current law about point C is

$$I_2 + I_3 = I_1 \qquad (2\text{-}3)$$

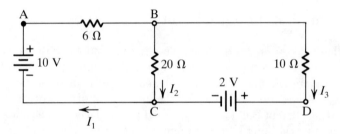

Fig. 2-1. Circuit in Example 2-1 analyzed by Kirchhoff's laws.

The simultaneous solutions of Eqs. (2-1) through (2-3) yields: $I_1 = 0.6842$ A, $I_2 = 0.2947$ A, and $I_3 = 0.3895$ A. The voltage at A, V_A, is arbitrarily set equal to zero; then $V_B = -4.10$ V, $V_C = -10$ V, and $V_D = -8.0$ V.

2-3 Thévenin Equivalent

A complicated circuit of sources and resistances serving a resistive load may be replaced by an equivalent circuit that will provide the same current through that load as the original circuit. Suppose, for example, that the circuit in Fig. 2-1 has the purpose of providing a current through the 10 Ω resistor, as shown in Fig. 2-2. The Thévenin equivalent of a circuit enclosed in the dashed lines is a voltage source in series with a resistance as shown in Fig. 2-3. The

2-3. Thévenin Equivalent

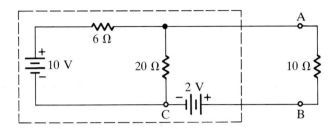

Fig. 2-2. Circuit providing current across load from point A to point B.

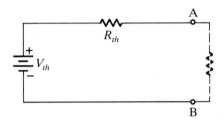

Fig. 2-3. Thévenin equivalent of circuit in Fig. 2-2.

Thévenin equivalent voltage V_{th} is the voltage that would prevail between A and B if the resistance between the two points were made infinite (an open circuit). The equivalent resistance R_{th} is the resistance looking back into the circuit with all the independent voltage sources removed (voltage sources replaced by a short circuit, and current sources by an open circuit).

Example 2-2. Determine the Thévenin equivalent of the circuit enclosed in the dashed lines of Fig. 2-2.

Solution. V_{th} is the voltage $V_A - V_B$ that would prevail if an infinite resistance existed between A and B. The $6\,\Omega$ and $20\,\Omega$ resistors divide the $10\,\text{V}$, resulting in

$$V_A - V_C = (20/26)(10) = 7.692\,\text{V}$$

V_B is 2 V higher than V_C, so

$$V_A - V_B = 7.692 - 2 = 5.692\,\text{V} = V_{th}$$

If the 10 V and 2 V sources were removed in Fig. 2-2, the circuit viewed from the A-B position would appear as shown in Fig. 2-4, which has an equivalent resistance

$$\frac{1}{R_{th}} = \frac{1}{6} + \frac{1}{20} = 0.2167$$

so $R_{th} = 4.615\,\Omega$. The values, then, in the Thévenin equivalent of Fig. 2-3 are $V_{th} = 5.692\,\text{V}$ and $R_{th} = 4.615\,\Omega$.

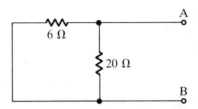

Fig. 2-4. Determining the resistance in the Thévenin equivalent.

As a check, a $10\,\Omega$ resistance could be placed across A-B and the Thévenin equivalent circuit would show the current through this resistance as $(5.692\,\text{V})/(4.615 + 10\,\Omega) = 0.3895\,\text{A}$, which checks I_3 in Example 2-1. The difference in voltage across A and B is $(10\,\Omega)(0.3895\,\text{A}) = 3.895\,\text{V}$.

What is the advantage of the Thévenin equivalent in view of the fact that the circuit could have been analyzed using Kirchhoff's laws? If the circuit is to be analyzed only once, there may be little advantage, but if numerous resistances in addition to the $10\,\Omega$ resistor between A and B are to be explored, the Thévenin equivalent provides the results quickly.

2-4 Norton Equivalent

A complicated circuit of voltage sources and resistances can also be replaced with an equivalent circuit that consists of a constant-current source and a resistance in parallel with it, as Fig. 2-5 shows. This circuit is called the Norton equivalent. The steps in determining the current I_N and the resistance R_N that characterize the Norton equivalent are: (1) Determine the voltage between positions A and B if there were an infinite resistance between A and B, (2) look back from A-B into the circuit and determine the equivalent resistance assuming no current source existed, and (3) determine I by dividing $V_{A\text{-}B}$ from Step 1 by the equivalent resistance from Step 2.

2-4. Norton Equivalent

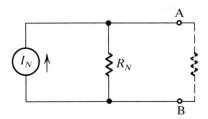

Fig. 2-5. The Norton equivalent.

Example 2-3. What is the Norton equivalent of the circuit enclosed in the dashed lines of Fig. 2-2?

Solution. The value of $V_{A\text{-}B}$ for the open circuit between A and B has already been found in Example 2-2 to be 5.692 V. Likewise the equivalent resistance from Example 2-2 is 4.615 Ω. Referring to Fig. 2-5 the magnitude of the constant-current source for the open-circuit case (and thus all situations) is $5.692/4.615 = 1.233$ A. The Norton equivalent is, then, as shown in Fig. 2-6.

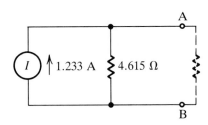

Fig. 2-6. The Norton equivalent of the circuit enclosed by dashed lines in Fig. 2-2.

If a resistance of 10 Ω is connected across A-B, the equivalent resistance of the two resistors is

$$R_{eq} = \frac{1}{\frac{1}{4.615} + \frac{1}{10}} = 3.158 \, \Omega$$

$$V_A - V_B = (1.233 \, \text{A})(3.158 \, \Omega) = 3.894 \, \text{V}$$

and the current through the 10 Ω resistor is 3.894/10 = 0.3894 A, which conforms to previous calculations.

2-5 RC Circuits

The analog circuits with which we deal will usually operate in a steady state. The voltage applied to them will be constant or will change slowly, and the components respond instantaneously. Components such as capacitors and inductors possess time response characteristics and require dynamic (changes with respect to time) analysis of the circuit. The capacitance C of a capacitor (Fig. 2-7) is defined by Eq. (2-4):

$$I = C \frac{dV}{dt} \tag{2-4}$$

where

$$
\begin{aligned}
I &= \text{current flowing into the capacitor, A} \\
dV/dt &= \text{rate of change of potential across capacitor, V/s} \\
C &= \text{capacitance, F}
\end{aligned}
$$

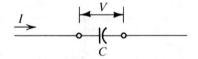

Fig. 2-7. A capacitor.

A basic combination of components is that of a capacitor and resistor in series with a voltage applied across the combination, as shown in Fig. 2-8. Of interest at the moment is the voltage across the capacitor as the applied voltage to the circuit changes.

Example 2-4. The applied voltage, $V_1 - V_3$, is zero until at time $t = 0$ the voltage makes a step change to 6 V. If $R = 2\,\text{M}\Omega$ and $C = 0.04\,\mu\text{F}$, develop the equation for $V_2 - V_3$ as a function of time.

Solution. The current I that flows through the resistor is the same magnitude as the current flowing into the capacitor, since

2-5. RC Circuits

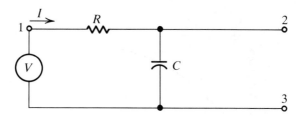

Fig. 2-8. An RC circuit.

no connection is made across points 2 and 3. Two equations for this current are

$$I = \frac{V_1 - V_2}{R} = \frac{V_1 - V_2}{2 \times 10^6 \Omega}$$

$$I = C\frac{dV}{dt} = (0.04\,\mu\text{F})\frac{d(V_2 - V_3)}{dt}$$

(2-5)

After the step change at $t = 0$,

$$I = \frac{V_1 - V_2}{2,000,000} = \frac{(V_1 - V_3) - (V_2 - V_3)}{2,000,000}$$

$$= \frac{6 - (V_2 - V_3)}{2,000,000}$$

(2-6)

Equating the two expressions for I from Eqs. (2-5) and (2-6) gives

$$\frac{6 - (V_2 - V_3)}{(2 \times 10^6)(0.04 \times 10^{-6})} = \frac{d(V_2 - V_3)}{dt}$$

$$= \frac{6}{0.08} - \frac{V_2 - V_3}{0.08}$$

(2-7)

Designate the expression on the right side of Eq. (2-7) as y; then its derivative

$$\frac{dy}{dt} = -\frac{1}{0.08}\frac{d(V_2 - V_3)}{dt}$$

and

$$\frac{d(V_2 - V_3)}{dt} = -0.08\frac{dy}{dt}$$

so Eq. (2-7) can be rewritten

$$-0.08\frac{dy}{dt} = y$$

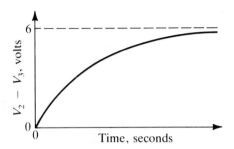

Fig. 2-9. Response of the RC circuit in Example 2-4.

for which the solution is

$$y = C_1 e^{-t/0.08}$$

or

$$V_2 - V_3 = 6 + 0.08 C_1 e^{-t/0.08}$$

The constant of integration C_1 can be determined by substituting the initial condition $V_2 - V_3 = 0$ when $t = 0$, yielding

$$C_1 = \frac{-6}{0.08}$$

so

$$V_2 - V_3 = 6(1 - e^{-t/0.08})$$

which takes the form shown in Fig. 2-9.

2-6   Resistors

Two important types of resistors are (1) carbon resistors and (2) power resistors. In instrumentation and control circuits the carbon resistor is most common, and several sizes are available, each with a different power rating. The sizes, shown to scale, are given in Fig. 2-10. The color coding of the resistor indicates both the magnitude of the resistance and the tolerance. As shown in Fig. 2-11, the first band indicates the first figure, the second band the second figure, and the third band the number of zeros to be added to the previous two digits. The last band indicates the tolerance; a gold band indicates 5% and a silver band 10%. Table 2-1 shows the numerical values associated with the color bands. A resistor banded with yellow-violet-orange-gold has a resistance of 47,000 Ω with a tolerance of 5%.

A power resistor, shown in Fig. 2-12, has the capability of sustaining a much higher power dissipation than the carbon resistor.

2-6. Resistors

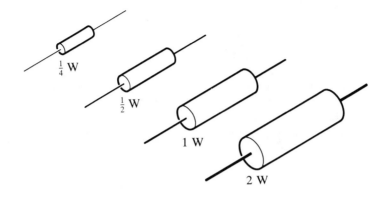

Fig. 2-10. Carbon resistors.

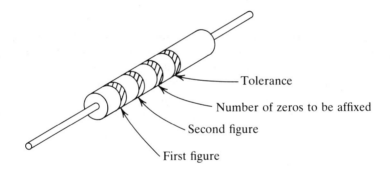

Fig. 2-11. Color code of resistors.

Table 2-1. Resistor Color Code

0	Black	4	Yellow	8	Gray
1	Brown	5	Green	9	White
2	Red	6	Blue		
3	Orange	7	Violet		

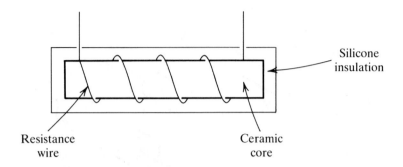

Fig. 2-12. A power resistor.

2-7 Diodes

A diode performs as a check valve for electricity, permitting flow in one direction only. The germanium diode begins to conduct current when the voltage in the forward direction is approximately 0.2 V, while the commonly used silicon diode begins to conduct when the voltage is about 0.6 V. The current-voltage characteristic of a silicon diode is shown in Fig. 2-13. The diode can hold against a high backward voltage of perhaps hundreds of volts (depending upon the characteristic of the diode) until it breaks down and lets the current flow freely. Whenever current is flowing through a diode, power is dissipated as heat, equal to the product of the voltage across the diode and the current through it, $(V)(I)$. Diodes for high current applications should therefore be mounted on a heat sink.

Figure 2-13 may suggest that the diode allows no current to pass in the forward direction until some characteristic voltage at which the current flows freely. A more precise representation of the current-ΔV characteristics is

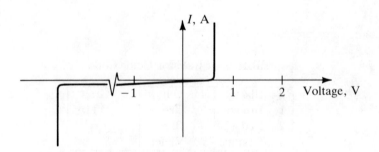

Fig. 2-13. Characteristics of a diode.

2-7. Diodes

given by the exponential relationship:

$$I = I_c(e^{q\Delta V/(AkT)} - 1) \tag{2-8}$$

where

I_c = a characteristic of the diode
q = 1.6×10^{-19} coulomb
A = a constant of the order of unity
k = Boltzmann's constant, 1.38×10^{-23} J/K
T = absolute temperature, K

When $A = 1$ and $T = 300$ K, Eq. (2-8) simplifies to

$$I = I_c(e^{39\Delta V} - 1) \tag{2-9}$$

With each increase in ΔV of 0.1 V, I increases by a factor of 50.

Example 2-5. A diode having a value of $I_c = 38 \times 10^{-12}$ A is rated at 1/2 W. What is the current at which this power rating is reached?

Solution. The 1 in the parentheses of Eq. (2-9) is insignificant, so

$$\text{Power} = 0.5\,\text{W} = (\Delta V)\left(38 \times 10^{-12}\right)\left(e^{39\Delta V}\right)$$

$$\Delta V = \frac{\ln[0.5/(38 \times 10^{-12})\Delta V]}{39}$$

which can be solved iteratively for ΔV. $\Delta V = 0.610$ V, so $I = (38 \times 10^{-12})(e^{39\Delta V} - 1) = 0.816$ A.

Many diodes are color coded to indicate their part number and also the directional bias. The symbol for the diode is shown in Fig. 2-14a and the color coding in Fig. 2-14b. One family of diodes has a prefix of 1N followed by the number indicated by the color code corresponding to Table 2-1. A diode with the color bands of red-yellow-gray-red is a 1N2482 diode.

Two important characteristics obtainable from the specification sheet of a diode are (1) the maximum forward current and (2) the breakdown voltage. Diodes are available that are capable of handling 1000 A. The maximum forward current limitation is usually a function of the extent of heating and the capability of keeping the diode cool. The breakdown voltage is the negative voltage the diode can hold against until it permits a heavy flow of current. The diode is not damaged by reverse current as long as the VI heating is not too high. The maximum breakdown voltage may be 5 to 10 V for some diodes and up to 2500 V for others.

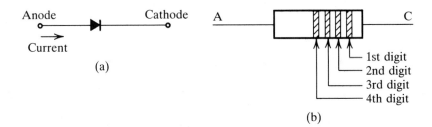

Fig. 2-14. Diode identification: (a) symbol, and (b) color code.

2-8 Rectifying Circuit

Most instrument circuits and the microcomputer are powered by low-voltage direct current, and the source of the dc is usually a transformed and rectified ac supply. Figure 2-15 shows a simple rectifying circuit. The 115 V, 60 Hz power feeds a step-down transformer, because the voltage supply in which we are interested is probably in the range of 5 to 15 V dc. The alternating current flows to a diode bridge which rectifies the alternating current to direct current. Without the capacitor C the voltage across the load R would be a "full-wave rectification" as shown in Fig. 2-16. It is easy to appreciate that the dc power supply suitable for the instrumentation circuits and microcomputers should have a uniform voltage (free of ripples) and maintain a constant voltage even as the load resistance R changes.

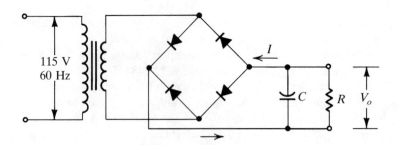

Fig. 2-15. Simple rectifying circuit.

2-9. Voltage Ripple

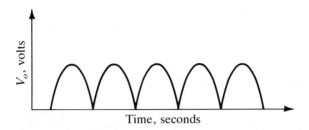

Fig. 2-16. Full-wave rectification.

2-9 Voltage Ripple

The capacitor placed across the output of the bridge circuit provides a significant improvement in the uniformity of voltage. The sequence of events of the rectifier-capacitor-load combination can be visualized as shown in Fig. 2-17. Suppose that the power is switched on and the voltage at the output of the bridge starts at A and builds up to B. During this portion of the cycle the bridge both feeds the load and charges the capacitor. At point B when the voltage supplied by the bridge begins to drop, the charge in the capacitor takes over the supply of the current. The diodes in the bridge prevent any current from flowing back out of the capacitor. During the BC portion of the cycle the capacitor discharges through the resistor and the voltage follows an exponential decay that is the inverse of the exponential buildup of Fig. 2-9. The discharge of the capacitor continues until point C, when the voltage out of the bridge in the next half-wave begins to replenish the capacitor and at the same time provides current for the load. The pattern of the output voltage is a repetition of the BCD profile.

The rate of voltage decay during the BC portion of the cycle can be kept low by maintaining a high value of the product of R and C. A high value of

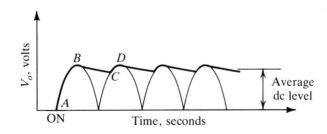

Fig. 2-17. Voltage from simple rectifying circuit in Fig. 2-15.

capacitance should thus be chosen, because the magnitude of R is dictated by the load. At high loads the resistance is low, causing a higher rate of decay and a droop in the average dc level.

2-10 Commercial Power Supplies

The components for the simple rectifying circuit of Fig. 2-15 can be purchased for a low cost, a fraction of the cost of a commercial power supply. Some of the extra cost of a commercial power supply is devoted to smoothing out the ripple of Fig. 2-17 and maintaining a nearly constant voltage regardless of the current draw on the rectifier. The difference in voltage between the peak and valley of the output is often no more than several millivolts in a commercial power supply. Many commercial power supplies have "sense terminals" that can be connected to the load to supply a feedback control circuit that adjusts a constant voltage. Connecting the sense terminals across the load rather than at the output terminals of the power supply further compensates for any voltage drop in the connecting leads between the power supply and the load. Commercial power supplies are also capable of holding the ripple within several tenths of a percent of the rated voltage over the entire load range.

2-11 Voltage Regulators

On a given project there may be the need for several different voltages. A convenient way to achieve this variety of supply voltages is to use a power supply for the highest of the voltages and step down to the other voltages by using a voltage regulator. The package style used by at least one manufacturer (Fairchild) is shown in Fig. 2-18, and the numbering pattern is presented in Table 2-2. The terminal assignment shown in Fig. 2-18 is applicable when the front with the letters is visible. The metal part of the regulator with the hole is for rejecting heat. The hole enables the use of a bolt to attach the voltage regulator to a heat sink.

The input voltage must be 3 or 4 V higher (more negative for the 7900 series) than the output voltage. The regulating capability might vary by several tenths of a volt from one regulator to another of the same type, but any one regulator usually maintains an output voltage within 50 mV. The current capacity of the regulators can be up to approximately 1 A, particularly if adequately cooled.

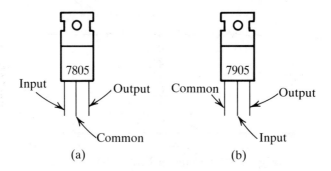

Fig. 2-18. Voltage regulator. (a) Positive, (b) negative voltage.

Table 2-2. Designation of Voltage Regulators

For Positive Voltage		For Negative Voltage	
Type number	Nominal voltage	Type number	Nominal voltage
7805	5	7905	−5
7806	6	7906	−6
7808	8	7908	−8
7812	12	7912	−12

Problems

2-1. (a) Determine the Thévenin equivalent of the circuit enclosed by the dashed lines in Fig. 2-19. (b) If a load of 20 Ω is placed across A-B, what will be the voltage drop across and the current through the resistance? **Ans.:** (b) 4.12 V.

2-2. (a) What is the Norton equivalent of the circuit in Fig. 2-19? (b) If a load of 20 Ω is placed across A-B, what will be the voltage drop across, and the current through the resistance? **Ans.:** (b) 0.207 A.

2-3. A 100 Ω potentiometer is set such that a high-impedance voltmeter reads 8 V between the wiper and ground when 12 V is applied, as shown in Fig. 2-20. If a 150 Ω resistance is attached between the wiper and ground, what current will flow through the resistance? **Ans.:** 0.0465 A.

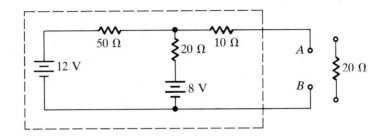

Fig. 2-19. Circuit in Probs. 2-1 and 2-2.

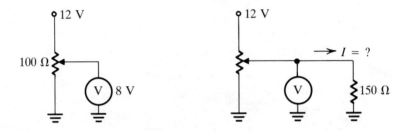

Fig. 2-20. Circuits in Prob. 2-3.

2-4. Using a combination of 1/4 W, 30 Ω resistors connected in series, parallel, or series-parallel, find the arrangement that will provide an equivalent resistance of 36 Ω capable of operating across a voltage difference of 15 V.

2-5. In the circuit shown in Fig. 2-8, the input voltage $V_1 - V_3$ is originally constant at 5 V and then at time $t = 0$ drops linearly to 0 V in 0.1 s. $V_1 - V_3$ then remains constant at zero thereafter. Using an expression for $V_1 - V_3$ of $5 - 50t$, the equation for $V_2 - V_3$ is

$$V_2 - V_3 = 5 - 50[t - RC(1 - e^{-t/RC})]$$

Sketch a graph of $V_1 - V_3$ and $V_2 - V_3$ as a function of time.

2-6. A diode and a 100 Ω resistor in series are supplied with 1 V. The diode has an I_c value of 12×10^{-12} A. What is the current that flows through the circuit? (Answer to check result: The voltage drop in the diode is 0.508 V.)

Problems

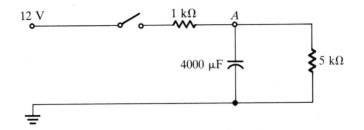

Fig. 2-21. Circuit in Problem 2-7.

2-7. In the circuit shown in Fig. 2-21 the capacitor is originally discharged, and then the switch is closed. How long does it take for the voltage at A to rise to 5 V?

2-8. In the rectifying circuit of Fig. 2-15 whose output is shown in Fig. 2-17, $V_B = 15\,\text{V}$, $C = 8000\,\mu\text{F}$, $R = 5\,\Omega$, and the ac supply frequency is 60 Hz. (a) Compute the time between B and C. (b) What is the voltage at C? **Ans.:** (a) 0.00683 s; (b) 12.65 V.

Chapter 3

Operational Amplifiers

3-1 Application of Operational Amplifiers

Operational amplifiers (op amps) provide output dc voltages that are related to the input dc voltages by an amplification factor. In combination with other electronic components they serve other purposes also, a few of which will be explained in this chapter. Measurement and control circuits use op amps widely, and the device is a building block of many circuits. The symbol of the op amp is shown in Fig. 3-1 with the inputs on the left and output on the right. Supply voltages of +12 and −12 V are shown leading to the op amp. Supply voltages are always required, and even though the symbol for the op amp may not show the supply voltages, it is assumed that they exist.

There are other types of amplifiers, and the op amp distinguishes itself from the power amplifier, for example, by its voltage and current range. The op amp usually operates with voltages between ±15 V and with currents less than perhaps 20 mA—values that are usually adequate for instrumentation and control.

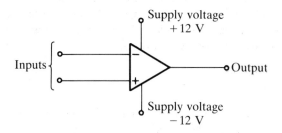

Fig. 3-1. An op amp.

3-2 Basic Characteristic of the Op Amp

In the connections to the op amp shown in Fig. 3-2a the + input is grounded and the − input is designated V_i. If V_i drops even a millivolt below the input voltage to the + connection, the output voltage V_o rises to the positive supply voltage. The −1 mV position is shown on Fig. 3-2b, but only to indicate that it takes a very small negative voltage, perhaps only a fraction of a millivolt, for the output to change to its maximum value. If V_i becomes slightly positive V_o falls sharply to the negative supply voltage.

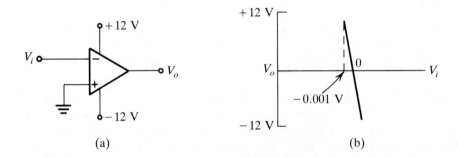

Fig. 3-2. Characteristic of an op amp: (a) connections, and (b) V_o vs. V_i.

3-3 Comparator

One application of the op amp is as a comparator where two signals, A and B, are checked and if A is higher than B a voltage of, for example, 5 V is passed on, otherwise zero voltage is transmitted. Figure 3-3 shows a possible circuit for a comparator. The connections through the diode to +5 V limits the voltage to 5 V, and more precisely to 5.6 V because of the 0.6 V needed for current to flow through the diode. The connection to ground through the diode prevents V_o from dropping below −0.6 V.

3-4 Inverting Amplifier

One of the frequently used connections for the op amp as an amplifier is as an inverting amplifier, shown in Fig. 3-4. Figure 3-2 indicates that a characteristic of the op amp is to increase V_o when there is even a slight reduction of the voltage at the inverting input (V_a in Fig. 3-4) compared to

3-4. Inverting Amplifier

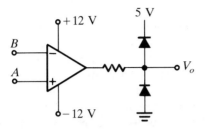

V_o, V	when
5	$A > B$
0	$A < B$

Fig. 3-3. A comparator.

the voltage at the non-inverting input (ground in Fig. 3-4). The op amp, therefore, adjusts V_o such that V_a is essentially the same as the + input. Another characteristic of the op amp is that very little current flows into or out of the op amp at the positive and negative input terminals (only currents of the order of several microamps). The current I that flows from V_o through R_1 to V_a is therefore essentially the same as that flowing through R_2 from V_a to V_i,

$$I = \frac{V_o - V_a}{R_1} = \frac{V_a - V_i}{R_2} \quad (3\text{-}1)$$

Since $V_a = 0$, Eq. (3-1) becomes

$$\frac{V_o}{-V_i} = \frac{R_1}{R_2} = \text{amplification ratio} \quad (3\text{-}2)$$

The amplification ratio (from the negative input V_i to the positive output V_o) is a ratio of the resistances, R_1/R_2.

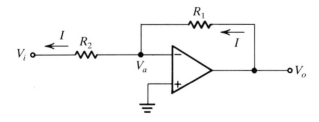

Fig. 3-4. An inverting amplifier.

3-5 Choice of Resistances

Equation (3-2) suggests that any combination of resistances R_1 and R_2 providing the same ratio of resistances would be equally acceptable. If $R_1 = 1000\,\Omega$ and $R_2 = 250\,\Omega$, for example, an input of -2 V for V_i would result in $V_o = 8$ V. The same amplification would result if $R_1 = 1\,\text{M}\Omega$ and $R_2 = 0.25\,\text{M}\Omega$. The latter choice of resistances may be preferable, however, because current conducts back from V_o to V_i, which in some cases may distort the source voltage V_i. It is usually good practice, then, to choose large resistances for R_1 and R_2, with a maximum of about $1\,\text{M}\Omega$.

3-6 Non-inverting Op Amp

The connection of the op amp in Fig. 3-4 providing the amplification of Eq. (3-2) results in a voltage inversion—a negative input to a positive output and vice versa. Many applications call for the amplification of a positive voltage to another positive voltage for an output. One possibility to meet this assignment is to use two op amps in series, with the first one providing the desired amplification ratio along with an inversion, and the second op amp providing an inversion with a 1:1 amplification. Another option is the non-inverting connection shown in Fig. 3-5. The op amp performing with the characteristics shown in Fig. 3-2b adjusts V_o so that V_a in Fig. 3-5 is essentially equal to V_i. If V_a starts to go negative relative to V_i, V_o increases. The current I is

$$I = \frac{V_o - V_a}{R_1} = \frac{V_o - V_i}{R_1}$$

and

$$I = \frac{V_a - 0}{R_2} = \frac{V_i - 0}{R_2}$$

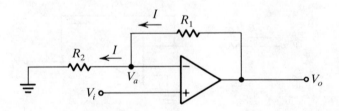

Fig. 3-5. Non-inverting op amp.

3-7. Buffer or Follower Amp

Equating the two expressions for I gives

$$\frac{V_o}{V_i} = \frac{R_1 + R_2}{R_2} \tag{3-3}$$

3-7 Buffer or Follower Amp

A special case of the non-inverting op amp is the one shown in Fig. 3-6. The connections in the circuits of Fig. 3-5 can be converted to Fig. 3-6 by making $R_1 = 0$ and $R_2 = \infty$, for which the amplification ratio from Eq. (3-3) reduces to unity. What is the purpose of a non-inverting amplifier with an amplification ratio of unity? The answer lies in the relative values of current inputs and outputs of the op amp. Suppose that the current drawn by a load is 10 mA, but the source is only capable of several microamps. A buffer amp can be placed between the source and the load to transmit the same voltage, but at a higher current.

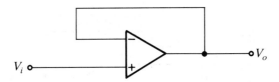

Fig. 3-6. Buffer or follower op amp.

3-8 Signal Conditioning

A frequent requirement in instrumentation and control is to convert a small voltage range to a larger one that extends upward from zero. Suppose, for example, that a pressure is being measured and the range of interest is 60 to 100 kPa. This particular pressure transducer converts those pressures into voltages that are 3.2 and 3.4 V, respectively, as shown in Fig. 3-7a. The overall conversion desired is that of spreading the range and adjusting the zero, as shown in Fig. 3-7b. The circuit shown in Fig. 3-8 will deliver a voltage between 0 and 3.4 V. The op amp is connected in a non-inverting mode, and the value of V_a is set at 3.4 V through the use of a voltage divider. To obtain 0 V output when the transducer output is 3.2 V requires the ratio $R_1/R_2 = 3.2/0.2 = 16$.

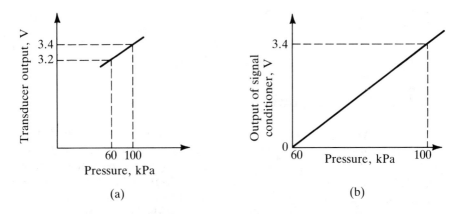

Fig. 3-7. Expanding the scale and adjusting the zero of (a) the output of a pressure transducer to (b) the desired conversion by means of a signal conditioner.

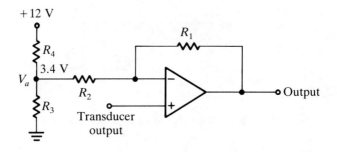

Fig. 3-8. Circuit to achieve signal conditioning indicated by Fig. 3-7.

3-9 Summing and Multiplying Amplifier

The op amp conventionally works as a multiplier, and it can function as a combination of multiplier and summer. Suppose that the input voltages are V_2 and V_3, and the output voltage V_1 is to be such that it satisfies the equation

$$V_1 = aV_2 + bV_3 \qquad (3\text{-}4)$$

where a and b are known constants. The circuit shown in Fig. 3-9 will perform the mathematical operation of Eq. (3-4) when the resistances are properly chosen. The negative value of V_1 is first obtained, and this voltage is passed through the inverting amplifier on the right. The currents I_2 and I_3 combine

3-9. Summing and Multiplying Amplifier

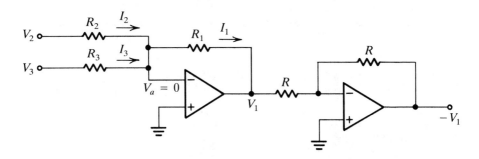

Fig. 3-9. Summing and multiplying amplifier.

to form I_1,

$$I_2 + I_3 = I_1$$

or

$$\frac{V_2 - 0}{R_2} + \frac{V_3 - 0}{R_3} = \frac{0 - V_1}{R_1}$$

so

$$-V_1 = \frac{R_1}{R_2}V_2 + \frac{R_1}{R_3}V_3 \tag{3-5}$$

so the resistances are chosen such that $R_1/R_2 = a$, and $R_1/R_3 = b$.

Example 3-1. During winter operation of many air-conditioning systems in large buildings, cold outdoor air is mixed with return air, as shown in Fig. 3-10, to provide ventilation air as well as to provide a source of cool air (perhaps at a temperature of mixed air t_m of 15°C) for removal of internal heat from the building.

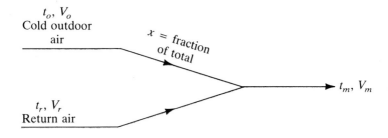

Fig. 3-10. Mixing of outdoor and return air.

At extremely low outdoor temperatures the fraction of outdoor air must be prevented from dropping lower than the minimum ventilation requirements.

In a given installation, the outdoor, return, and mixed air temperatures t_o, t_r, and t_m, respectively, are being sensed and transduced linearly to voltages V_o, V_r, and V_m, respectively. The temperature t_m is controlled at 15°C, and there is a desire to know when the controller has reduced the value of the fraction of cold air, x, less than 0.2.

Devise an op amp circuit to send a +5 V signal when x drops to 0.2; otherwise the signal is zero.

Solution. The energy balance at the junction in Fig. 3-10 is

$$xt_o + (1-x)t_r = t_m$$

which in terms of transduced voltages is

$$xV_o + (1-x)V_r = V_m$$

The regulation of x as the outdoor temperature changes is shown in Fig. 3-11. At outdoor temperatures above 15°C, the mixed air temperature of 15°C cannot be maintained. For a given combination of V_o and V_r, the V_m that occurs at $x = 0.2$, called $V_{0.2}$, is

$$V_{0.2} = 0.2V_o + 0.8V_r$$

If $V_{0.2}$ drops lower than the V_m being controlled, it indicates that x is less than 0.2.

The op amp circuit to provide the desired function is shown in Fig. 3-12 and consists of a summing-multiplying circuit that feeds into an inverter and then into a comparator.

3-10 Generalized Circuit for an Op Amp

A generalized op amp circuit of which most of the previously described circuits are special cases is shown in Fig. 3-13. The expression for the output voltage V_o is given by Eq. (3-6):

$$V_o = \frac{(R_1 + R_2)}{(R_3 + R_4)} \frac{R_4}{R_2} V_{i_2} - \frac{R_1}{R_2} V_{i_1} \qquad (3\text{-}6)$$

3-11. Integrator

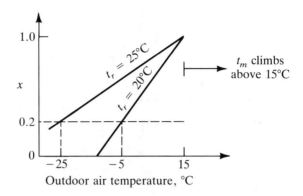

Fig. 3-11. Fraction of outdoor air needed to control a mixed air temperature of 15°C.

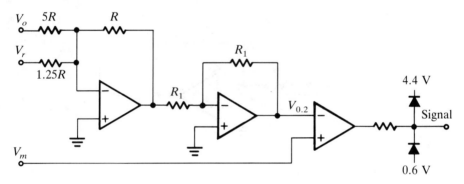

Fig. 3-12. Op amp circuit to give a 5 V signal if the fraction of outdoor air drops below 0.2.

3-11 Integrator

By utilizing a capacitor in the op amp circuit as illustrated in Fig. 3-14, an integrating circuit can be constructed. Two expressions for the current I are

$$I = \frac{V_a - V_i}{R} \quad \text{and} \quad I = C\frac{d(V_o - V_a)}{dt}$$

where C = capacitance, farads. Since $V_a = 0$,

$$\frac{dV_o}{dt} = -\frac{V_i}{RC} \tag{3-7}$$

As shown in Fig. 3-14b, the profile of V_o is a straight line with a slope of $-V_i/RC$, which is the integral of Eq. (3-7).

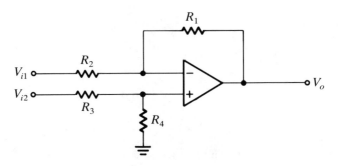

Fig. 3-13. Generalized op amp circuit.

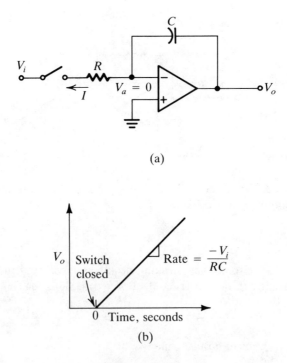

(a)

(b)

Fig. 3-14. (a) Integrating circuit, and (b) output voltage as a function of time.

3-12 Pin Diagram of 741 Op Amp

One general-purpose op amp bears the number SN72741 (Texas Instruments, Inc.) and has the eight-pin DIP (dual in-line package) configuration shown in Fig. 3-15a with the designation of pins illustrated in Fig. 3-15b. Pin 2 is the inverting input which is the negative input on the op amp symbol.

Pin 3 is the non-inverting input, and pin 6 is the output. The supply voltages are provided to pins 4 and 7, $-15\,\text{V}$ and $+15\,\text{V}$, for example. Pin 8 is not used and need not be connected.

Pins 1 and 5 are marked "offset null" and are often not connected but are used when precise amplification is required. The slight mismatch of the components that compose the op amp could result in an output that is non-zero when a zero input is imposed at the input. In such a case a potentiometer (perhaps $10\,\text{k}\Omega$) can be connected across the null offsets as shown in Fig. 3-16 and a negative voltage supplied to the center tap as needed in order to bring the output voltage to zero.

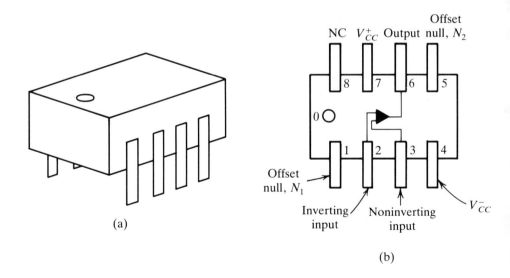

Fig. 3-15. (a) Sketch of 741 op amp, and (b) pin diagram.

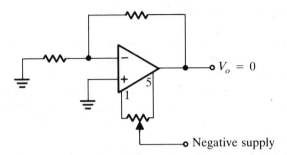

Fig. 3-16. Use of the null offsets.

3-13 Limitations and Ratings of the Op Amp

The 741 op amp has current and voltage characteristics such that its application is primarily limited to control and it is not used to provide power to motors, heaters, or other devices requiring high current. These limitations, as well as other rating information, are stated on the specification sheet of the op amp. Typical restrictions and characteristics are:

Voltages. Supply voltages should not exceed ±18 V. The output voltage will not be higher than positive supply voltage, V_{CC+}, nor lower than the negative supply, V_{CC-}. Input voltages should be between +15 and −15 V.

Power and current output. Continuous power dissipation is limited to 500 mW. Drawing too much current at the output of the op amp may result in a dropoff of the output voltage and/or the chip becoming hot and burning out. With a 30 V difference in the two power supplies, the 500 mW limitation corresponds to a current of 17 mA output current.

Input resistance. We stated in Sec. 3-4 that very little current flows into or out of the op amp at the input voltage terminals. The input resistance is 2 MΩ between the inverting and non-inverting input connections.

Dynamic response. The op amp responds rapidly, but not infinitely fast. If an ac voltage is applied at the input of the op amp, the response of the output voltage as a function of frequency is shown in Fig. 3-17. The peak-to-peak output voltage holds constant until the frequency of the input signal climbs above approximately 10 kHz.

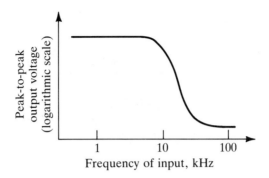

Fig. 3-17. Dynamic response of an op amp.

General References

1. W. G. Jung, *IC Op-amp Cookbook*, Howard W. Sams & Company, Indianapolis, IN, 1974.

2. B. D. Wedlock and J. K. Roberge, *Electronic Components and Measurements*, Prentice-Hall, Englewood Cliffs, NJ, 1969.

Problems

3-1. Show the circuit that utilizes only one op amp and designate the magnitude of resistances that performs the following (V_i, V_o) amplifications: $(-1, -4), (-1.5, -6)$.

3-2. Derive Eq. (3-7) for the generalized op amp circuit.

3-3. If $V_{\text{input}} = 4\,\text{V}$ in the circuit shown in Fig. 3-18, what is V_{output}?

3-4. The summing amp circuit shown in Fig. 3-19 is to provide the following voltage relationship:
$$-V_1 = 3V_2 + 2V_3 + 3$$
Determine the values of R_2, R_3, and V_{bias}.

3-5. In the differential amplifier shown in Fig. 3-20, determine the values of R_1 and R_2 such that $V_o = 0\,\text{V}$ when V_{i1} and V_{i2} are equal, and $V_o = 2\,\text{V}$ when $V_{i2} - V_{i1} = 0.2\,\text{V}$.

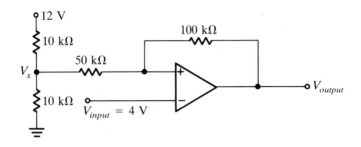

Fig. 3-18. Op amp circuit in Prob. 3-3.

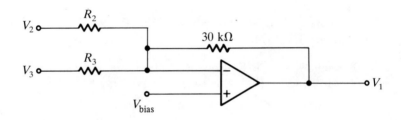

Fig. 3-19. Summing amp in Prob. 3-4.

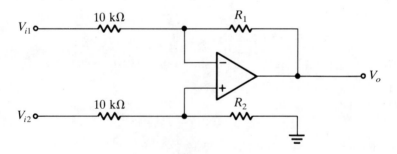

Fig. 3-20. Differential amplifier in Prob. 3-5.

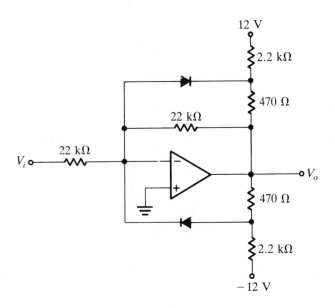

Fig. 3-21. Circuit in Prob. 3-6. Assume that the diodes permit full current flow when the voltage difference across them is 0.6 V.

3-6. For the circuit shown in Fig. 3-21, sketch V_o as V_i varies from -10 V to $+10$ V.

3-7. If the diode in the circuit of Fig. 3-22 for small forward-biased voltages has a current–voltage relationship of

$$I = k_1 e^{k_2(V_a - V_o)}$$

show that $V_o = c_1 + c_2 \ln(V_i)$.

3-8. There is a frequent need for a circuit that provides a time delay. An example is the need to bypass the low-pressure cutout of the oil pump during startup when the oil pump serves a compressor and is driven directly off the compressor shaft. Design a circuit using op amp(s) and other basic components to provide 5 V immediately upon closure of a switch and then a sudden drop to 0 V after 3 s. Specify the magnitudes of the components.

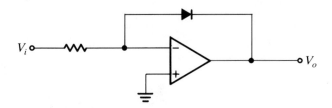

Fig. 3-22. Logarithmic converter in Prob. 3-7.

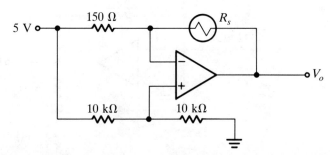

Fig. 3-23. Op amp circuit with a temperature sensor as one resistance in Prob. 3-9.

3-9. The op amp circuit in Fig. 3-23 converts a varying resistance of a temperature sensor into a corresponding voltage. The resistance of the sensor R_s is $100\,\Omega$ when its temperature is $0°C$ and $125\,\Omega$ when the temperature is $50°C$. What is V_o when $t = 0$ and when $t = 50°C$?

Chapter 4

Transistors

4-1 Impact of the Transistor

The invention of the transistor generated a revolution during the past several decades, not only in the technical world, but in social and personal lives as well. The transistor replaced the vacuum tube by virtue of its lower cost, smaller size, lower power requirement, cooler operation, and faster startup time. The transistor is now a common component in instrument and control circuits. In comparison to the op amp, which is fundamentally a voltage amplifier, the transistor is a current amplifier. Just as the op amp does not generate a higher voltage on its own but only controls the voltage, the transistor does not generate higher currents but only controls them. The op amp, while basically a voltage multiplier, is a building block in circuits that perform a wide variety of functions. In a similar manner the basic function of the transistor may be combined with other components in a circuit to achieve one of a number of assignments.

This chapter will concentrate on only a few applications of transistors—those directly applicable to instrumentation and control. These applications include transistor circuits that serve as switches, current amplifiers, and voltage amplifiers. The Zener diode will be explained and combined with a transistor in a circuit to develop a constant-current source and a constant-voltage source.

4-2 Symbols and Terminology

Two classes of transistors are the *npn* type and the *pnp* type. The symbol for the *npn* transistor is shown in Fig. 4-1a and derives its name from the arrangement of the *n*- and *p*-type semiconductor materials as shown in Fig. 4-1b.

Figures 4-2a and 4-2b are corresponding diagrams for the *pnp* transistor. The leads of a transistor are called the base, emitter, and collector, as shown in Fig. 4-3. The middle semiconductor material is the base, and the outer materials are the emitter and collector. In the transistor symbol the branch with the arrow is the emitter.

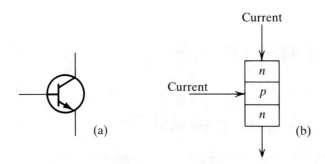

Fig. 4-1. An *npn* transistor. (a) The symbol, and (b) the arrangement of the semiconductor materials.

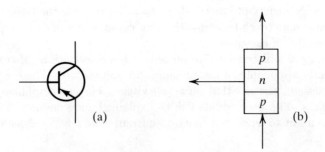

Fig. 4-2. A *pnp* transistor. (a) The symbol, and (b) the arrangement of the semiconductor materials.

4-3. Current Characteristics

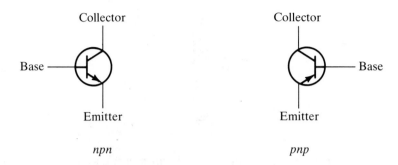

Fig. 4-3. The base, emitter, and collector of a transistor.

4-3 Current Characteristics

If a transistor, such as the *npn* one in Fig. 4-4, is connected with a voltage source applied to the collector, the relationship of the collector current I_c to the base current I_b is as shown in Fig. 4-5. When a small current is permitted to flow into the base, a much larger current is permitted to flow into the collector (I_c) and out the emitter (I_e). Once the voltage difference across the collector-emitter rises above a fraction of a volt, the collector current I_c is almost independent of the voltage. For a given value of V_{ce} there is a constant ratio of I_c to I_b,

$$I_c = \beta I_b \tag{4-1}$$

where β is a constant that could vary from perhaps 40 to 200 from one transistor model to another.

The transistor functions as a *current amplifier*. It does not generate the current I_c but only controls the current. There must be a voltage source in order for the current to flow.

Kirchhoff's current law applies around the transistor such that

$$I_e = I_b + I_c \tag{4-2}$$

but since I_b is small relative to I_c, I_e is almost equal to I_c.

The shape of the curves in Fig. 4-5 applies equally well to a *pnp* transistor, shown in Fig. 4-6, but the base current I_b and collector current I_c flow out of the transistor, and the voltage V_{ce} is reversed. Equation (4-2) continues to apply.

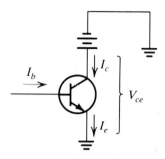

Fig. 4-4. Flow of current in an *npn* transistor.

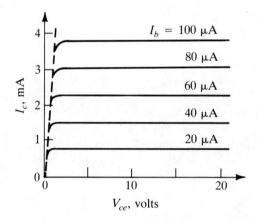

Fig. 4-5. Transistor characteristics.

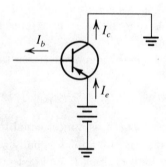

Fig. 4-6. Flow of current in a *pnp* transistor.

4-4 Bipolar-Junction and Field-Effect Transistors

Two major types of transistors are the bipolar-junction transistor (BJT) and the field-effect transistor (FET). The FET, like the BJT, is a three-terminal device, but the terminals are called the gate, source, and drain. One of the characteristics of the FET is that the controlling terminal, the gate, draws virtually no current. The source-drain current is controlled by a field produced by the applied gate voltage. While the BJT is predominantly a linear device as indicated by Eq. (4-1), the source-drain current in the FET bears a squared relationship to the gate voltage.[1] This chapter concentrates on the BJT, which is the common type used in discrete (not integrated) circuits.

4-5 Voltages at the Transistor Terminals

In the *npn* transistor of Fig. 4-3 the connection through the transistor from the base to the emitter is the same as in a diode. Thus, even a drop in base voltage below the emitter voltage does not result in current flowing out of the base. On the other hand, when the voltage at the base terminal rises 0.6 to 0.7 V above the voltage of the emitter terminal, current will flow into the base.

A comparable relationship of terminal voltages prevails for the *pnp* transistor of Fig. 4-6. As current flows into the transistor at the emitter, there is a voltage drop of 0.7 V from the emitter to the base. The diode behavior of the transistor prohibits current from flowing into the transistor at the base.

Further relationships of the voltages and currents in a transistor are illustrated in Fig. 4-7 for the *npn* type. As the base voltage V_b increases above the emitter voltage V_e, the base current I_b follows the diode characteristic as represented in the exponential equations, Eqs. (2-8) and (2-9). The collector current I_c increases as a multiple of I_b by the factor β. The collector voltage V_c is the same as the supply voltage, V_{CC}, when $I_c = 0$ and progressively drops as I_c increases. When the abrupt rise in I_c takes place, $V_c - V_e$ drops until it reaches its minimum value of 0.2 V. Further increases in $V_b - V_e$ will not increase I_c further, because 0.2 V is the minimum that can occur between the collector and emitter.

4-6 Voltage Amplifier

The transistors in Figs. 4-4 and 4-5 are current amplifiers, because a low control current I_b regulates much higher collector and emitter currents, I_c and I_e, respectively. The current-amplifying property of the transistor can be employed to form a voltage amplifier by insertion of several resistors, as in

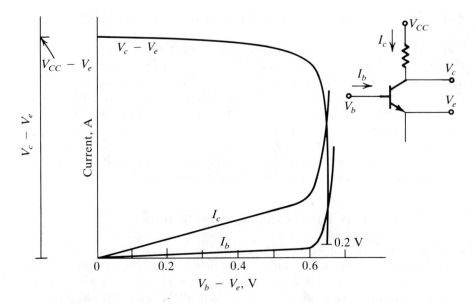

Fig. 4-7. Voltage and current relationships in an *npn* transistor.

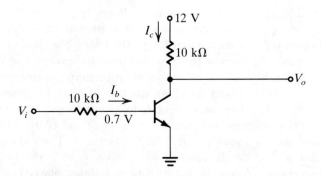

Fig. 4-8. A transistor as a voltage amplifier.

Fig. 4-8. A 10 kΩ resistor is connected in the line to the base, and another is connected between a 12 V source and the collector. The base current is

$$I_b = \frac{V_i - 0.7}{10,000}$$

and if β of the transistor is 100, for example,

$$I_c = 100\frac{V_i - 0.7}{10,000}$$

4-7. Transistor as a Switch

Also,
$$I_c = \frac{12 - V_o}{10,000}$$

For several values of V_i the corresponding values of V_o are shown in Table 4-1. When V_i changes from 0.79 to 0.81 V, V_o changes from 3 to 1 V. The transistor circuit in Fig. 4-8 is thus an inverting amplifier with a ratio of 100.

Table 4-1. Voltage Amplification of Transistor Circuit in Fig. 4-8

V_i, volts	I_b, μA	I_c, mA	V_o, volts
0.79	9	0.9	3
0.80	10	1.0	2
0.81	11	1.1	1

Example 4-1. If in the circuit of Fig. 4-8 the resistance in the line to the base is changed to 20 kΩ, what is V_o when $V_i = 0.8$ V?

Solution.
$$I_b = \frac{0.8 - 0.7}{20,000} = 5\,\mu\text{A}$$
$$I_c = \beta(5\,\mu\text{A}) = (100)(5) = 0.5\,\text{mA}$$
$$V_o = 12 - I_c(10,000\,\Omega) = 7\,\text{V}.$$

4-7 Transistor as a Switch, and Saturating the Transistor

If, in the circuit of Fig. 4-8, the value of V_i is increased further, Table 4-1 continues as follows:

V_i, volts	I_b, μA	I_c, mA	V_o, volts
0.82	12	1.2	0
0.83	13	1.3	−1

The condition where $V_o = -1$ V is not possible, because there is no voltage supply available below ground. Even the $V_o = 0$ V condition when $V_i = 0.82$ V is not possible because, as Fig. 4-7 shows, the drop of approximately 0.2 V is the minimum that can prevail when current flows between the collector and emitter.

Any magnitude of V_i greater than 0.82 V permits a free flow of current between the collector and emitter (except for the 0.2 V drop). The opposite to the free flow status is when V_i is less than 0.7 V and the current flow between the collector and emitter is blocked. These two conditions show the two states of a switch, as indicated in Fig. 4-9.

When $V_i = 0$ the switch is open, but it is closed when $V_i = 1$ V. At any V_i above 0.82 V the transistor is "saturated" and the equation $I_c = \beta I_b$ no longer holds. At $V_i = 1$ V, for example, $V_o = 0.2$ V, $I_c = 1.18$ mA, and $I_b = 30\,\mu$A, indicating β equals 39.3 and is thus no longer equal to 100.

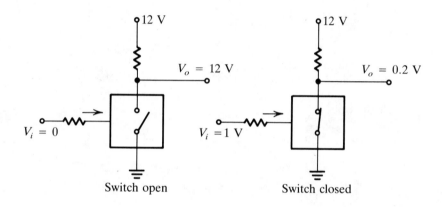

Fig. 4-9. A transistor used as a switch.

4-8 Common Emitter and Common Collector Circuits

One of the ways of connecting a transistor in a circuit, shown in Fig. 4-8, is called a common-emitter (CE) connection. This circuit is shown again in Fig. 4-10a along with a common-collector (CC) connection in Fig. 4-10b. The differences in the two circuits lie in the location of the output voltage tap and in whether the resistance is located in the collector or emitter line. Table 4-1 showed values of V_i slightly above 0.7 V for the common-emitter connection. With the common-collector arrangement V_i must be 0.7 V or more higher than V_o for current to flow. The common-collector arrangement is called an "emitter-follower" because V_o tracks V_i (except for the 0.7 V differential).

4-9. Zener Diode

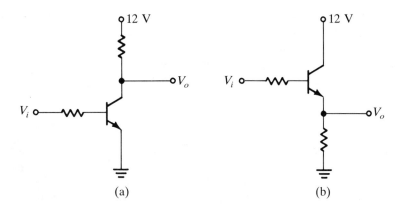

Fig. 4-10. (a) Common-emitter and (b) common-collector connections.

4-9 Zener Diode

A constant-current source is often needed in instrumentation circuits, and one possible circuit for this source will now be presented. A component of that constant-current source is a Zener diode. The Zener diode has characteristics similar to the ones discussed in Sec. 2-7, except that the voltage at which reverse current flows is sharply defined. Figure 4-11a shows the symbol of the Zener diode, and Fig. 4-11b the current-voltage characteristics of a Zener diode that has a nominal voltage V_D of 5 V for the flow of reverse current. Zener diodes are available in a variety of nominal voltages, for example, 2.8, 3.3, 3.9, ..., 9.1, 10.0, 11.0, ..., 75.0, 82.0, 90.0, etc.

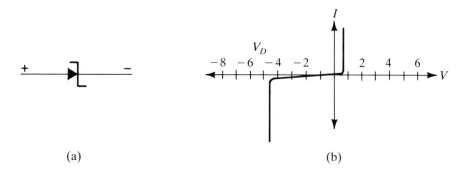

Fig. 4-11. A Zener diode: (a) symbol, and (b) current-voltage relationship.

Like most semiconductor devices, the Zener diode is temperature dependent such that V_D may vary by about 0.005 V per °C change in temperature. In some types of Zener diodes V_D may increase and in other types decrease with an increase in temperature. The availability of these two types of Zener diodes is an advantage, because the two can be combined into a circuit that is self-compensating for temperature.

4-10 Constant-Current Source

There are occasions when a current source is needed that maintains a constant current through a changeable resistance. Such a need will be described in Chapter 5 in discussing temperature measurement. The Zener diode can be incorporated in the transistor circuit of Fig. 4-12 to form such a source. The Zener diode maintains a constant voltage V_D, and thus the voltage at the emitter is $V_D - 0.7$. The current passing out the emitter through R_E to ground, I_e, thus remains fixed. Since $I_c = I_e - I_b$ and the change in I_b is small relative to the other currents, I_c is also constant, even if the load resistance R_L changes. If the circuit in Fig. 4-12 is stabilized at some operating condition and R_L decreases, the transistor resists the tendency of I_c to increase by closing off the collector-emitter path slightly to raise V_c.

Often one of the load terminals is grounded, so a *pnp* transistor must be used as in Fig. 4-13. The emitter voltage is $12 - V_D + 0.7$ V and I_e is constant at $(V_D - 0.7)/R_E$, so the current through the load R_L remains essentially constant.

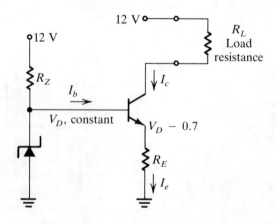

Fig. 4-12. Constant-current source.

4-11. Designing a Constant-Current Source

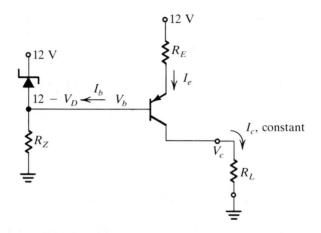

Fig. 4-13. A constant-current source when one of the load terminals must be grounded.

4-11 Designing a Constant-Current Source

The specifications for a constant-current source are usually (1) the desired current, (2) the maximum resistance of the load, and (3) the supply voltage. The choices to be made are V_D through the selection of the Zener diode, R_E, and R_Z. If in the circuit of Fig. 4-13 R_L is zero, V_c will be zero and still provide the specified current. As R_L increases V_c must increase proportionally, which it can do until the transistor becomes saturated. At the saturated condition the difference between V_e and V_c is 0.2 V, so the highest possible value of V_c is $V_e - 0.2 = V_b + 0.7 - 0.2 = 12.5 - V_D$. The upper limit of R_L can be increased by choosing a Zener diode with a lower V_D and then reducing R_E to achieve the desired current. The limitation of this progression is that when R_E becomes small, slight perturbations in V_D cause large percentage changes in the current through R_E and thus R_L as well.

Example 4-2. Design a constant-current source to supply 1 mA to a load that varies from 0 Ω to 5 kΩ. Power the current source with a 12 V supply, and use the circuit shown in Fig. 4-13.

Solution. The maximum voltage across the load is equal to $(5000\,\Omega)(0.001\,\text{A}) = 5\,\text{V}$. The range of V_C is thus between 0 and 5 V. The voltage difference $V_E - V_C$ ranges from 0.2 V upward, so V_E must be above 5.2 V. Since

$$V_E = V_b + 0.7$$

V_b must be above 4.5 V. If the nominal voltage of the Zener diode is chosen as 5.0 V, then $V_b = 7$ V and $V_E = 7.7$ V, which allows V_C to range between 0 and 5 (actually up to 7.5 V). Next, select the resistances: $R_E = (12 - 7.7\,\text{V})/(0.001\,\text{A}) = 4.3\,\text{k}\Omega$, and $R_Z = 700\,\Omega$ to draw a nominal 10 mA through the Zener diode. Too little current results in a fluctuation of V_D, and too much current causes excessive power dissipation and possible overheating of the diode.

4-12 Operating Limits of a Transistor

It is reasonable to assume that the voltage and current applied to a transistor cannot be increased indefinitely. If the voltage between the collector and emitter is increased beyond the range shown in Fig. 4-5, a breakdown occurs, resulting in high collector and emitter currents even though the base current is small. The distorted characteristic curves are shown in Fig. 4-14. Most transistors have a breakdown voltage in excess of 30 V.

The power dissipated in the transistor is the product of the current and the difference in voltage across the collector-emitter. A transistor with a maximum rated power dissipation of 0.5 W has the permitted operating region bounded by a hyperbola, as shown in Fig. 4-15.

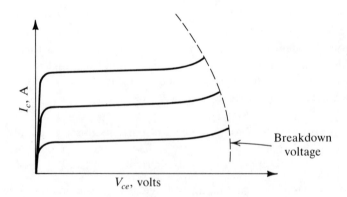

Fig. 4-14. Breakdown voltage.

4-13. Transistor Packages

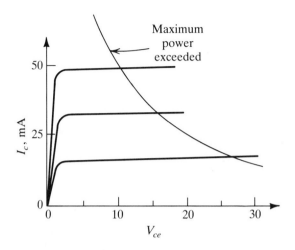

Fig. 4-15. Maximum power dissipation.

4-13 Transistor Packages

Two of the frequently used transistor packages are shown in Fig. 4-16. The package with the flat shape in Fig. 4-16a might be a TO-3 and is the usual package for high current and/or voltage. The one in Fig. 4-16b may be a TO-5 to TO-72. The designation of the pins is also shown in Fig. 4-16, but the specification sheet for the transistor being used should be consulted.

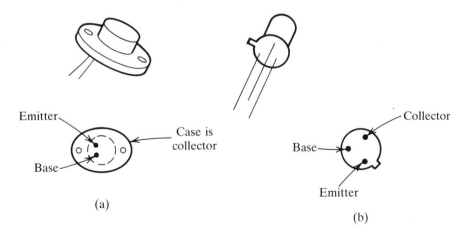

Fig. 4-16. Transistor packages.

References

1. P. Horowitz and W. Hill, *The Art of Electronics*, Cambridge University Press, Cambridge and New York, 1980.

Problems

4-1. When $V_i = 0.9\,\text{V}$ the voltage amplifier circuit shown in Fig. 4-17 provides $V_o = 6.0\,\text{V}$. What is V_o when $V_i = 0.92\,\text{V}$? **Ans.:** $5.4\,\text{V}$.

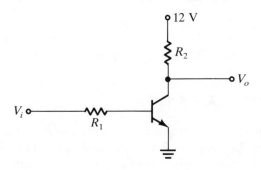

Fig. 4-17. Voltage amplifier in Prob. 4-1.

4-2. The *pnp* transistor with the common emitter in Fig. 4-18 has a value of $\beta = 80$, the sink voltage is $-12\,\text{V}$, and $R_L = 500\,\Omega$. If the power dissipated in R_L is to be $32\,\text{mW}$, what should be the resistance R_b? **Ans.:** $113\,\text{k}\Omega$.

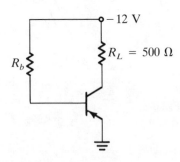

Fig. 4-18. Common-emitter transistor in Prob. 4-2.

Problems

4-3. A Zener diode is installed in the dc power supply shown in Fig. 4-19 in order to limit the output voltage to 5.1 V.

At point A in the circuit the voltage follows the pattern of Fig. 2-17, having a peak of 8 V and a time average of 7.2 V. The Zener diode has a V_D value of 5.1 V, can tolerate a maximum current of 70 mA, and can dissipate up to 1 W. What is the resistance R_1 to protect this Zener diode at the most demanding case of infinite load resistance? **Ans.:** 41 Ω.

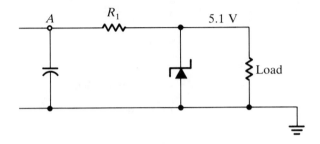

Fig. 4-19. Zener diode as voltage limiter in Prob. 4-3.

4-4. An *npn* transistor with a β of 80 is connected in the circuit shown in Fig. 4-20. At what value of I_b does the transistor experience the greatest power dissipation? **Ans.:** 150 μA.

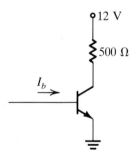

Fig. 4-20. Transistor circuit in Prob. 4-4.

4-5. The constant-current source shown in Fig. 4-21 incorporates a Zener diode that maintains 5 V across itself. (a) What is the magnitude of the constant current, and (b) what is the maximum R_{load} for which the device can maintain the constant current? **Ans.:** (b) 2616 Ω.

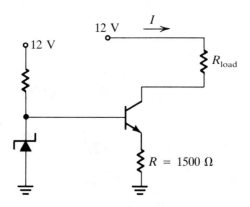

Fig. 4-21. Constant-current source in Prob. 4-5.

4-6. Design a constant-current source conforming to the circuit of Fig. 4-12 to maintain a constant current I_c of 2 mA through the load that may vary from 0 to 2 kΩ. (a) Specifically select R_E and V_D when the voltage source is 12 V. (b) If $V_D = 6$ V, compute the power dissipated in the transistor when the load is 500 Ω. **Ans.:** (b) 11.4 mW.

Chapter 5

Transducers

5-1 Importance of Good Instrumentation

This chapter focuses attention on instrumentation, including measurement of temperatures, pressures, velocities, flow rates, motion, position, electric current, and some other variables. The instrumentation portion of a computer control system must be of high quality. There is no purpose served in choosing an elaborate computer to control an expensive mechanical or thermal system if the instrumentation is faulty. Perhaps instrumentation does not excite many engineers, but there can be no shortcut around careful choices and proper applications of instruments.

Not only should the instrumentation be of good quality, but the computer control system introduces an additional specification. In almost every case the measured variable must be translated into a *voltage*. Thus, a visual reading of a thermometer or a pressure gauge is not adequate, because the temperature or pressure must be converted into a corresponding voltage that in turn translates to a digital signal. Transducers are therefore an integral part of a computer control system.

The previous statement about the inadequacy of a visual reading should be properly interpreted. Installation of a thermometer well or a pressure gauge for duplicating a temperature or pressure transducer is highly recommended, because the transducer and other components in the circuit have a way of going out of calibration. Providing a visual indication of the magnitude of the variable for calibration for periodic checks of the automated instrumentation should be a standard practice.

This chapter first concentrates on three important types of temperature sensors—thermocouples, resistance temperature devices, and integrated circuits. Those three types of sensors are widely used, although new concepts are continually emerging as well.

5-2 Thermocouples

The thermocouple circuit, as shown in Fig. 5-1, combines the function of a sensor and transducer, because as the sensed temperature changes the voltage output changes. Two dissimilar metals—copper and constantan in Fig. 5-1—are connected such that one junction senses the temperature and the temperature of the other junction is held at some reference value. The voltage developed is almost linearly proportional to the difference in temperature between the sensing and reference junctions. The characteristics of three of the dozen or more available thermocouple pairs are shown in Table 5-1.

Table 5-1. Characteristics of Several Thermocouple Pairs

Type	Symbol	Maximum Usable Temperature, °C	mV/°C (approximate)
Copper-constantan	T	400	0.0428
Iron-constantan	J	900	0.0527
Chromel-alumel	K	1250	0.0410

The voltage developed is of the order of 0.05 mV per °C temperature difference, so a 20°C temperature difference between the junctions would, for example, develop 1 mV. The desired magnitude of the voltage needed for conversion to a digital signal is of the order of 0 to 10 V, so an amplification of more than 10,000 is required. A multistage amplifier is required with components and a circuit that resists drift at these high amplification ratios.

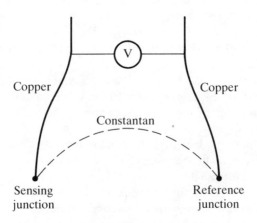

Fig. 5-1. A thermocouple circuit.

5-3 Thermocouple Reference Junction

The traditional technique for applying thermocouples has been to place the reference junction of Fig. 5-1 in an ice bath. For unattended computer control this approach is unsatisfactory, so automatic-compensating reference junctions are used. Two of the several approaches to reference junction compensation are: (1) using a small automatic ice generator[1] and (2) applying a voltage corrector that utilizes a resistance-sensitive temperature sensor.[1,2] The circuit of one type of compensator is shown in Fig. 5-2, where one leg of a bridge is a temperature-sensitive resistance R_T that is thermally integrated with the reference junction. If the reference temperature is to be 0°C, for example, the bridge is adjusted so that there is zero voltage between A and B when the reference junction and R_T are at 0°C. At reference junction temperatures other than 0°C the bridge provides a compensating voltage for the deviation of R_T from 0°C.

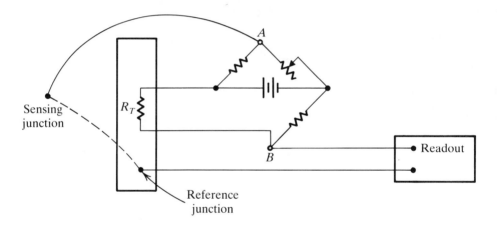

Fig. 5-2. Reference junction compensator for a thermocouple circuit.

5-4 Metal and Thermistor Resistance-Temperature Devices

While the thermocouple circuit accomplishes a direct translation of temperature to voltage, another class of sensors called resistance-temperature devices, RTDs, converts a temperature change into a resistance change that in turn is translated into a change in voltage. Two types of RTDs are thermistors, which are formed from semiconductor materials, and metals. Figure 5-3 shows relative resistances as a function of temperature for several materials.[3,4] The metals, tungsten and platinum, experience an increase in resistance, while on the other hand the resistance of the thermistor drops as the temperature increases.

The percentage change in resistance per °C change in temperature is higher for the thermistor than for metals, but the resistance-to-temperature variation is much closer to linear for the metal than for the thermistor.

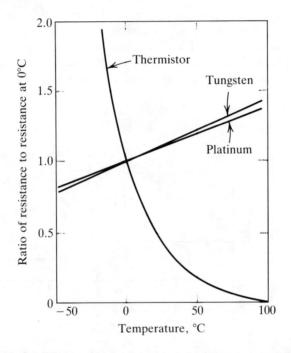

Fig. 5-3. Ratios of resistances to the resistance at 0°C.

5-5 Series Circuit

The next several sections explore how the change in resistance of the RTD is converted to a change in voltage. Three techniques that will be described are (1) series circuit, (2) bridge circuit, and (3) three- and four-wire constant-current arrangements. In the series circuit, Fig. 5-4, a constant voltage is applied between A and C to the two resistances in series—the sensor resis-

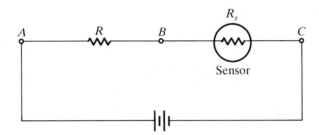

Fig. 5-4. A series circuit for converting a change in resistance to a change in voltage.

tance R_s and another resistance R. After the sensor has been selected, several additional decisions to be made are: (1) the magnitude of the applied voltage, (2) the magnitude of the resistance R, and (3) whether the voltage should be sensed from A to B or from B to C. Some of the factors that influence the above decisions are the I^2R heating in the sensor, which distorts the temperature, and the linearity of the temperature-to-voltage relation. A qualitative approach to these decisions is that the I^2R heating in the sensor should be kept low, so the applied voltage should be low and/or the resistance R kept high relative to R_s. With either option the need for amplifying the voltage or the change in voltage is likely. If the low voltage and high R approach is used, the magnitude of change in either V_{AB} or V_{BC} will be low. The usual choice is to make the applied voltage high (of the order of 5 to 15 V) and use a high R relative to R_s. The final decision is whether to sense voltage V_{AB} or V_{BC}. When R is high relative to R_s, the percentage change in the voltage across the sensor, V_{BC}, will be greater than that across the resistance R. The calculations in Prob. 5-2 illustrate some of the magnitudes of voltage changes and I^2R heating. Example 5-1 investigates the influence of the non-linearity of temperature versus resistance for a thermistor sensor.

Example 5-1. In the series circuit of Fig. 5-4 the resistance R is 5 kΩ, the applied voltage is 10 V, and the resistance of the thermistor at three different temperatures are 1120 Ω at 0°C, 560 Ω at 20°C, and 280 Ω at 40°C. (a) What are the values of V_{BC} at the foregoing three temperatures, (b) what is the I^2R heating in the sensor at 0, 20, and 40°C, and (c) what are the average values of $\Delta V/\Delta$°C between 0 and 20°C and between 20 and 40°C?

Solution.

t, °C	R_s, Ω	I, mA	I^2R_s, mW	V_{BC}, V
0	1120	1.634	2.99	1.830
20	560	1.799	1.811	1.007
40	280	1.894	1.004	0.530

Average $\Delta V/\Delta$ °C between 0 and 20°C
$= (1.830 - 1.007)/20 = 0.0412$
Average $\Delta V/\Delta$°C between 20 and 40°C
$= (1.007 - 0.53)/20 = 0.0239$

The choice of the magnitude of R, for a given set of resistances R_s, has several conflicting influences on the magnitude of $\Delta V/\Delta$°C and the I^2R heating in the sensor. A high value of R decreases $\Delta V/\Delta$°C, which is undesirable, but decreases the I^2R heating, which is desirable. One maker of thermistors states that for its sensors 1 mW of I^2R heating results in a 1°C increase in temperature for the thermistor in still air.

In Example 5-1 the average $\Delta V/\Delta$°C between 0 and 20°C was much greater than that between 20 and 40°C. Had R been chosen as 570 Ω, however, the average $\Delta V/\Delta$°C for both temperature ranges would have been equal, a technique that will be illustrated in Example 5-2. This lowered value of R would have increased the I^2R heating, which perhaps could be compensated for by reducing the imposed voltage from 10 V.

The non-linearity of the resistance-temperature characteristics of the thermistor may at first seem to be a drawback, but by making a proper choice of R in Fig. 5-4, a linear relationship of V_{AB} to temperature for three different points in the range of interest can be developed.[5] The voltage-temperature curve will have a slight S shape, so there is a small deviation from linearity between the points.

Example 5-2. Choose the resistance R in the series circuit of Fig. 5-4 so that V_{AB} is linear with temperature at the three points for which the thermistor resistance is given in Table 5-2. V_{AC} is constant.

Solution. To provide linearity of the three points, the difference in V_{AB} from 0°C to 15°C must equal the difference from 15°C to 30°C. Thus,

$$V_{AC}\left(\frac{R}{R+7855} - \frac{R}{R+16325}\right) = V_{AC}\left(\frac{R}{R+4028} - \frac{R}{R+7855}\right)$$

Table 5-2. Thermistor Resistance

t, °C	R_s, Ω
0	16,325
15	7,855
30	4,028

which yields $R = 6108\,\Omega$. With that resistance the V_{AB} voltages expressed as fractions of V_{AC} are 0.272, 0.437, and 0.603 at 0, 15, and 30°C, respectively.

5-6 Bridge Circuits

An alternative arrangement to the series circuit of Fig. 5-4 for converting a change in resistance to a change in voltage is the bridge circuit of Fig. 5-5. The RTD having a resistance R_s forms one leg of the bridge, and the adjacent legs are adjustable resistances. An advantage of the bridge circuit is that by proper adjustment of R_1 and R_2 the voltage output V_o can be zeroed for one specified value of R_s, and another specified value of V_o (within limits) can be obtained for another certain value of R_s.

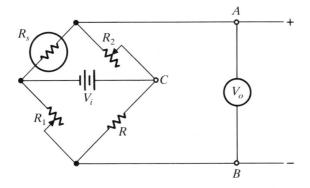

Fig. 5-5. Bridge circuit.

Example 5-3. The value of R_s for a certain platinum RTD is 800 Ω at 0°C and 980 Ω at 50°C. The fixed resistance R in the bridge of Fig. 5-5 is 1 kΩ. The applied voltage $V_i = 15$ V. Determine R_1 and R_2 such that V_o is 0 when the sensor temperature is 0°C and 0.5 V when the temperature is 50°C.

Solution. At 0°C when $R_s = 800\,\Omega$ and $V_o = 0$,

$$0 = (15\text{ V})\left(\frac{800}{800 + R_2} - \frac{R_1}{R_1 + 1000}\right) \quad (5\text{-}1)$$

At 50°C when $R_s = 980\,\Omega$ and $V_o = 0.5$,

$$0.5\text{ V} = (15\text{ V})\left(\frac{980}{980 + R_2} - \frac{R_1}{R_1 + 1000}\right) \quad (5\text{-}2)$$

Subtracting Eq. (5-1) from Eq. (5-2) gives

$$\frac{0.5}{15} = \frac{980}{980 + R_2} - \frac{800}{800 + R_2}$$

which is a quadratic equation satisfied by

$$R_2 = 231.4\,\Omega \text{ and } 3388.6\,\Omega$$

The corresponding R_1 values are

$$R_1 = 3458\,\Omega \text{ and } 236.1\,\Omega, \text{ respectively.}$$

The same concern about $I^2 R$ heating of the sensing element prevails in the bridge circuit as was true in the series circuit. Of the two (R_1-R_2) resistance combinations in Example 5-3, the preferred choice would be 236.1-3388.6 Ω, because then the high R_2 resistance would be in series with the sensor, such that low current would pass through the sensor.

In Example 5-3 the output voltage of 0.5 V was specified for a sensor temperature of 50°C. What would have happened if 1 V had been requested at that sensor temperature? The answer is that the equations for R_1 and R_2 would have developed complex numbers, indicating a physical impossibility. A consequence of Eq. (5-1) is that to achieve 0 V at 0°C, $R_2 = 800{,}000/R_1$. When this expression for R_2 is substituted into the equation for V_o [of the form of Eq. (5-2)], V_o at 50°C is

$$V_o \text{ at } 50°\text{C} = (15\text{ V})\left(\frac{980}{980 + 800{,}000/R_1} - \frac{R_1}{R_1 + 1000}\right) \quad (5\text{-}3)$$

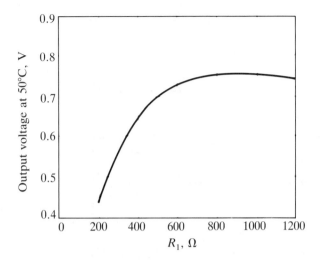

Fig. 5-6. Output voltage of bridge in Example 5-3 at a sensor temperature of 50°C as a function of R_1.

When the V_o at 50°C is plotted in Fig. 5-6 against R_1, a maximum appears, and the location of this maximum can be determined by differentiation. The optimum R_1 is 903.5 Ω, yielding a maximum V_o of 0.7604 V. The value of R_2 at this condition is 885.4 Ω.

5-7 Amplification of a Bridge Output

Section 3-8 discussed using an op amp circuit to zero and amplify a small voltage range. The example cited in Fig. 3-7 was to translate a 3.2 to 3.4 V span to a 0 to 3.4 V output. The amplification of the bridge circuit in Fig. 5-5 poses a special problem since V_o represents a voltage difference that changes with respect to ground voltage as R_s changes. In Example 5-3 $V_A - V_B$ in Fig. 5-5 varies from 0 to 0.5 V as the sensor temperature changes from 0 to 50°C, but V_A does not change from 0 to 0.5 V because V_B changes as R_s changes. One side of the power supply in Fig. 5-5 is usually grounded, such as at point C. An amplification circuit such as the one shown in Fig. 5-7 provides an output voltage proportional to the difference of input voltages,

$$V_o = k(V_B - V_A) \qquad (5\text{-}4)$$

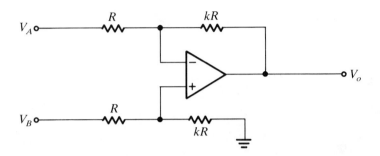

Fig. 5-7. Amplification circuit for a bridge output.

5-8 RTD Circuits Supplied with Constant Current

The circuit in Fig. 5-4 places a resistance in series with the sensor and supplies the combination with a constant voltage. The bridge circuit of Fig. 5-5 is also powered with a constant voltage. Another widely used concept is to measure the voltage drop across the sensor when it is supplied with a constant current. Section 4-10 showed the circuit for a rudimentary constant-current source, and sources that hold a constant current quite accurately are commercially available. Several approaches to sensing the voltage are illustrated in Fig. 5-8 as the two-wire, three-wire, and four-wire circuits.[6] In all cases a constant current is supplied, and the voltage indicated by a high-impedance voltmeter. The two-wire circuit in Fig. 5-8a is the least expensive but has the disadvantage of reading the voltage drop through the lead wires as well as through the sensor. On the other end of the spectrum is the four-wire circuit in Fig. 5-8c where only the voltage drop through the sensor is measured. A compromise between the two concepts is the three-wire circuit where the three leads become a part of a bridge circuit. If the resistances of the lead wires are identical, these resistances cancel out.

One of the techniques programmed into some computer controllers is to apply the constant current for only a brief time (several milliseconds, for example) and measure the voltage during that short time span. This technique avoids most of the I^2R heating occurring with continuous current that could distort the sensed temperature.

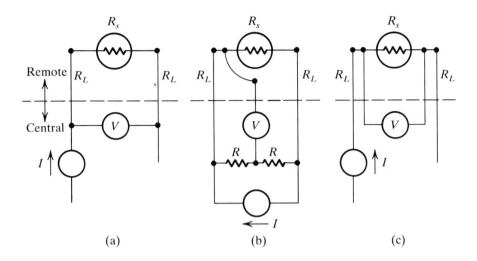

Fig. 5-8. Constant-current source with voltage drop through the sensor measured in (a) a two-wire, (b) a three-wire, and (c) a four-wire circuit.

5-9 Temperature-Dependent Integrated Circuits

Recent entries into the transducer market are temperature-sensitive integrated circuits—one class that provides a voltage proportional to the temperature (10 mV/K, for example), and another class that permits a current to flow that is proportional to the temperature. A voltage is supplied to the two leads of the current-regulating transducer, and the current in μA permitted to pass is equal to the absolute temperature of the sensor, as shown in Fig. 5-9. The nominal voltage supplied to the transducer is 5 V, but the device is not sensitive to variations in the supply voltage. If the supply voltage were 10 V rather than 5 V, an error of about 1 μA would result—thus an error of 1°C. The current can be transduced to a voltage by sensing the voltage drop across a fixed resistance in the supply line, as shown in Fig. 5-10a. Two of the many possible combinations of these transducers are shown in Figs. 5-10b and 5-10c.

5-10 Application of Sensors—Liquid Temperature

Four of the various arrangements for measuring the temperature of a liquid flowing through a pipe or tube are: (1) direct insertion of the sensor, (2) a probe, (3) a well, and (4) bonding the sensor to the surface of the tube. These methods are illustrated in Figs. 5-11a through 5-11d. Errors in measurement

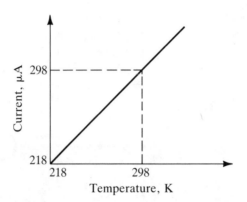

Fig. 5-9. Current-temperature characteristics of an integrated circuit temperature transducer (Analog Devices, Inc.).

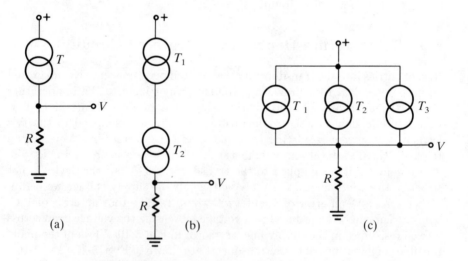

Fig. 5-10. Temperature-dependent integrated circuits with a current output connected to indicate (a) a single temperature, (b) the minimum of two temperatures, and (c) the average of three temperatures (when the current is divided by 3).

5-10. Application of Sensors—Liquid Temperature

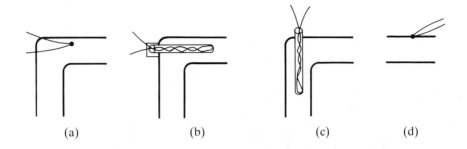

Fig. 5-11. Methods of applying sensors to measure the temperature of liquids: (a) direct insertion, (b) probe, (c) well, and (d) surface temperature measurement.

are least likely to result when the temperature profile of the flowing fluid is uniform and this temperature is the same as the tube. The direct insertion method of Fig. 5-11a has the advantage of no heat transfer barrier between the fluid and the sensor, but sometimes the seal is prone to leak. Also, the sensor should be supported so that it remains at or near the center of the stream.

One of the popular methods is the use of a manufactured probe where the sensor is bonded to the inside end of the probe. Mechanical seals are available that permit convenient removal of the probe.

A well, as in Fig. 5-11c, has been a traditional receptacle for thermometers. The well is mounted vertically, and a conducting fluid that will not evaporate, such as oil, is added after inserting the thermometer. Thermocouples can be inserted in wells, but the junction of the thermocouple should be in good thermal contact with the tip of the well.

A very simple procedure for measuring the temperature of liquid flowing through a pipe is to bond a sensor to the outside surface of the pipe. Of the methods mentioned so far, this technique is probably the one most fraught with error, since the pipe may not be at the same temperature as the fluid. Reasons for the discrepancy are convection between the ambient air and the pipe, conduction through the pipe to the pipe supports, and conduction through the lead wires of the sensor. Several steps that improve the accuracy are to run the lead wires in thermal contact with the pipe and to insulate the pipe and lead wires for some distance upstream and downstream from the point of measurement.

The frequently encountered situation of the tube being at a different temperature than the liquid is one of the basic sources of error in measuring the temperature of liquids. A probe may be considered a fin, as in Fig. 5-12, where

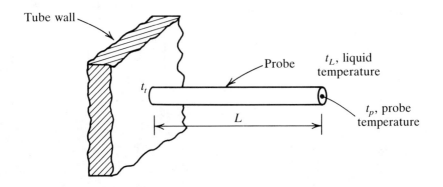

Fig. 5-12. Heat transfer in a temperature probe treated as a circular fin.

the root temperature is that of the tube. The probe reads the tip temperature of the fin, which is somewhere between that of the liquid and the tube. The equation for the tip temperature of a protruding rod[7] is

$$t_p - t_L = (t_t - t_L)\left(\frac{1}{\cosh mL + (h/mk)\sinh mL}\right) \qquad (5\text{-}5)$$

where

t_p = temperature of probe tip, °C
t_L = liquid temperature, °C
t_t = temperature of tube, °C
L = length of tube, m
k = thermal conductivity of probe metal, W/(m-K)
h = convection coefficient (liquid to probe), W/(m²-K)
m = $\sqrt{hP/kA}$ m^{-1}
P = perimeter of probe, m
A = cross-sectional area of probe metal, m²

Example 5-4. What is the probe temperature when $t_L = 80°\text{C}$ and $t_t = 50°\text{C}$ in a fluid where $h = 120$ W/(m²-K) for a probe 150 mm long, having a perimeter of 40 mm, a cross-sectional area of 0.00009 m², and a conductivity of 105 W/(m-K)?

Solution.

$$m = \sqrt{(120)(0.040)/(105)(0.00009)} = 22.54\,\text{m}^{-1}$$
$$mL = (22.54)(0.150) = 3.381$$
$$\cosh mL = 14.72$$
$$\sinh mL = 14.68$$
$$h/mk = 120/((22.54)(105)) = 0.0507$$
$$t = 80 + (50-80)\left(\frac{1}{14.72 + (0.0507)(14.68)}\right)$$
$$= 78.1°\text{C}$$

so conduction through the probe results in an error of $1.9°$C.

5-11 Application of Sensors—Temperature of Air and Other Gases

The conduction through the probe discussed in the previous section is also a concern when measuring the temperature of air or other gases flowing in a duct or tube. In addition, three more sources of error must also be considered: compressibility, radiation, and stratification. In high-velocity flow of a compressible fluid the temperature of the flowing gas rises to the stagnation temperature when the gas is brought to rest at the temperature sensor. This rise in temperature is normally neglected at low velocities (for example, air velocities less than 50 m/s).

A serious source of error in the measurement of gas temperatures is radiation to surrounding surfaces when these surfaces are at a different temperature than the gas. The sensor, as in Fig. 5-13, "sees" surfaces and exchanges thermal radiation with them.

Under equilibrium conditions a balance of heat flow prevails such that the rate of heat transfer at the sensor by radiation to (from) the surroundings equals the rate by convection from (to) the gas. The rate of heat transfer by radiation from the sensor to the surroundings is given by the Stefan-Boltzmann equation

$$q_{\text{rad}} = A_s \sigma F_e (T_s^4 - T_t^4) \tag{5-6}$$

where

q_{rad} = rate of heat transfer by radiation, W
A_s = area of sensor, m^2
σ = Stefan-Boltzmann constant, 56.7×10^{-9} W/(m^2-K^4)

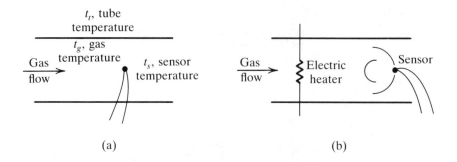

Fig. 5-13. Radiation affecting the measurement of a gas temperature: (a) sensor exposed to tube temperature different than gas temperature, and (b) a radiation shield to reduce radiation from an electric heater.

F_e = radiant heat-transfer factor, which is a function of the geometry and the emissivities of the sensor and the surroundings, dimensionless

T_s and T_t = absolute temperatures of sensor and tube, respectively, K

Example 5-5. Air at a temperature of 90°C flows through a tube that has a temperature of 70°C. What will be the sensor temperature if the convection coefficient between the air and the sensor is 45 W/(m²-K) and the sensor is considered completely enclosed by the tube for which case F_e equals the emissivity of the sensor, which is 0.7?

Solution. The two expressions for rates of heat transfer are
Radiation: $q_{rad} = A_s(56.7 \times 10^{-9})(0.7)(T_s^4 - 343.15^4)$
Convection: $q_{con} = A_s(45)(363.15 - T_s)$

Equating q_{rad} and q_{con} and solving the non-linear equation for T_s yields
$$T_s = 360.48 \text{ K} \quad \text{or} \quad t_s = 87.3°C$$

Control engineers in the air-conditioning field are afflicted with a problem of stratification where the temperature of air varies several °C over the cross section of the duct. Thus the point at which the temperature sensor is located may not be representative of the mean temperature of the air. The condition of stratification is particularly pronounced immediately after mixing streams of different temperatures; this is illustrated in Fig. 5-14, where cold outdoor ventilation air mixes with warm recirculated air. Instead of the mixture assuming a uniform temperature, cold air will migrate to the lower portion of the mixed-air duct and warm air to the upper section.

Some approaches to correcting stratification include (1) rearranging the air entrances so that the cold air enters at the top of the duct rather than at the bottom as it does in Fig. 5-14, (2) locating the temperature sensor following a fan, and (3) placing in the air stream a coil through which water or antifreeze is recirculated. One type of temperature sensor available for pneumatic control systems consists of a sensing element 1 to 2 m long that is distributed over the cross section of the duct.

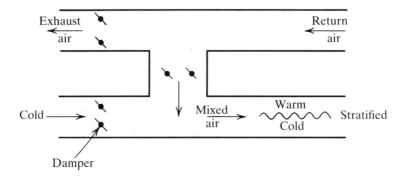

Fig. 5-14. Mixing of air streams of different temperatures resulting in stratification.

5-12 An Overview of Temperature Sensors and Transducers

The fact that several types of temperature transducers are in wide use suggests that no one sensor is best for all applications. The trends, however, seem to be as follows: Thermocouples continue to dominate in laboratory installations, particularly when a data-acquisition instrument is available that provides the reference junction and stable amplification. For industrial control, metal RTDs are widely used, and the 100 Ω platinum sensor has achieved considerable popularity. The class of transducers that is seeming to gain in acceptance is the integrated circuit type.

An important category of transducers used in industrial practice follows the 4 to 20 mA convention in which the transducer is supplied with a constant voltage and the combination of a sensor and circuit provides an output current of 4 to 20 mA that is proportional to the specific temperature range of the sensor. An advantage of transmitting an analog value in the form of an electric current rather than as a voltage is that the resistance in the wires does not reduce the magnitude of the signal.

Because the market for temperature sensors is so massive, the search goes on for new materials and concepts in temperature measurement. Fiber optics, for example, is the basis for some new types of sensors,[8,9] which can be so extremely small that they are unaffected by the electrical environment.

Of equal importance to the proper choice of temperature sensor is the installation of the sensor to be sure that the temperature being sensed is truly the one intended.

5-13 Flow Rate and Velocity Measurement

Measuring velocities and flow rates is important in many computer control systems. In some process industries the flow rate of one or more streams must be regulated precisely, so a measurement is required. Flow rates may also be regulated without being measured, when, for example, the flow rate is adjusted to hold a certain temperature, pressure, or another variable to a desired value.

The emergence of computer control systems is bringing with it the desirability of measuring flow rates to monitor energy consumption. A fundamental requirement for conserving energy in a complex system is to know how the energy is being used and whether various individual processes are taking place with high efficiency. In large industrial compressed air systems, for example, the motors driving the multiple compressors might require several thousands of kW. In many such plants the operators have no accurate data on the compression efficiencies of individual compressors at full and partial load. In order to operate the entire plant efficiently, it is necessary to choose which compressors should be in service and at what fraction of their full capacity they should be operating. Furthermore, a compressor that was efficient when new might decline in efficiency as it wears. To conserve energy in this compressed air facility, the flow rate delivered and the power required by each compressor should be measured. In heating and cooling systems for buildings and industrial plants, air, gas, and/or water sometimes flow continuously at higher rates than necessary. Measurement and reduction of these flow rates can conserve energy.

The next several sections cover the measurement of both velocity and flow rate, but the major interest in measuring the velocity is as a step in computing the flow rate. There are dozens of different devices for measuring flow rates, each having various advantages and disadvantages with respect to accuracy, need or ease of calibration, cost, and penalty with respect to pressure drop. Several very reliable types of meters are the nutating-disk type (wobble plate) for water that is used to measure water consumption in residences and commercial buildings. Another standard type of meter for air flow is the rotating-vane anemometer in which the number of rotations in a given time can be converted to an air velocity. Both of these types of meters are of the integrating type, and to adapt these instruments to computer control they would have to be equipped to provide an electric pulse for each revolution of one of their rotating elements.

For a computer control system the types of velocity and flow meters capable of providing a voltage are preferred. Several important types to be examined are venturis, orifices, pitot tubes, hot-wire anemometers, and turbines. Two recently developed types that will be explored utilize ultrasound, and another employs the vortex-shedding principle.

5-14 Venturi Tubes—Liquid Flow Measurement

The most straightforward calculations of differential-pressure type flow meters (venturi, orifice, pitot tube) prevail for liquid flow through a venturi tube. The calculation is essentially an application of Bernoulli's theorem. More complicated equations for orifices and for compressible fluids will be studied later. The standard reference book for these meters, which includes data on other types of meters as well, is *Fluid Meters*, published by ASME.[10] The classical Herschel venturi tube, shown in Fig. 5-15, has circular cross sections; one pressure tap is located in the entrance section at the left. The convergent section has an included angle of 21° tapering to the throat, where the other pressure tap is located. The length of the throat is one-third the throat diameter. After exiting the throat the fluid passes through the divergent section, which does not participate in the measurement but has the purpose of recovering as much pressure as possible. The diverging angle is usually about 7° to 8°.

As a general rule the flow in a converging section where pressure converts to velocity is an efficient process, while the opposite process of converting velocity to pressure often entails losses. Since the measurement portion of the venturi is in the converging section, the process is almost thermodynamically

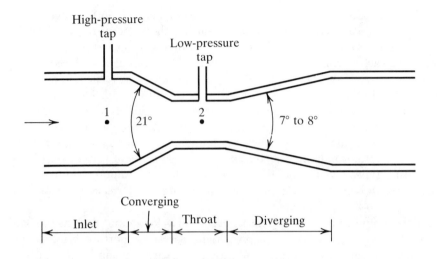

Fig. 5-15. Venturi tube.

reversible, so the Bernoulli equation applies:

$$\frac{V_1^2}{2} + \frac{p_1}{\rho} = \frac{V_2^2}{2} + \frac{p_2}{\rho} \tag{5-7}$$

where

V_1 and V_2 = velocity at points 1 and 2, respectively, in Fig. 5-15, m/s

p_1 and p_2 = pressure at points 1 and 2, respectively, in Fig. 5-15, Pa

ρ = density of liquid, kg/m^3

For liquid the density is assumed constant.

Another relationship between points 1 and 2 derives from the continuity equation, which equates the mass rate of flow w at both positions,

$$w, \text{kg/s} = V_1 A_1 \rho = V_2 A_2 \rho$$

where A_1 and A_2 are the cross-sectional areas at 1 and 2, respectively, in m^2. Then

$$V_1 A_1 = V_2 A_2$$

and

$$V_1 = V_2 \left(\frac{D_2^2}{D_1^2}\right) \tag{5-8}$$

5-15. Orifice—Liquid Flow Measurement

The ratio of diameters, D_2/D_1, appears so frequently in venturi and orifice relations that it is designated by its own symbol, β, where $\beta = D_2/D_1$.

The combination of Eqs. (5-7) and (5-8) yields an expression for V_2:

$$V_2 = \sqrt{\frac{2(p_1 - p_2)}{\rho(1 - \beta^4)}} \tag{5-9}$$

Because the process of accelerating the fluid between points 1 and 2 is so efficient, the theoretical equation, Eq. (5-9), would serve adequately as an expression for V_2 in practical calculations. To compensate for the slight inefficiency of the process, a factor C_v, the discharge coefficient, may be introduced to represent the fact that not all of the change in pressure, $p_1 - p_2$, is converted into velocity,

$$V_2 = C_v \sqrt{\frac{2(p_1 - p_2)}{\rho(1 - \beta^4)}} \tag{5-10}$$

For the venturi the factor C_v depends upon the roughness of the inlet and the converging sections and on the diameters. Typical values[10] of C_v fall between 0.985 and 0.995.

5-15 Orifice—Liquid Flow Measurement

Another instrument that measures flow rates by sensing a pressure difference resulting from a change in velocity is the orifice meter, as shown in Fig. 5-16. The orifice plate is a thin (3 to 10 mm thick) plate with a circular opening. The fluid accelerates as it passes through the orifice and continues to accelerate for a short distance beyond the orifice until the minimum cross-sectional area is reached at the "vena contracta."

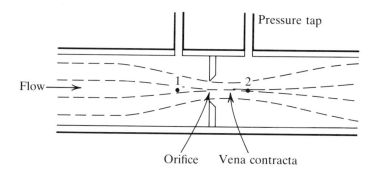

Fig. 5-16. The orifice flow meter.

The ideal situation would be to measure the low pressure at the vena contracta, but the distance from the orifice to the vena contracta is not always constant for a given installation. The location of the vena contracta may range from 0.3 to 0.9 pipe diameters downstream from the inlet face of the orifice. Since the downstream pressure tap is at a fixed position, the shift of the vena contracta is accommodated by adjusting the value of the factor C_o in an equation comparable to Eq. (5-10),

$$V_o = C_o \sqrt{\frac{2(p_1 - p_2)}{\rho(1 - \beta^4)}} \qquad (5\text{-}11)$$

where V_o is the mean velocity at the orifice diameter, and β is the ratio of the orifice diameter to the upstream pipe diameter.

The value of C_o for most applications is approximately 0.6; it is a function of β, the Reynolds number at the orifice, and the upstream pipe diameter. Tables of values of C_o are available in Reference 10 and are excerpted for one pipe size in Fig. 5-17. The coefficients in Fig. 5-17 are applicable to "flange taps" where the centerline of the upstream tap is located 25.4 mm ahead of the inlet face of the orifice, and the downstream tap 25.4 mm after the outlet face of the orifice. The refined calculation incorporates an iteration, since the Reynolds number is initially unknown.

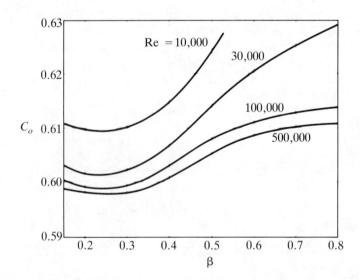

Fig. 5-17. Values of orifice coefficient C_o for 50 mm pipe when using flange taps.[10]

5-16. Flow Measurement of a Compressible Fluid

Example 5-6. What is the flow rate of 25°C water indicated by an orifice flow meter when the pipe diameter is 50 mm, the orifice diameter is 35 mm, and the pressure difference measured at the flange taps is 40 kPa? At 25°C the density of water is 996.9 kg/m³ and the viscosity is 0.905×10^{-3} Pa-s.

Solution. An iterative process is necessary because the flow coefficient is initially unknown. Substituting a trial value of C_o of 0.60 into Eq. (5-11) yields a trial V_o:

$$V_{o,\text{trial}} = 0.60 \sqrt{\frac{2(40,000)}{996.9\,(1 - (35/50)^4)}} = 6.17 \text{ m/s}$$

The Reynolds number at the orifice is

$$\text{Re} = \frac{VD\rho}{\mu} = \frac{(6.17)(0.035 \text{ m})(996.9)}{0.905 \times 10^{-3}} = 237,900$$

At this Reynolds number and $\beta = 0.7$, Fig. 5-17 shows $C_o = 0.612$, so

$$V_o = 0.612 \sqrt{\frac{2(40,000)}{(996.9)\,(1 - (0.7)^4)}} = 6.29 \text{ m/s}$$

so

$$\text{Flow rate} = (6.29 \text{ m/s}) \left(\frac{\pi (0.035)^2}{4}\right) (996.9 \text{ kg/m}^3)$$

$$= 6.03 \text{ kg/s}$$

5-16 Flow Measurement of a Compressible Fluid in a Venturi or Orifice

The Bernoulli equation, Eq. (5-7), is a special case of a more general expression of the steady-flow energy equation

$$h_1 + \frac{V_1^2}{2} = h_2 + \frac{V_2^2}{2} \tag{5-12}$$

where h_1 and h_2 are the enthalpies in J/kg at points 1 and 2, respectively. Equation (5-12) is applicable to points 1 and 2 in the converging section of the venturi as shown in Fig. 5-18. If the fluid is assumed to be a perfect gas,

$$h_1 - h_2 = c_p(T_1 - T_2) = c_p T_1 (1 - T_2/T_1) \tag{5-13}$$

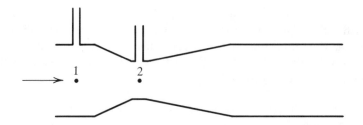

Fig. 5-18. A venturi as a converging nozzle.

where

c_p = specific heat at constant pressure, J/(kg-K)
T_1, T_2 = absolute temperature, K

The process in a converging nozzle is highly efficient in converting enthalpy to kinetic energy, so the assumption of an isentropic process is valid. Then

$$\frac{T_1}{T_2} = \left(\frac{p_2}{p_1}\right)^{(k-1)/k}$$

where k = ratio of specific heats, c_p/c_v. When this temperature-pressure relation is substituted into Eq. (5-13) and then into Eq. (5-12), the result is

$$\frac{V_2^2 - V_1^2}{2} = c_p T_1 \left[1 - \left(\frac{p_2}{p_1}\right)^{(k-1)/k}\right] \qquad (5\text{-}14)$$

To express the upstream velocity V_1 in terms of V_2, the continuity equation,

$$\frac{V_1 A_1}{v_1} = \frac{V_2 A_2}{v_2} \qquad (5\text{-}15)$$

applies, and the specific volumes for the isentropic process are related to the pressures through the equation

$$p_1 v_1^k = p_2 v_2^k \qquad (5\text{-}16)$$

Substituting Eq. (5-16) into Eq. (5-15) and then into Eq. (5-14) yields the expression for V_2:

$$V_2 = \sqrt{\frac{2 c_p T_1 \left[1 - (p_2/p_1)^{(k-1)/k}\right]}{1 - \beta^4 (p_2/p_1)^{2/k}}} \qquad (5\text{-}17)$$

When an orifice is used to measure the flow rate of a compressible fluid, the correction for the compressibility may be accommodated by the introduction of an "expansion factor," as explained in Reference 10.

5-17 Pitot Tubes

A pitot tube is a small-diameter tube that has the end directed upstream to the flow, as shown in Fig. 5-19. The purpose of the pitot tube is to measure the pressure rise generated by decelerating the fluid from its velocity V_1 to zero at point 2. The pressure at point 1 in one popular design of pitot tube is sensed in the annular tube surrounding the inner tube that measures the total pressure. For an incompressible fluid the Bernoulli equation, Eq. (5-15), may be applied with $V_2 = 0$, yielding an expression for V_1:

$$V_1 = \sqrt{\frac{2(p_2 - p_1)}{\rho}} \qquad (5\text{-}18)$$

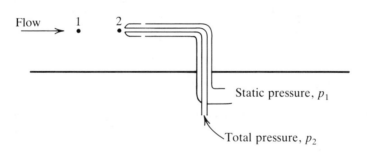

Fig. 5-19. A pitot tube.

5-18 Hot-Wire Anemometer

The key element of a hot-wire anemometer is a small sensing wire whose electrical resistance is dependent on its temperature. When an electric current is passed through the sensing wire the temperature is a function of the cooling provided by the stream whose velocity is to be measured. The heat-transfer coefficient over the wire is roughly proportional to the square root of the velocity of the fluid. Since the temperature of the sensing wire is to be used as the indicator of velocity, the temperature of the fluid must be measured directly or eliminated as a factor through electronic circuitry.

Small-diameter hot-wire anemometers are so rapid in their response that they can detect even high-frequency turbulence in the fluid flow. Hot-wire anemometers are used for turbulence measurements, but they are also suitable for steady-flow velocity indications.

A companion device to the above-mentioned laboratory instrument is a commercial device consisting of two temperature probes, as in Fig. 5-20. One of the probes is heated to a specified temperature while the other is unheated. The probes are RTDs, so the current needed to maintain the temperature of the heated probe is an indicator of the velocity.

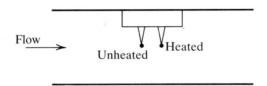

Fig. 5-20. A heated-element air-velocity sensor.

5-19 Turbine Flow Meter

The velocity of a fluid can be measured by inserting a propeller-type rotating element in the stream. The greater the velocity of the fluid the more rapidly the propeller or turbine rotates. The turbine has a magnet on one of its rotating elements that generates an electric pulse each time it passes a pickup. The pulse is transmitted to a frequency-to-voltage transducer, or the pulses may be counted by the computer for a specified time interval to indicate the flow rate. The turbine flow meter may occupy the entire cross-sectional area in a small-diameter pipe, or in large pipes a small tube containing the turbine may sample the velocity at one location. At very low flow rates the turbine may lock because of the magnetic force, so turbine meters are beginning to appear that sense the passage of the rotation element with an integrated circuit.

5-20 Ultrasonic Flow Meters

The ultrasonic flow meter takes advantage of the fact that high-frequency sound will pass faster in the direction of flow than opposite to the flow. The velocity of the sound is the sum of the fluid velocity which is to be measured and the sonic velocity. One means of separating the influence of the sonic velocity is the "sing-around" concept[11] shown in Fig. 5-21. A pulse is transmitted upstream and picked up by its receiver, which triggers another pulse. A series of pulses thus develops. Another loop passes pulses in the downstream direction, and the difference in the frequencies of these two pulse sequences indicates the velocity of the fluid.

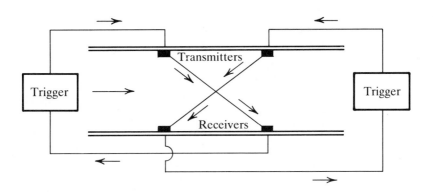

Fig. 5-21. Ultrasonic method velocity measurement.

5-21 Vortex-Shedding Flow Meters

A blunt object in a flow stream causes vortices to form that shed alternately on either side of the object, as shown in Fig. 5-22. The volumetric flow rate is proportional to the rate of vortex passage, and a sensor detects the passage of each vortex. The sensing of a vortex is accomplished either by using a very sensitive pressure pickup or by sensing the force impulses on the object. The pulse rate can be determined by the computer counting the number in a given time.

Two advantages of the vortex meter are that the vortex rate is independent of changes in density and viscosity of the fluid, and that the ratio of the high to low velocities capable of being sensed is of the order of 15. This velocity ratio, called the turn-down ratio, is of the order of 7 for an orifice or venturi. The reason for the independence of the fluid viscosity and density lies in the fundamental behavior of vortices.[12] Figure 5-23 shows a graph of the Strouhal number for vortices shed from a circular cylinder.

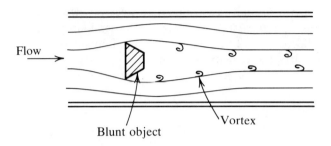

Fig. 5-22. Vortex-shedding flow meter.

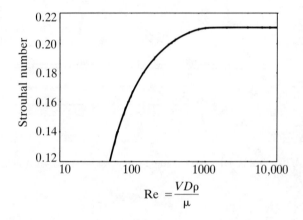

Fig. 5-23. Strouhal number relating vortex frequency for flow over a cylinder.

$$\text{Strouhal number} = nD/v$$

where

n = frequency of vortices, s^{-1}
D = diameter of the cylinder, m
V = velocity of the fluid, m/s

Figure 5-23 applies to a cylinder while trapezoidal obstructions are typical of commercial meters, but the key idea is that for Reynolds numbers above about 1000 the vortex frequency is proportional to the velocity and is independent of density and viscosity.

5-22 Evaluation of Flow-Measuring Devices

The fact that no one flow or velocity transducer dominates over all the others suggests that all have advantages and disadvantages. One advantage of the meters employing velocity-to-pressure conversion (orifices, venturis, and pitot tubes) is that if they are manufactured and installed according to specifications they do not need additional calibration. Users of measuring instruments have experienced drifts away from calibration with most flow meters, but little can go wrong with venturis, orifices, and pitot tubes. The pressure difference that they develop would normally be picked up by a pressure transducer, and in-place gauges provide the possibility of periodic checks of pressure.

Orifices and venturis introduce a pressure drop into the system that may be of some concern when attempting to conserve energy. The pitot tube introduces very little pressure drop in the system, but the pressure signal that it gives is often small and difficult to transduce. A velocity of 20 m/s is a fairly high velocity in an air-conditioning duct, for example, and the pitot tube in this duct would read 250 Pa, which would require a sensitive transducer. The system pressure losses of orifices and venturis are shown in Fig. 5-24, represented as a fraction of the measured pressure differential.

The other methods of measuring flow described in this chapter generally introduce very little pressure drop, but most of these meters require some calibration. The computer control system designer should not overlook meth-

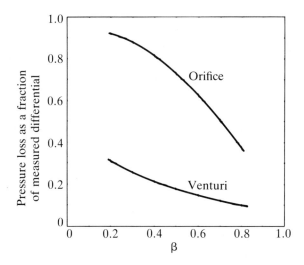

Fig. 5-24. System pressure losses introduced by orifices and venturis.

ods of determining the flow rate by means other than those covered in this chapter. A two-stage air compressor normally employs an intercooler, for example, and by measuring the temperature in and out of the cooling water and the flow rate of water, the flow rate of the air can be calculated. The energy loss associated with a venturi in the cooling water line could be an order of magnitude less than that caused by a venturi in the compressed air line.

Since energy conservation is one major reason for installing a computer control system, and determining energy flow rates often requires measuring a flow rate of fluid, it is expected that strong activity in developing new and improved techniques of flow and velocity transducing will continue.

5-23 Pressure Transducers

Probably the most widely used transducers in engineering systems are temperature transducers, but the next most frequently encountered type is for sensing and transducing pressure. There are various principles on which different pressure transducers work, and there are probably several hundred manufacturers of pressure transducers. Two broad classifications are "mechanical" and "physical." In the mechanical type a change in pressure results in a physical motion of an element such as a diaphragm, bellows, C-shaped tube, or helical tube. This motion is then transduced into a voltage by a motion or position transducer. A number of these position and motion transducers will be explored further in Sec. 5-31.

Pressure transducers that use the change in a physical property include the silicon sensor and piezoelectric elements. Quartz is an example of a piezoelectric element that develops a voltage when subjected to pressure. The output of the quartz crystal is passed to a charge amplifier. An advantage of the piezoelectric principle is its high-frequency response.

5-24 Evaluation of Types of Pressure Transducers

Some of the considerations in selecting a transducer include cost, accuracy, durability, and frequency response. Also, the materials of the transducer must be chemically compatible with the fluid whose pressure is being measured. Certain of the transducers that employ a mechanical motion are subject to some hysteresis, which reduces accuracy. The response time of some mechanical transducers may be inadequate for measuring pressures inside the cylinder of an engine, for example, but quite adequate for the response of most thermal systems. The rapid-response transducers are often, but not always, more expensive than the slower types. Measurement of low pressures (several hundred

Pa, for example) is difficult, and those transducers are often large in order to accommodate a diaphragm of sufficient size.

Installation of a pressure gauge along with the transducer is highly recommended for calibration purposes and also so that errors in the transducer and its circuitry can be detected.

5-25 Force

Measurement of a force is usually achieved by conversion of that force into a proportional deflection. The deflection can then be sensed by a motion transducer.

5-26 Torque

A torque measurement is often valuable, because when the rotative speed is known, the power delivered by a shaft can be determined. Just as was true of a linear force, torque can be translated to an angular deflection that can be sensed by a motion detector.

5-27 Electric Current

One important reason for interest in computer control is to conserve energy, so measurement of electric power is a frequent assignment. One component of the electric power is the current, which can be measured by a current transformer that surrounds a conductor as shown in Fig. 5-25. A current transformer might have a transformation ratio of 200 A to 5 A; this secondary current is passed through a low resistance of perhaps $0.1\,\Omega$, and the voltage

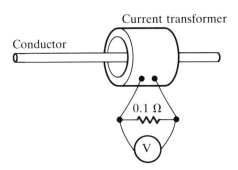

Fig. 5-25. A current transducer.

across this resistance is then sensed. The single phase power P is

$$P = IV \cos\theta$$

where

$$
\begin{aligned}
I &= \text{current, A} \\
V &= \text{voltage, V} \\
\cos\theta &= \text{power factor} \\
\theta &= \text{phase angle}
\end{aligned}
$$

The voltage remains nearly constant, but the power factor is drawn down from unity if the load has an inductive component that could be caused by electric motors on the line. Power or watt transducers are available that sense the voltage and current and the power factor as well.

5-28 Humidity Sensors

For sensing and transducing moisture content of air, the most convenient device is one whose electrical properties change with humidity. One type of humidity sensor, for example, has carbon particles impregnated in a material that expands and contracts as the humidity changes. The change of spacing of the carbon particles changes the electrical resistance of the sensor. Certain solid-state and polymer materials change their electrical properties in response to changes in relative humidity. The humidity-sensitive material must be combined with electrical circuitry to translate the change in property to a voltage change. Alternative means of humidity detection sense the dewpoint of the air. It is fairly easy to find humidity transducers that are accurate in a limited range of humidity and temperature. The challenge is when humidities are to be sensed near 0% and 100% relative humidity and/or over a wide range of temperature.

5-29 Chemical Composition

Particularly in the chemical and process industries the need arises to monitor continuously the fraction of a certain substance in a stream. There are hundreds of different types of these sensors, and the basic objective is to find some property or characteristic that changes as the composition of the substance in question changes.

Certain materials, such as some semiconductors, will absorb a number of different vapors, resulting in a change of a property such as the electrical conductivity. A substance might, for example, absorb ammonia, carbon

5-30. Liquid Level

monoxide, and methane. This detector might be used in one application to measure the concentration of ammonia if CO and CH_4 do not normally appear. In another application where NH_3 and CH_4 are not present the sensor could be used as a CO detector.

An example of sensing a chemical composition is the use of infrared radiation to detect the composition of CO_2. Carbon dioxide will absorb certain infrared wavelengths, so the infrared light intensity passing through a gas sample can indicate the CO_2 content.

5-30 Liquid Level

Next to measuring pressure and temperature, the measurement of liquid level is the most frequent sensing task in the chemical and process industries.[13] A particularly widespread assignment is to sense whether the liquid level is at a certain point and take action if it is not. We are more interested, however, in the task of transducing the level into a voltage that is proportional to the level. Several of the many available techniques for level sensing are (1) capacitance measurement, (2) measuring pressure at the bottom of the vessel, and (3) ultrasonic reflection. Using the capacitance principle, two plates are immersed in the vessel and the level of liquid between the plates influences the capacitance, which is the measured variable. A pressure transducer at the bottom of the tank can be used to measure the pressure, which can be translated into the level. In the ultrasonic measurement, as shown schematically in Fig. 5-26, a sound frequency of tens of kHz is directed from the emitter to the surface of the liquid. The time delay of the signal passing from the emitter and reflected back to the detector is converted into a distance to indicate the liquid level.

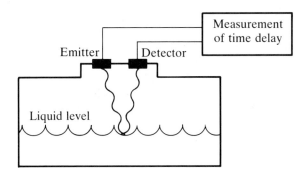

Fig. 5-26. Ultrasonic indicator of liquid level.

5-31 Position and Motion Sensors

Transducers that measure the closeness to a given point (proximity), the distance moved from a given reference point (displacement), velocity, and acceleration are widely used as self-contained transducers, but may also be a part of a transducer serving another purpose. Some types of pressure transducers employ movable elements whose displacement must be measured. A few of the principles used in position and motion sensing include potentiometric action, strain gauge measurement, capacitance, reluctance, differential transformer action, and the Hall effect.

One of the most straightforward position or motion transducers is formed by connecting the moving element to the wiper of a potentiometer, as in Fig. 5-27. This transducer is inexpensive and consists of a simple circuit. Several deficiencies of the potentiometer are that force is required to move the wiper, the wires on the coil eventually wear, and the output changes in small steps as the wiper slides from one wire to the next, rather than providing a smooth variation in voltage.

A device capable of measuring very small displacements is the strain gauge, illustrated in Fig. 5-28. An increase in the length increases the resistance. Strain gauges are often configured in a bridge circuit arranged to compensate for changes in resistance due to variations in temperature.

A capacitive pickup works by virtue of a change in the capacitance as the dimension of a gap between two plates changes. The magnitude of the capacitance must ultimately be translated into a voltage, and that progression can be accomplished by incorporating the capacitance in an RC circuit whose changing frequency can be sensed and converted to voltage.

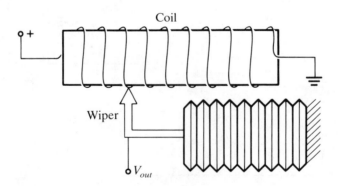

Fig. 5-27. Potentiometric motion transducer.

5-31. Position and Motion Sensors

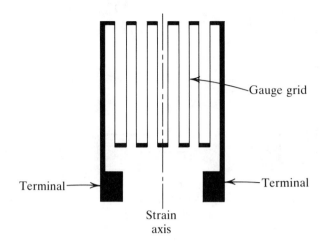

Fig. 5-28. Elongation of the strain gauge along the axis increases its resistance.

Proximity transducers operating on the reluctance principle respond to the closeness of a ferromagnetic object. These transducers can measure the distance between a ferromagnetic surface and the sensitive face of the transducer without physical contact.[14] The variable-reluctance proximity transducer is shown schematically in Fig. 5-29a, and the trend in ac output voltage is presented in Fig. 5-29b. The primary of a transformer is supplied with constant

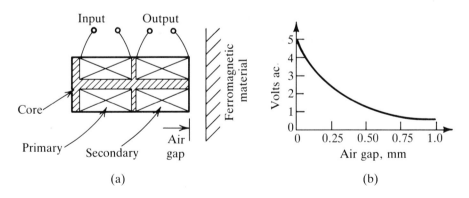

Fig. 5-29. (a) A variable reluctance proximity transducer, and (b) its output voltage.

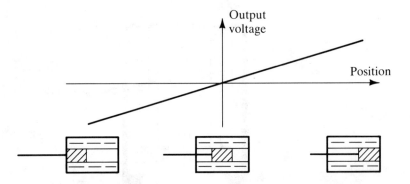

Fig. 5-30. A differential transformer used as a linear variable displacement transformer (LVDT).

ac voltage at a frequency usually between 400 and 5000 Hz. The output of the secondary is influenced by the reluctance received from the ferromagnetic material, which is additive to the reluctance between the primary and secondary in the device.

The differential transformer is a transducer with a movable core whose position regulates the output voltage of the transformer, as illustrated in Fig. 5-30.

A principle that is the basis for a number of commercial devices that transduce position is the *Hall effect*. If a strip of conductor or semiconductor, as in Fig. 5-31, is supplied with a current and a magnetic field normal to the strip, a voltage will develop across the other axis.[15] The applicable equation is

$$V_H = k i_c B / d \qquad (5\text{-}19)$$

where

V_H = Hall voltage
k = Hall coefficient (a function of the material and temperature)
i_c = current
B = field strength, gauss
d = thickness of the strip

The Hall effect can be applied in numerous ways, for example, by allowing the position or motion to control B, which has a linear influence on V_H.

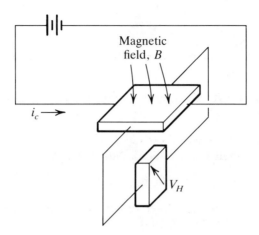

Fig. 5-31. The Hall effect voltage resulting from a current flowing normal to a magnetic field.

5-32 Rotative Speed

The most popular principle of transducing rotative speed is to sense the passage of a magnet on the shaft, which provides an electric pulse with each passage. The frequency of these pulses is then converted to a voltage through a frequency-to-voltage transducer. With the assistance of the microcomputer controller a frequency-to-voltage converter may not be needed, but instead the microcomputer can count the number of pulses occurring in a specified time interval.

5-33 How to Choose Transducers

The variety of types of transducers for a given assignment is often overwhelming, so the designer or operator of a computer control system could spend enormous amounts of time just studying and selecting transducers. As a practical means of moving on to other tasks, many engineers find one type of transducer that serves the purpose and continue to use that type. The user can then benefit from becoming familiar with the characteristics of that transducer. A shift to another type can come after trying a sample and comparing characteristics with the familiar one.

References

1. *Temperature Measurement Handbook*, Omega Engineering, Inc., Stamford, CT, 1984.

2. *Two-Terminal IC Temperature Transducer*, Analog Devices, Norwood, MA, 1979.

3. R. P. Benedict, *Fundamentals of Temperature, Pressure, and Flow Measurements*, John Wiley & Sons, New York, 2d ed., 1977.

4. Fenwal Electronics, Framingham, MA.

5. A. Burke, "Linearizing Thermistors with a Single Resistor," *Electronics*, pp. 151–154, June 2, 1981.

6. F. Kreith, *Principles of Heat Transfer*, Intext Educational Publishers, New York, 3d ed., 1973.

7. R. H. Kennedy, "Selecting Temperature Sensors," *Chemical Engineering*, pp. 54–71, Aug. 8, 1983.

8. "Fiber-Optic Sensors," *Electronics Week*, p. 38, Oct. 22, 1984.

9. "Shrinking for a Bigger CPI Role," *Chemical Engineering*, p. 18, Feb. 21, 1983.

10. *Fluid Meters—Their Theory and Application*, 6th ed., American Society of Mechanical Engineers, New York, 1971.

11. P. Harrison, "Flow Measurement—A State of the Art Review," *Chemical Engineering*, vol. 87, no. 1, pp. 97–104, Jan. 14, 1980.

12. H. Schlichting, *Boundary-Layer Theory*, trans. J. Kestin, McGraw-Hill Book Company, New York, 7th ed., 1979.

13. B. Lazenby, "Level Monitoring and Control," *Chemical Engineering.*, vol. 87, no. 1, pp. 88–96, Jan. 14, 1980.

14. *Handbook of Measurement and Control*, Schaevitz Engineering, Pennsauken, NJ, 1976.

15. H. V. Malmstadt, C. G. Enke, and S. R. Crouch, *Electronic Measurements for Scientists*, W. A. Benjamin, Inc., Menlo Park, CA, 1974.

General References

1. J. R. Mannion, "Water System Instrumentation," *ASHRAE Transactions*, vol. 82, part I, pp. 505–519, 1976.

2. V. Cavaseno, "Flowmeter Choices Widen," *Chemical Engineering*, vol. 85, no. 2, pp. 55–57, Jan. 30, 1978.

3. J. McDermott, "Sensors and Transducers," *EDN*, pp. 123–142, Mar. 20, 1980.

Problems

5-1. In Fig. 5-2, should the resistance R_T increase or decrease with an increase in temperature in order for the circuit to function properly?

5-2. A platinum sensor, as in Fig. 5-32, is placed in series with a resistance, and 12 V is applied across the combination.

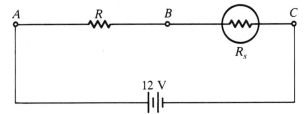

Fig. 5-32. Sensor in a series circuit, Prob. 5-2.

Complete the following table and observe magnitudes of voltage that could be picked up and amplified. Observe also the linearity (or lack of it) and the relative amounts of I^2R heating.

R	Temp., °C	R_s	V_{A-B}, V	V_{B-C}, V	I^2R Heating in Sensor, mW
1 kΩ	10	60			
	20	67			
	30	74			
10 kΩ	10	60			
	20	67			
	30	74			

5-3. A thermistor sensor is placed in a bridge circuit, Fig. 5-33, across which 12 V is applied. The resistance of the thermistor is shown in Table 5-2, Sec. 5-5. (a) Choose R_1 and R_2 such that $V_o = 0$ V when $t = 0°$C, and $V_o = 2$ V when $t = 30°$C. (b) What is V_o when $t = 15°$C? **Ans.:** (a) $R_1 = 2503$ and $R_2 = 52{,}170\,\Omega$.

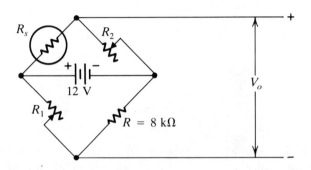

Fig. 5-33. Bridge circuit in Prob. 5-3.

5-4. In the circuit shown in Fig. 5-34, where the sensor is a thermistor with resistances given in Table 5-2, Sec. 5-5, choose R so that the voltage is linear to the temperature at the three temperatures in Table 5-2.

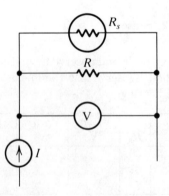

Fig. 5-34. Resistance in parallel with thermistor in Prob. 5-4.

5-5. Section 5-6 discussed the choice of the adjustable resistances in a bridge circuit to achieve the maximum output voltage. What is the combination of resistances (R, R_1, and R_2) that provides the maximum V_o at 50°C and 0 V at 0°C in Example 5-3? **Ans.:** $R_2 = 885.4\,\Omega$ and $R = 1.107 R_1$.

5-6. Prove Eq. (5-4).

5-7. In the three-wire circuit of Fig. 5-8b, at a certain temperature of the sensor the following resistances prevail: $R_L = 8\,\Omega$, $R_s = 110\,\Omega$, and $R = 10{,}000\,\Omega$. If $I = 15\,\text{mA}$, what is the reading of the voltmeter?

5-8. Two integrated-circuit temperature transducers are arranged in a circuit with an op amp as shown in Fig. 5-35. One transducer is at a temperature of T_1 K and the other at T_2 K. What is the equation for the voltage, V_o, in terms of T_1 and T_2?

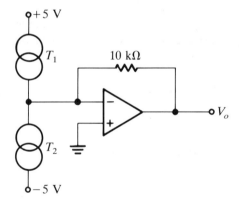

Fig. 5-35. Integrated circuit transducers in Prob. 5-8.

5-9. Select the orifice diameter in a pipe that has an internal diameter of 50 mm. The desired pressure difference measured at the flange taps is 70 kPa with a flow rate of water of $2\,\text{L/s}$. Assume an orifice coefficient $C_o = 0.61$. **Ans.:** 18.7 mm.

5-10. An orifice with a diameter of 30 mm installed in a 50 mm inside-diameter pipe indicates a pressure difference of 25 kPa when water at 25°C flows through the meter. (a) What is the value of C_o, and (b) what is the volume flow rate indicated by this measurement? **Ans.:** (b) 3.28 L/s.

5-11. The flow of air is to be measured by a venturi that has an upstream diameter of 37 mm and a throat diameter of 25 mm. Air enters the venturi at an absolute pressure of 325 kPa and a temperature of 30°C. The throat pressure is 290 kPa absolute. For air: $c_p = 1.0\,\text{kJ/(kg-K)}$, $R = 0.287\,\text{kJ/(kg-K)}$, and $k = 1.4$. What is the flow rate in kg/s indicated by these measurements? **Ans.:** 0.26 kg/s.

5-12. Water is pumped through a system where the measuring station is in a pipe 100 mm ID. The pump has an efficiency of 60% and the motor 90%. The system operates 4000 hours per year, the average water velocity in the 100 mm pipe is 2 m/s, and electricity costs $0.05 per kWh. Compute the annual pumping cost attributable to the flow meter (a) if the meter is an orifice with $\beta = 0.5$ and $C_o = 0.6$, and (b) if the meter is a venturi with $\beta = 0.4$ (which gives comparable $p_1 - p_2$ to the orifice) and $C_v = 0.99$. **Ans.:** (a) $354.

5-13. Water flows over a vortex-shedding flow meter that is cylindrical and has a diameter of 20 mm. When the rate of vortex shedding is 10 per second, what is the water velocity?

5-14. In the ultrasonic liquid level indicator of Fig. 5-26, if the signal passes through air at 20°C, what is the change in time delay when the liquid level changes 100 mm?

5-15. Since the resistance R of wire in a strain gauge $= rL/A$, where

R = resistance of section, Ω
r = resistivity of material, Ω-m
L = length of section, m
A = cross-sectional area, m^2

why is dR/R for most strain gauges greater than dL/L, where dL/L = strain?

5-16. Equation (5-19) indicates that V_H can be increased by increasing i_c and/or decreasing the thickness of the strip, d. The value of k decreases, however, with an increase in temperature, which could be caused by I^2R

heating. If the surface temperatures of the strip are held constant, which of the approaches illustrated in Fig. 5-36 results in the lowest peak temperature, doubling i_c as in Fig. 5-36a or reducing d by 50% as in Fig. 5-36b?

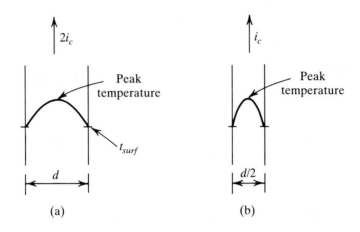

Fig. 5-36. Two approaches to increasing the Hall voltage V_H in Prob. 5-16 with least penalty of high temperature.

Chapter 6

Actuators

6-1 Actuators for Computer Control Systems

After the controlled variables have been sensed, transduced, and converted to digital signals, the microcomputer makes decisions on what action should be taken based on the user-supplied program. Ultimately a switch will be turned on or off or the position of a valve, damper, variable-speed motor drive, or other modulating regulator will be changed. This chapter concentrates on these "actuators" which exist at the extremity of the control system.

The computer can provide signals in digital form (0 V or 5 V) only, and a single line can be the control for opening and closing a switch, as in Fig. 6-1a. If an analog signal is ultimately needed, a number of digital lines in parallel are required from the computer, as in Fig. 6-1b. In Fig. 6-1b all eight lines from the computer pass through a digital-to-analog converter (which will be discussed in Chapter 8) to convert the signal to a voltage that is proportional to the magnitude of the binary number represented by the eight input lines. This voltage, which might be available at a very low current, must ultimately

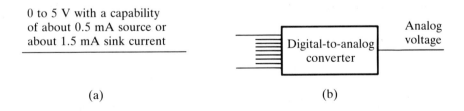

Fig. 6-1. Signals that the microcomputer can provide to prompt an action.

modulate the position of some component that perhaps requires an appreciable mechanical force.

This chapter presents techniques of stepping up the current to take two-position action with the signal shown in Fig. 6-1a, and also describes several types of devices that convert an analog voltage to a proportional mechanical position. Finally, this chapter explores the characteristics of stepping motors.

6-2 Two-Position DC Electric Switch

A standard requirement in computer control systems is to translate the voltage (0 or 5 V) from one line of the microcomputer output into the opening or closing of a switch that starts or stops a motor or other electrical equipment. The task is to step up the current from about 1 mA available from the microcomputer to perhaps 100 mA to operate a relay. The transistor, Fig. 6-2a, is a natural current amplifier, although two stages of current amplification, as in Fig. 6-2b, may be required.

Certain integrated circuit chips can be used as switches such as the 7405 inverter shown in Fig. 6-3. When the line from the microcomputer is high, the inverter sets a low voltage at the output that is capable of sinking a current of the order of 16 mA through the relay coil to ground. When the line from the microcomputer is low, the output of the inverter is high and no current flows through the relay coil.

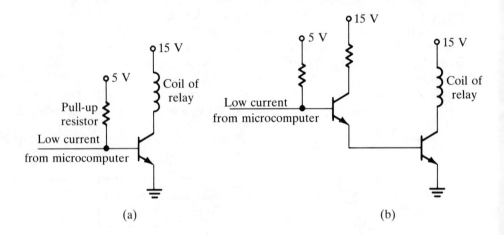

Fig. 6-2. Amplifying the current of a signal from the microcomputer to operate a switch: (a) single transistor, and (b) two-stage transistor.

6-3. Silicon-Controlled Rectifier (SCR) for DC Switching

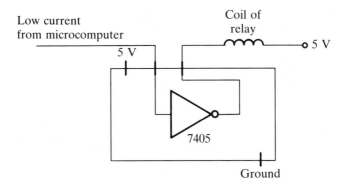

Fig. 6-3. An inverter chip used as a switch.

6-3 Silicon-Controlled Rectifier (SCR) for DC Switching

A class of switches called thyristors, are assemblies of four semiconductor materials that form a p-n-p-n combination. Two groups of thyristors are those that permit current to flow in one direction only and those that permit current to flow in either direction. The unidirectional group includes the silicon-controlled rectifier (SCR), the gate turn-off (GTO), and the programmable unijunction transistor (PUT). In the bidirectional group appear, among others, the triac, and the silicon bilateral switch. We shall concentrate on the most widely used switches, the SCR and the triac.[1]

The SCR consists of an anode (positive terminal), a cathode (negative terminal), and a connection for the gate current. The symbol for the SCR is shown in Fig. 6-4. When a voltage is imposed on the SCR, line current will not flow until current is provided to the gate. When gate current is applied, the SCR permits line current to flow and to continue even though the gate current is removed. The line current does not stop until the applied voltage between the anode and cathode is removed. No current flows when voltage is reapplied to the anode/cathode until the SCR is once again "fired" by the gate current.

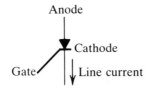

Fig. 6-4. A silicon-controlled rectifier.

A sample application of an SCR in a control circuit is the heater control shown in Fig. 6-5. The adjustment of the 10 kΩ variable resistor regulates the position in the sine wave at which the gate current reaches a value sufficient to fire the SCR. Once the SCR current is flowing it continues until the forward voltage across the SCR has dropped to zero. During the negative voltage portion of the sine wave, the unidirectional characteristic of the SCR prevents the flow of any reverse current.

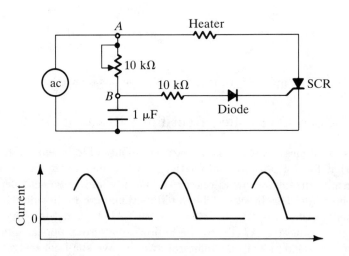

Fig. 6-5. A heater control circuit utilizing an SCR.

6-4 Triac—Alternating Current Switching

In contrast to the SCR, the triac will permit current to flow in either direction when the triac is fired by providing the gate current. A positive gate current permits the line current to flow in one direction, and a negative gate current permits flow in the other direction. The symbol of the triac is shown in Fig. 6-6, and the voltage-current characteristics are shown in Fig. 6-7.

If a voltage difference is imposed on the triac and a gate current in either direction is applied, the triac will conduct current in the direction of voltage drop. When the triac is fired and the gate current is removed, the line current can continue to flow until it drops to the holding current value I_m; then the triac shuts off. Another characteristic of the triac is V_{BD}, the breakdown voltage. If a voltage difference exceeding the magnitude of V_{BD} is impressed

6-4. Triac—Alternating Current Switching

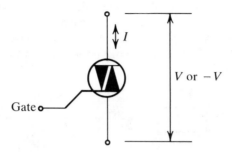

Fig. 6-6. Symbol for a triac, a switching device that permits current to flow in either direction.

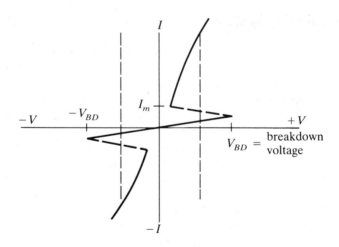

Fig. 6-7. Voltage-current characteristics of a triac.

on the triac, it will conduct current even though there is no gate current. Exceeding the magnitude of V_{BD} does not normally damage the triac, so the device can be subjected to reasonable voltage transients.

A circuit where a triac chops an ac current is shown in Fig. 6-8. If the load is inductive (such as a motor), the RC network shown in dashed lines is necessary to protect the triac from high countervoltage resulting from the abrupt switching.

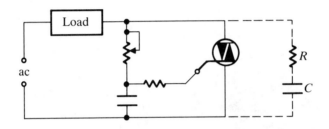

Fig. 6-8. Full-wave power controller using a triac.

6-5 Optically Isolated Switch

The triac is adaptable to alternating current, but the current must be retriggered with each half cycle, and devices are available to perform this function. A frequent assignment in computer control systems is to translate a change from 0 to 5 V at the output port of the microcomputer to the activation of an ac relay. Thus, a very low direct current (several microamperes) must switch alternating current. One possibility for performing this assignment is shown schematically in Fig. 6-9. It incorporates an optically isolated triac driver (one model being the MOC-3010). The dc signal from the microcomputer is here amplified in a two-stage amplifier, and the output of this amplifier is the control current for the triac driver. The optically isolated triac is limited to about 100 mA, which is large enough to serve as the control current for many relays. As long as the minimum direct current flows through the triac driver, the triac conducts current and operates the relay.

6-6 Solid-State Relays

Just as is true of electromechanical relays, solid-state relays (SSR) open or close a circuit carrying line current in response to a relatively low control

6-6. Solid-State Relays

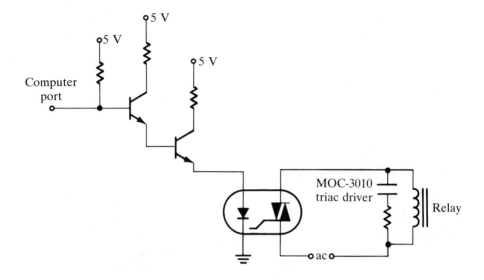

Fig. 6-9. Using an optically isolated triac to switch alternating current.

current. SSRs are available to switch alternating currents of 40 A or more at voltages of 250 V or higher. These relays are also available to switch direct current, although they are not generally available in current ratings as high as ac relays. Triacs are central elements of most SSRs, so the characteristics of triacs help explain the relative time sequences of control and line currents, as shown in Fig. 6-10. Following the onset of control current,

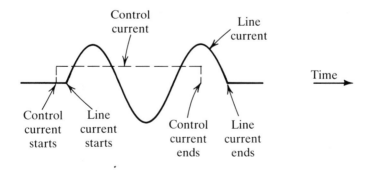

Fig. 6-10. Turn-on and turn-off of line current in an SSR in response to the control current.

the line current waits until the next zero-voltage point to begin. The minimum interval between these two turn-on times is approximately 10 µs. When the control current ceases, the line current continues to the next zero-voltage point, usually reacting within about 40 µs. The zero-voltage switching is often an advantage—to the relay itself, as well as to many ac sources and loads.

SSRs often require a heat sink, because the voltage drop through the relay may be of the order of 1.5 V. This voltage multiplied by the current handled by the relay indicates the magnitude of heat to be dissipated.

6-7 Electric-Motor Actuators

Two of the several types of electric-motor actuators are shown in Figs. 6-11 and 6-12. In the arrangement of Fig. 6-11 a variable voltage or variable resistance is translated into a modulating position of a mechanical arm. The variable resistance or voltage is developed from a translation of the computer output. An interface then converts the voltage or resistance to a dc voltage of a range required for the motor.

The motor is reversible so that it can move the arm in either direction called for by the controller. One of the required characteristics of many controllers is that they revert to fail-safe positions if there is a loss of power. The straightforward motor drive would simply remain motionless on loss of power, so one option available is a spring return that is activated upon loss of power. This spring return drives the arm to the fail-safe position.

Another type of electric-motor actuator is a reversible motor as shown in Fig. 6-12a. When fed with, for example, 24 V ac through lines A and the common, the motor operates in one direction, and it reverses its direction when powered through lines B and the common. To provide a varying rate of change of position, some actuators are controlled through pulse-width modulation as shown in Fig. 6-12b, where pulses of power are provided at regular time intervals. The duration of the power supply determines the net rate of change of position.

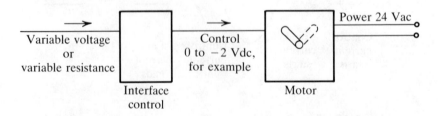

Fig. 6-11. A modulator driven by an electric motor.

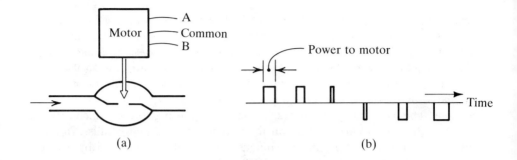

Fig. 6-12. (a) A reversible motor operating a valve, and (b) a method of supplying power to the motor using pulse-width modulation.

6-8 Magnetic Operator

One concept for actuators of the modulating type is the use of the magnitude of the input voltage to develop a magnetic field of varying strength. This magnetic field drives the core against a spring, as shown in Fig. 6-13. When applied to a normally closed valve, the spring force keeps the valve closed until the magnetic force opens it against the spring. One manufacturer of this type of controller has chosen 0 to 20 V dc as the powering voltage range. The maximum power required to completely open a nominal 25 mm valve is 16 W, and to open a 100 mm valve fully requires 120 W.

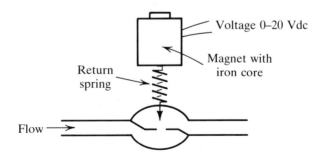

Fig. 6-13. Magnetic modulating valve.

6-9 Hydraulic Actuator

One of the challenges confronting electric motor drives is to develop the required force for operating a valve, damper, or other element. The difficulty is at least partially surmounted by gearing down the rotation of the motor shaft. Another approach to achieving the force amplification is to use the electric motor to drive a small hydraulic pump. The hydraulic pressure is regulated by a vent to the reservoir, which in turn is adjusted by the control voltage. The hydraulic piston works against an opposing spring, which drives the operator to its normal position if the power to the motor fails. In one design the electric motor runs continuously and could be powered by one of a number of different ac voltages (24, 120, 208, etc.).

6-10 Pneumatic Valve and Damper Operators

The mating of computer control systems with pneumatic operators to achieve the final application of force will probably be standard practice for at least the near future. Pneumatic piston-cylinders, diaphragms, or bellows are simple in construction, straightforward to maintain, usually enjoy a long life, and can develop large forces.

The designer of a computer control system should be aware that many actuators that provide mechanical motion are subject to hysteresis. Figure 6-14 shows the stroke as a function of the applied pressure in a pneumatic operator. When the actuator has been moving in one direction, a step change of pressure is needed to begin the progression in the opposite direction. The pneumatic, hydraulic, and magnetic operators are all subject to hysteresis. It is possible to reduce the magnitude of hysteresis by resorting to high-precision and expensive hardware. An opportunity for the computer control system is to use lower-cost hardware and compensate for the hysteresis in software.

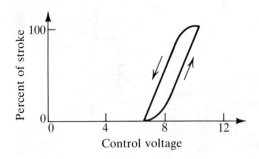

Fig. 6-14. Hysteresis in many types of mechanical actuators.

6-11 Electric-to-Pneumatic Transducer

When pneumatic actuators are chosen, it is sometimes necessary to use an interface to translate an electric signal, such as a voltage, into a modulating pneumatic pressure. One such interface, shown schematically in Fig. 6-15, is called an electric-to-pneumatic transducer. There may be some hysteresis in the electric-to-pneumatic transducer, but the magnitude is usually minimal.

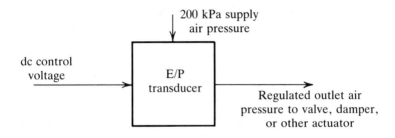

Fig. 6-15. An electric-to-pneumatic transducer.

6-12 Stepping Motors

A multipole motor that advances a fraction of a revolution (one step) for each time a prescribed sequence of pulses is applied to it is called a stepping or stepper motor.[2] Stepping motors are good for digitally controlled positioning mechanisms (printers, disk drives, metal cutting machines, packaging machines, etc.), because of their good acceleration and because of the predictable shifts in rotor position in response to given pulse cycles. In addition, they are finding a role in positioning modulating actuators, such as valves. Some typical step sizes available in commercial stepping motors are:

Degrees	1.8	2.0	2.5	5.0	15	30
Steps per revolution	200	180	144	72	24	12

The approximate maximum speed of most stepping motors is about 1000 steps/s if no delivery of power is demanded from the shaft, and the typical maximum speed under load is several hundred steps per second.

Two different classes of stepping motors are distinguished by the rotor style—the permanent magnet type and the variable reluctance type.[3] The

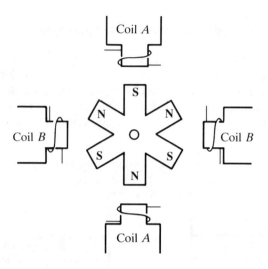

Fig. 6-16. The poles of a stepping motor.

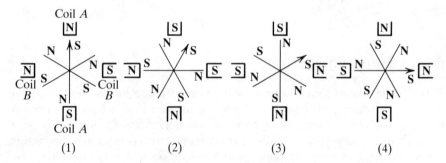

Fig. 6-17. A sequence for clockwise rotation of a stepping motor.

stepping motor shown in Fig. 6-16 has a rotor of the permanent-magnet type consisting of a number of teeth, each of which is a pole of a magnet. The rotor is bounded by pairs of stator magnet coils whose polarity is determined by the direction of the current through the coil. The sequence of polarity changes to achieve a clockwise rotation is shown in Fig. 6-17. At position 1, because of the proximity of coil A to its rotor teeth, the attraction of opposite poles holds the rotor fixed despite the attempt of coil B to move the rotor in a clockwise direction. To shift to position 2 the polarity of coil A is reversed, and momentarily the **S** polarity of coil A is adjacent to the **S** of the rotor, which provides a repulsive force that is neutral with respect to the direction

6-12. Stepping Motors

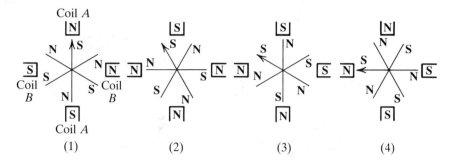

Fig. 6-18. A sequence for counterclockwise rotation of a stepping motor.

Table 6-1. Sequence in Four-Step Cycles of a Stepping Motor

(Coil A in the + status corresponds to **N** at the top pole of Fig. 6-18; coil B in the + status corresponds to **N** of the right pole.)

	Step	Coil A	Coil B	
CW	1	+	−	
↓	2	−	−	↑
	3	−	+	
	4	+	+	CCW
	1	+	−	

of motion. The force exerted by coil B on its two nearest teeth, however, provides the clockwise force. To take the next step to position 3, the polarity of coil B is reversed, and finally the polarity of coil A is reversed to move to position 4 and complete the cycle, which results in a 90° clockwise rotation.

The three-step sequence providing a 90° counterclockwise rotation (30° per step) is shown in Fig. 6-18. A summary of the cycles is given in Table 6-1. With other combinations of numbers of stator and rotor poles, other step sizes result. For example, if the rotor has five pairs of poles with two pairs of stator poles, the step size would be 18°.

In the stepping motor with a variable-reluctance rotor, the rotor has nonpolarized teeth, as shown in Fig. 6-19. Only one set of stator poles is energized at a time, and the rotor turns in such a way that the nearest teeth align them-

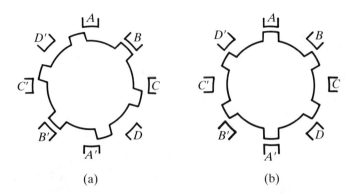

Fig. 6-19. A variable-reluctance stepping motor. The motor steps 15° counterclockwise when the stator polarization changes from (a) where pole A is energized to (b) where pole B is energized.

selves with the active stator poles. The stepping motor in Fig. 6-19 is called a four-phase motor, because only one of the four sets of poles is energized at a time. In Fig. 6-19a, pole A has drawn a pair of rotor teeth into alignment. If pole A is switched off and pole B energized, the teeth closest to pole B move into alignment with it, and the rotor turns counterclockwise 15°.

6-13 Performance of Stepping Motors

While there are various types of stepping motors exhibiting a spectrum of behavior, there are certain common characteristics in their performances. The rotation of one step requires a finite amount of time, and, in fact, the response is like that shown in Fig. 6-20, where the rotor begins turning slowly, picks up speed, and then overshoots the new steady-state position. The rotor oscillates several times about its new position until the oscillations damp out.

The programmer of a microcomputer that controls a stepping motor is faced with a dilemma. One of the advantages of the stepping motor is that after a series of single pulses has been sent to the motor, the shaft position is still precisely known. In order to achieve this exact response the changes of the driving pulses must be approximately synchronized with the changes of teeth positions of the rotor. In order to rotate the shaft at a high speed, it is possible to call for the next step about the time the rotor has reached the shaft position designated as region A in Fig. 6-20. This type of operation,

6-13. Performance of Stepping Motors

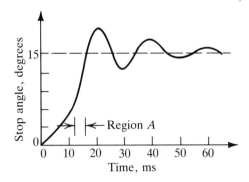

Fig. 6-20. Single-step response.

called slewing, achieves high rotative speeds and if carefully done can still maintain precise registration.

When the motor starts from a motionless state and is to progress through a large number of steps to a new motionless state, the entire translation should normally take place as rapidly as possible. The motor must therefore be accelerated to a high speed quickly and then slow down rapidly to a stop. The permissible acceleration and deceleration rates are influenced by the inertia and friction of its own rotor as well as the inertia and friction of the load the motor is driving. Stepping motors possess torque-speed characteristics typical[4] of those shown in Fig. 6-21. The graph shows the "start-without-error" torque and the "running" torque. The start-without-error torque is the external torque the motor can provide at constant pulse rates during starting and stopping operation without loss of a step. The running curve is the torque that can be provided at constant speed or with small accelerations. The difference between the two curves is the torque devoted to motor inertia during acceleration or deceleration.

For design calculations it may be convenient to approximate the torque-speed curves for a given motor by straight lines, such as in Fig. 6-22.

Example 6-1. A 5° stepping motor with torque-speed curves shown in Fig. 6-22 is to accelerate a load from a motionless state to 200 steps/s with a constant acceleration. The load is essentially frictionless with an inertia of 0.6 g-m². How many steps are required to develop the speed of 200 steps/s?

Solutions. The critical condition is the highest velocity reached during the process, which is 200 steps/s, because here the start-

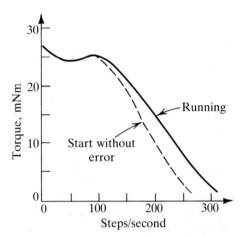

Fig. 6-21. Torque-speed characteristics of a stepping motor.

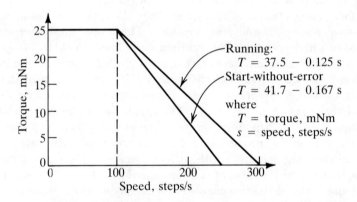

Fig. 6-22. Linear approximation of torque-speed curves of Fig. 6-21.

without-error torque is at its lowest value, 8.3 mN-m. Since the torque T in millinewton-meters equals the product of the inertia J in g-m² and the acceleration α in rad/s²,

$$T = J\alpha$$

$$\alpha = \frac{8.3 \text{ mN-m}}{0.6 \text{ g-m}^2} = 13.83 \text{ rad/s}^2$$

$$\alpha = 2.20 \text{ rev/s}^2 = 158.5 \text{ steps/s}^2$$

The time required to bring the assembly up to speed is

$$\frac{200}{158.5} = 1.262\,\text{s}$$

During this time interval the stepping motor turns through

$$(200/2)(1.262\,\text{s}) = 126\,\text{steps}$$

The potential for applying a microcomputer to control a stepping motor is attractive, because it is possible to program the microcomputer to provide the required sequence of pulses with a timing that corresponds to desired rotative speeds and accelerations. Stepping motors are appearing on the market with progressively higher power, which opens new opportunities for the stepping motor to serve as an actuator in a control assignment.

References

1. H. V. Malmstadt, C. G. Enke, and S. R. Crouch, *Electronic Measurement for Scientists*, W. A. Benjamin, Menlo Park, CA, 1974.

2. P. Giacomo, "A Stepping Motor Primer," *Byte*, vol. 4, no. 2, pp. 90–105, February 1979, and vol. 4, no. 3, pp. 142–149, March 1979.

3. B. C. Kuo, *Theory and Application of Step Motors*, West Publishing Company, St. Paul, MN, 1974.

4. Application Data, AIRPAX/North American Philips Controls Corporation, Cheshire, CT, 1980.

Problems

6-1. In the heater control circuit of Fig. 6-5 the power supply is 115 V rms and the adjustable resistor is set at 5 kΩ. The gate current required to turn on the SCR is 5 mA. (a) What is the peak voltage at A? (b) To what value must the voltage at B rise in order to turn on the SCR? (c) Sketch one cycle of V_A and V_B in their proper relative positions and indicate the point at which the SCR fires.

6-2. A 5° stepping motor with torque-speed curves shown in Fig. 6-22 drives a load with an inertia of 0.5 g-m^2 and a torque in mN-m to overcome friction of (0.018)(speed, steps/s). The operating mode is to bring the rotative speed of the assembly up to 150 steps/s at constant acceleration, then return to rest with a constant deceleration. What is the minimum total time for the acceleration-deceleration abiding by the start-without-error curve? **Ans.:** 0.81 s.

6-3. A load with an inertia of 0.8 g-m^2 and negligible friction load is to be accelerated to a speed of 200 steps/s by a 15° stepping motor whose speed-torque curves are shown in Fig. 6-22. The full start-without-error torque is to be used throughout the acceleration in order to use minimum time. (a) Sketch a graph of the acceleration and speed as a function of time from the start to full speed. (b) Compute the time elapsed and the number of steps executed in the process. **Ans.:** total steps = 261.

Chapter 7

Binary Numbers and Digital Electronics

7-1 Transition to Digital Electronics

The subject of the previous six chapters can be classified as "analog electronics." The emphasis was on analog voltages that were obtained from sensors and transducers, on analog voltages that adjust the position of actuators and circuits that modify these voltages. Generally these voltages could vary continuously between −15 and +15 V.

This chapter is the first one in this book devoted to digital electronics. Digital circuits have application apart from computer controllers but are an appropriate interface to computers as well. First there will be a brief description of binary numbers, but only one arithmetic operation will be explained—addition. Other arithmetic operations will be examined in Chapter 10, Binary Arithmetic. The first components to be explored are logic gates, which are often the building block of digital circuits. This chapter is also the appropriate place to study what are often called sequence controllers, which are important adjuncts for mechanical engineering systems. Sequence controllers built around electromechanical relays have been used for decades. The way they will fit into this chapter is that the logic can be represented in the form of logic gates, and low-cost controllers using gates instead of electromechanical relays can be constructed. For a short span of years sequence controllers were built using gates, but these devices were rapidly supplanted by "programmable controllers." Today's programmable controllers are sometimes indistinguishable from what we would call computer controllers. One of their characteristics, however, is that among other tasks they execute ladder diagrams, which because of their wide use in mechanical engineering and thermal applications will be treated in this chapter. The chapter then explains several characteristics and requirements of outputs, namely the three types of output signals encountered from integrated circuits—(1) totem pole, (2) open collector, and (3) three-state. Some specific chips with those output signals will be identified and examples of their uses presented. Additional chips to be explained are clock oscillators, flip-flops, digital counters, Schmitt triggers, multivibrators, latches, comparators, and analog switches. A problem that frequently arises when interfacing a mechanical switch with the microcomputer is that the switch contacts bounce when the switch is closed. A debouncing circuit will be described. Next follows an explanation of seven-segment LEDs and binary-coded decimal.

7-2 Binary Numbers

The computer does its work through the rapid processing of a mass of zeros and ones. The designation of a 0 or 1 is a convenient arithmetic abstraction of voltages of 0 or 5 V, respectively, prevailing in a conductor. Even these voltage values are nominal, because most digital circuitry recognizes, as Fig. 7-1 shows, a range of voltages as a logical zero and another as a logical 1. Furthermore, the ranges are different for the input and output of a device. Those shown in Fig. 7-1 apply to transistor-transistor logic (TTL), in contrast, for example, to complementary metal oxide semiconductors (CMOS). The voltages interpreted as logical 0 and logical 1 need not be precisely 0 and 5 V, respectively, but bands of voltages are interpreted as shown in Fig. 7-1. Between the voltage ranges representing logical 0 and logical 1, there is a band of voltages which provides a margin of safety between the two logical levels.

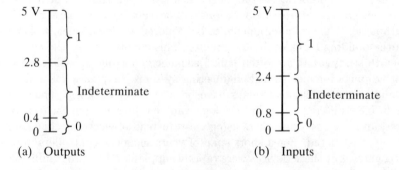

Fig. 7-1. Voltage ranges that represent a logical 1 and a logical 0 for (a) outputs and (b) inputs, applicable to the transistor-transistor logic (TTL) family.

7-3 Conversion between Binary and Decimal Numbers

The binary system, since it is composed of only zeros and ones, exhausts its possibilities in the rightmost column with these two digits. A further increase in magnitude shifts over to the next column on the left. A comparison of a few base ten and base two (binary) numbers is shown in Table 7-1.

Each digit in a binary number is called a bit. The rightmost bit is called the least-significant bit (LSB) and the leftmost bit is called the most-significant bit (MSB).

To convert a binary number to a decimal number, multiply each of the digits by 2 to an exponent starting at zero for the LSB and increasing by one for each shift to the left, then sum the result. In Fig. 7-2, for example, the binary number 101101 is shown to equal 45_{10}.

To convert in the opposite direction, divide successively by 2, using the remainder as the binary digit, as shown in Fig. 7-3 for the decimal number 471.

Table 7-1. Some Numbers in the Binary System and Their Decimal Equivalents

Binary (Base 2)	Decimal (Base 10)
0	0
1	1
10	2
11	3
100	4
101	5

$$1 \quad 0 \quad 1 \quad 1 \quad 0 \quad 1$$
$$1 \times 2^5 + 0 \times 2^4 + 1 \times 2^3 + 1 \times 2^2 + 0 \times 2^1 + 1 \times 2^0$$
$$32 + 0 + 8 + 4 + 0 + 1 = 45$$

Fig. 7-2. Conversion of binary number 101101 to decimal 45.

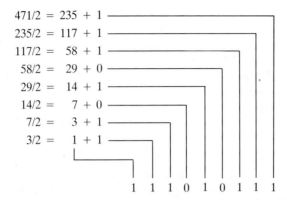

Fig. 7-3. Conversion of decimal number 471 to binary 111010111.

7-4 Addition of Binary Numbers

Chapter 14 will present a more extensive treatment of binary arithmetic, and only one operation will be discussed here—addition. The rules for the addition of two digits are as follows:

$$0 + 0 = 0$$
$$0 + 1 = 1$$
$$1 + 0 = 1$$
$$1 + 1 = 10, \quad \text{or 0 with a carry of 1}$$

Putting these rules into operation for adding 1101 (decimal 13) to 110 (decimal 6) gives the following result:

```
     1 1 0 1
   + 1 1 0
   ─────────
   1 0 0 1 1  =   decimal 19
```

7-5 Basic Logic Operations

We turn now to logic operations that are based on whether the input values to the operation are 0 or 1. The operations are used directly in digital electronics and also are the building blocks for more-complicated logic. Six basic logic operations are shown in Table 7-2. These operations will next be explained individually in Secs. 7-6 through 7-11.

7-6. OR Gate

Table 7-2. Basic Logic Operations

Abbreviation	Name	Operation Indicator
OR	or	+
AND	and	· or concatenate
NOT	not	bar over expression
NOR	not or	bar over OR expression
NAND	not and	bar over AND expression
XOR	exclusive or	⊕

7-6 OR Gate

A logic gate provides an output of 0 or 1 depending upon the states (0 or 1) of the inputs. The OR gate that performs the logical OR operation provides an output of 1 if either or both of the inputs are 1; otherwise the output is zero. This characteristic is often shown in a "truth table" that indicates the possible combinations of inputs and the outputs. For the OR gate the truth table and the symbol are shown in Fig. 7-4.

I	J	$I+J$
0	0	0
0	1	1
1	0	1
1	1	1

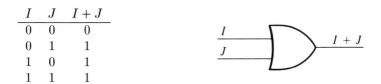

Fig. 7-4. Truth table and symbol of the OR gate.

7-7 AND Gate

In the AND gate, if I and J are both 1, then the output is 1, otherwise the output is zero, as illustrated in Fig. 7-5.

I	J	I · J = IJ
0	0	0
0	1	0
1	0	0
1	1	1

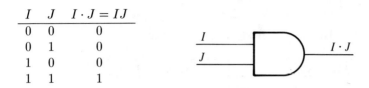

Fig. 7-5. Truth table and symbol of the AND gate.

7-8 Inverter

The inverter is a device such that if the input is 1 the output is 0 and vice versa, as shown in Fig. 7-6.

NOT 1 = $\bar{1}$ = 0
NOT 0 = $\bar{0}$ = 1

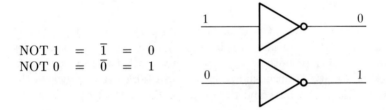

Fig. 7-6. The inverter.

7-9 NOT-OR (NOR) Gate

If the output of an OR gate is sent into an inverter the combination is a NOT-OR or NOR gate as shown in Fig. 7-7. In the combined symbol the small circle is placed at the output of the OR gate to indicate the negating operation.

I	J	$\overline{I+J}$
0	0	1
0	1	0
1	0	0
1	1	0

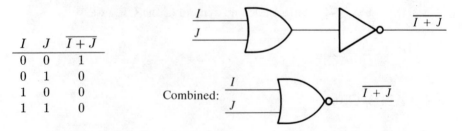

Fig. 7-7. Truth table and symbol of a NOR gate.

7-10 NOT-AND (NAND) Gate

The counterpart of the NOR gate is the NAND gate, whose truth table and symbol are shown in Fig. 7-8.

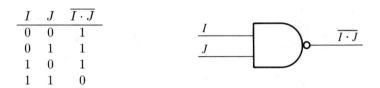

I	J	$\overline{I \cdot J}$
0	0	1
0	1	1
1	0	1
1	1	0

Fig. 7-8. Truth table and symbol of a NAND gate.

7-11 Exclusive-OR (XOR) Gate

The XOR gate is one that provides an output of zero if both the inputs are the same, but provides an output of 1 if the inputs are different. Figure 7-9 shows the truth table and symbol of the XOR gate.

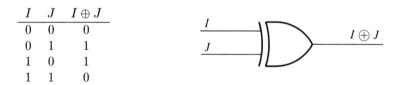

I	J	$I \oplus J$
0	0	0
0	1	1
1	0	1
1	1	0

Fig. 7-9. The exclusive-OR (XOR) gate and truth table.

7-12 Combining and Cascading Gates

Frequently AND and OR gates will be shown with more than two inputs. An AND gate with three inputs requires all the inputs to be 1 before the output is 1. The output of an OR gate with three inputs will be 1 if one or more of the three inputs is 1.

Various logic arrangements can be developed by cascading gates. Figure 7-10 shows a gate circuit that provides the XOR logic described in Fig. 7-9.

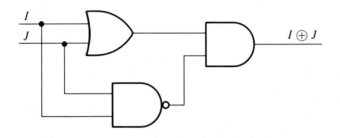

Fig. 7-10. Cascaded gates to form XOR logic.

7-13 De Morgan's Laws

De Morgan's laws are two theorems that show the equivalence of several logic functions:

$$\overline{A+B} = \overline{A} \cdot \overline{B} \tag{7-1}$$

and

$$\overline{AB} = \overline{A} + \overline{B} \tag{7-2}$$

The validity of De Morgan's laws can be shown by developing truth tables for each of the combinations in Eqs. (7-1) and (7-2). De Morgan's laws may be useful in developing alternative logic that uses available OR or NOR gates rather than AND or NAND gates, or vice versa.

Example 7-1. The gate circuit shown in Fig. 7-11a provides the desired response to the various combinations of the C, D, and E inputs. The circuit would require three chips, so an equivalent circuit is sought that would use two multiple-input chips.

Solution. The output X is high for the $C = 0$, $D = 0$, $E = 1$ combination and is low for all other input combinations. The first

7-14. Gate Chips

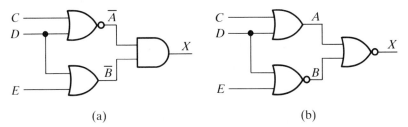

Fig. 7-11. (a) Original gate circuit in Example 7-1, and (b) equivalent circuit.

De Morgan law, Eq. (7-1), can be applied by defining A as $C + D$ in Fig. 7-11a and B as $\overline{D + E}$. Then

$$X = \overline{A} \cdot \overline{B} = \overline{A + B} = \overline{C + D + \overline{D + E}} \tag{7-3}$$

The circuit suggested by the rightmost expression of Eq. (7-3) is shown in Fig. 7-11b and can be developed with two chips—a multiple-input NOR gate and an OR gate.

7-14 Gate Chips

Gates are usually assembled in dual-in-line packages (DIP), and one chip often incorporates several gates. Some commonly used gate chips are shown in Fig. 7-12. The user of gate chips will quickly observe that the inputs need not be precisely 0 V and 5 V in order for the gates to function properly. Instead, the ranges of input voltages shown in Fig. 7-1 will be adequate for the gates to operate. The output voltage of the gate is normally not 5 V but is usually approximately 4 V. If an input is to be 0, it is not sufficient to leave the input as an open circuit, but it must be set to ground or at least less than 0.8 V. If an input is left as an open circuit, the gate is likely to treat it as a high. If not all of the four gates on a chip in Fig. 7-12 are used, the inputs of the unused input gates should be tied high through a pull-up resistor. Chips are also available that incorporate gates with three inputs.

7-15 Ladder Diagrams for Conditional and Sequential Control

The focus of this chapter is logic gates, and the next two sections present several subjects that may seem unrelated to gates. There is a relation, however, between gates and sequential/conditional controllers. For decades some of

132   Chapter 7. Binary Numbers and Digital Electronics

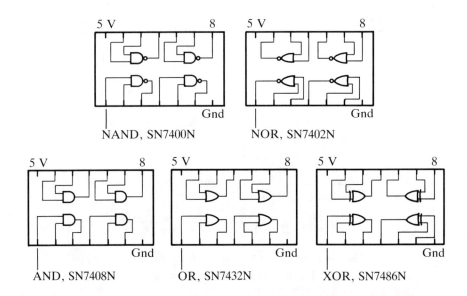

Fig. 7-12. Several gate chips (TI number designations).

the logic for controlling mechanical and electrical devices has been provided by electromechanical relays where one contact must be closed, for example, before another action can take place. Perhaps a time delay in the sequence is also a requirement. The graphic presentation of this logic has traditionally been provided by ladder diagrams. The designer constructs a ladder diagram, and the electrician chooses relays, cutouts, and interlocks that when properly connected execute the sequences of the ladder diagram.

Much of the logic in a ladder diagram can be represented in the terminology of gates. A further merging of the technologies is that programmable controllers are now used to execute the logic of ladder diagrams, eliminating electromechanical relays. Most programmable controllers are now microprocessor-based and may be difficult to distinguish from what we would call a computer controller. Ladder diagrams continue to serve as the means of communicating the control sequence, so they will be briefly explained.

Some symbols that appear in ladder diagrams are shown in Fig. 7-13: (a) contact normally open, (b) contact normally closed, (c) normally closed push button (push buttons are subject to momentary changes of position), (d) normally open push button, (e) coil that changes a contact from its normal position, (f) line switch for a motor or other device, (g) mechanical link-

7-15. Ladder Diagrams for Conditional and Sequential Control

age (usually connecting push buttons), (h) special function that should be described in the box, (i) time delay, (j) switch normally open, and (k) switch normally closed.

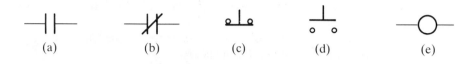

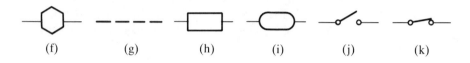

Fig. 7-13. Symbols used in ladder diagrams

The construction and typical appearance of a ladder diagram will be illustrated through an example.

Example 7-2. The temperature in a room is regulated by a warm-air system that uses an electric heater, as shown in Fig. 7-14. A push button is to start the fan and heater, and another push button stops the fan and heater. On startup the fan is to be allowed

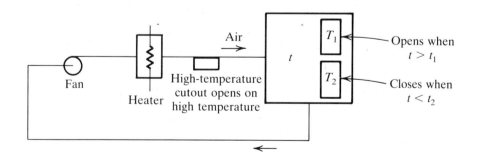

Fig. 7-14. A warm-air heating system whose ladder diagram is shown in Fig. 7-15.

to run for 10 s before the heater is activated. During operation the fan is to run continuously, but the heater cycles on and off under the control of the room temperature as sensed by the temperature switches T_1 and T_2. The heater switches off when t rises to t_1 and remains off until t drops to t_2. When the room temperature drops to t_2 the heater switches on and remains on until t rises to t_1. A high-temperature cutout in the air stream turns off the heaters when the temperature exceeds a specified setting. Construct the ladder diagram for this control sequence.

Solution. The ladder diagram appears in Fig. 7-15. The basic structure of the ladder diagram is that a 115 V line runs down the left side of the diagram, and a bus tied to ground runs down the right side. To start and run the equipment (in this case the fan and heater), 115 V must be able to reach the respective switches for the equipment.

The starting and operating sequences beginning from the idle state are as follows. A momentary depression of the start push button energizes coil A in line 1, which in turn closes the three contacts marked A. In line 1 the closure of contact A that parallels the start push button permits the start button to be released but coil

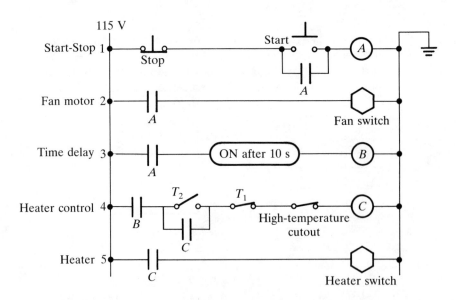

Fig. 7-15. Ladder diagram for warm-air heating system.

A remains energized. In line 2 the closure of A starts the fan motor. In line 3 the closure of A activates the time delay, which after 10 s transmits the voltage to coil B. Contact B then closes in line 4, but if temperature switch T_2 is open (because the room temperature is above t_2), the voltage passes no further. When the room temperature drops to t_2, however, the voltage passes on to switch T_1, which will be closed because T_2 is set less than T_1. Coil C is energized, which closes contact C in line 5, which in turn switches the heater on.

When the heater is on for a time, the temperature in the space will presumably rise until t_1 is reached, at which time switch T_1 opens, coil C is deactivated, and contact C opens, interrupting power to the heater. When the room temperature drops below t_1, switch T_1 closes, but not until the room temperature once again drops to t_2 does the heater resume operation. The opening of the high-temperature switch deactivates coil C and shuts down the heater.

A momentary depression of the stop button deactivates coil A in line 1, which opens all the A contacts to stop the fan and deactivate the B coil, shutting off the heater.

7-16 Ladder Diagram Using Gates

The logic that is displayed on the ladder diagram can be represented in gate symbolism, and in fact can be duplicated physically through the use of gate circuits. The "off" status of a switch can be replaced by a logical 0 and the "on" status by a logical 1. To activate a line power switch requires a logical 1, and the switch is off when a logical 0 is provided. The equivalence of the AND and OR gates to ladder diagram combinations is shown in Fig. 7-16. In Fig. 7-16a two switches in series can be replaced by an AND gate, and in Fig. 7-16b the operation of two switches in parallel is accomplished by an OR gate. Combinations of gates can often be used to duplicate other commonly used ladder diagram logic. The coupled NOR gates of Fig. 7-17a can perform the function of the push-button switch arrangement of Fig. 7-17b.

Suppose that the idle position in Fig. 7-17a is that $P = 1$ and $Q = 0$. Then if the start button is depressed, the upper NOR gate switches P to 0, activates A, and switches Q to 1. Release of the start button results in no change of P and Q. When the stop button is pressed, Q goes to 0, P to 1, and A is deactivated. Release of the stop button causes no change in the status.

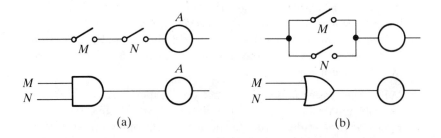

Fig. 7-16. (a) Switches in series, and (b) switches in parallel.

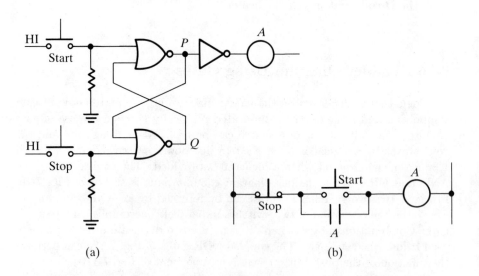

Fig. 7-17. Interlocks shown (a) with coupled gates, and (b) in ladder diagram symbolism.

7-17 Sequential Logic Circuits

Another example of a gate circuit performing a ladder diagram control function is a sequential operation. There are situations in which a certain sequence of steps must be taken before the desired action occurs. Suppose, for example, that push button A must be pressed momentarily before the momentary pressing of push button B in order to start the equipment. Figure 7-18 shows a circuit in which a momentary depression of A results in a high leaving both gates 1 and 2. Those signals remain high even when A is released. Now when B is pressed, the outputs of gates 3 and 4 go high and remain high even when B is released. A momentary depression of B first and subsequent pressing of A does not raise the final signal. In order to prepare for the next sequence after an operation, the RESET switch is momentarily pressed to drop the output of gate 2.

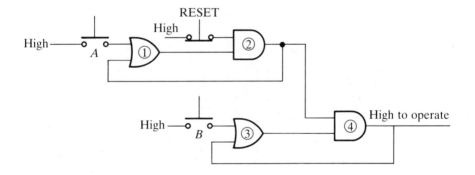

Fig. 7-18. Gate circuit that requires a momentary depression of A prior to pressing B in order to operate the equipment.

7-18 Binary Addition with Gates

This chapter opened with a discussion of binary numbers and will close with a description of addition using gates. Since it is not our intent to build a computer, we will drop the subject quickly after introducing it. We will show the gate circuit for adding two 2-bit numbers called A and B. Provision will also be made for a carry. The identification of the bits involved in the addition are shown in Fig. 7-19.

138   Chapter 7. Binary Numbers and Digital Electronics

The gate circuit for adding the LSBs is shown in Fig. 7-20a and the circuit for the MSBs in Fig. 7-20b. For a two-bit adder the two circuits are combined to take inputs A0, B0, A1, and B1 to provide outputs of S0, S1, and C1.

```
     C1   C0            carry
          A1   A0       A-number
          B1   B0       B-number
         ─────────
          S1   S0
```

Fig. 7-19. Addition of two 2-bit numbers.

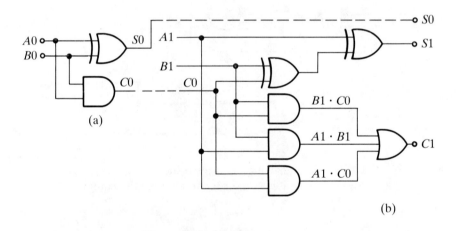

Fig. 7-20. Gate circuit for addition: (a) LSB, and (b) MSB.

7-19 Pull-Up Resistor

Most integrated-circuit chips can sink more current than they can supply. For example, the chip shown in Fig. 7-21 may have the capability of supplying 4 mA when it is attempting to provide an output of 5 V and can sink 16 mA

7-20. Three Classes of Outputs

when its output is to be 0 V. The circuit the chip supplies may, for example, require 6 mA in one direction for a logical high and 6 mA in the opposite direction for a logical low. Suppose that a pull-up resistor R is chosen with a resistance of 470 Ω. When the output signal attempts to go high it will supply 4 mA, which leaves 2 mA to be provided through the pull-up resistor to provide the desired 6 mA. To deliver 2 mA, a voltage drop of 0.94 V is required through the pull-up resistor so 4.06 V is provided to the load, which (as explained in Fig. 7-1) is sufficiently high to be treated as a logical high. When the output signal reverses and sinks 16 mA, 10 mA flows through the pull-up resistor, resulting in a drop of 4.7 V through the resistor and leaving 0.3 V at the output, which is sensed as a logical low.

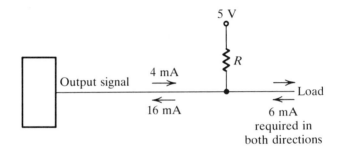

Fig. 7-21. A pull-up resistor.

7-20 Three Classes of Outputs Found on Inverters and Buffer Gates

Three possible output signals available from different chips are (1) totem pole, (2) open collector, and (3) three-state. The totem pole circuit, shown in Fig. 7-22, consists of two transistors with base currents provided by inputs A and B. If A is high and B is low, high voltage is available at the output. If, on the other hand, A is low and B high, then the output is low. If A and B are both high or both low, the output voltage does not take on a definite high or low. The 7404 inverter chip is an example of a chip with a totem pole circuit.

The open collector output is basically the circuit shown in Fig. 7-23. If the input is high the transistor pulls the output to ground, and if the input is low

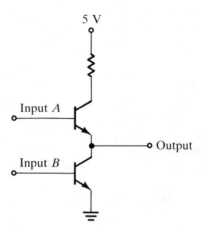

Fig. 7-22. Totem pole circuit.

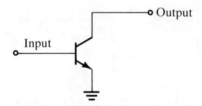

Fig. 7-23. Open-collector output.

the collector of the transistor is isolated from the ground so the output sees a high resistance. The 7405 and 7406 chips utilize open-collector outputs.

The third class of output is the three-state or tristate output, which is shown schematically in Fig. 7-24. Chips that provide this type of output

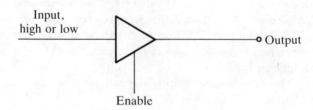

Fig. 7-24. Three-state output.

7-21. Debounced Switch

include the 74125 and 74126 three-state buffers. If the chip is enabled, the high or low input passes directly to the output. If the chip is disabled, the output of the chip is essentially disconnected from the external circuitry. The 74125 chip is enabled by a low signal at the enable pin, and the 74126 by a high signal at the enable. Both the three-state and open-collector output structures are able to assume a "disconnected" mode, but the three-state devices are able to reach the high level faster than open-collector chips. When a three-state output is supplying a high output, it is actively driving current, charging any external capacitance to a high voltage. In contrast, the open-collector relies on the external pull-up resistor to exponentially charge any capacitance toward the supply voltage.

7-21 Debounced Switch

Signals generated by the microcomputer or by logic gates make a transition cleanly from one logic level to the other, and if the interconnecting wires are kept reasonably short, the signals will arrive at their destinations undistorted. On the other hand, signals generated by a mechanical switch suffer from the characteristic of switch bounce, as shown in Fig. 7-25. The bounce of a switch occurs as the contacts physically bounce apart momentarily as the switch is moved to its new position. The ragged start of the switch closure shown in the figure would cause erratic operation if the circuitry to which it is connected were monitoring transitions. A typical "debounce" circuit incorporating two NAND gates is shown in Fig. 7-26.

In the process of moving the switch from position X to position Y, the sequence of events is as shown in Fig. 7-27. In Interval 1, $A = 0$, $B = 1$, $C = 1$, and $Q = 0$. In crossing the boundary from Interval 1 to Interval 2 the switch first leaves X so that A rises to 1, but B is still 1 so there is no change of C and Q. The first time the switch touches Y, however, B drops to 0, which changes Q to 1 and C to 0. Even as B bounces between 0 and 1 there

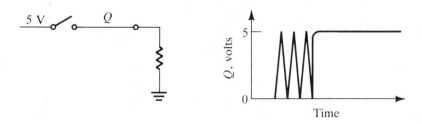

Fig. 7-25. Bounce of a switch.

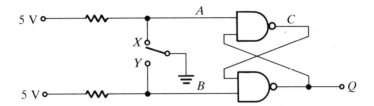

Fig. 7-26. Circuit to debounce a switch.

is no change in Q. During Interval 3 the switch leaves Y, resulting in a rise of B to 1, but since A remains 1 there is no change in C or Q. Not until A drops to zero for the first time does C go to 1 and Q to 0. The output Q thus changes cleanly.

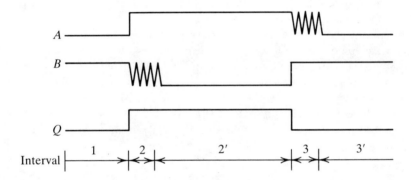

Fig. 7-27. Sequence of events in switching from X to Y and back.

7-22 Clocks and Oscillators

In electronic circuitry a clock provides a series of regular pulses as shown in Fig. 7-28. The frequencies available in different models of clocks vary from about 0.01 to higher than 2 MHz. Certain chips require clock pulses to sequence their operation, and microprocessors also are driven by a clock. Circuits, generally based on a resistance-capacitance combination, can be devised to provide clock pulses. The more common practice is to choose a crystal oscillator, as shown in Fig. 7-29, whose pins are 1, 7, 8, and 14 on a DIP configuration. A clock oscillator is tuned for a given frequency, and the deviation of frequency may be 1% in low-cost oscillators but of the order of 0.0025% in precise models.

7-23. Flip-Flops

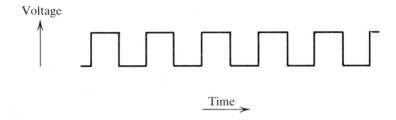

Fig. 7-28. A train of clock pulses.

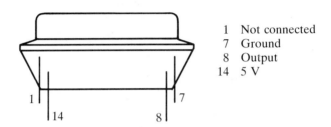

1	Not connected
7	Ground
8	Output
14	5 V

Fig. 7-29. A crystal clock oscillator.

7-23 Flip-Flops

A flip-flop changes state upon either a falling or rising transition of an input signal. Some flip-flops chips are:

- 7473 Dual J-K with clear

- 7474 Dual-D-type positive edge-triggered with preset and clear

- 75112, 74113, and 74114 dual J-K edge-triggered

In the 7473 J-K flip-flop, for example, shown in Fig. 7-30, the input lines are J, K, the clock, and the clear. The outputs are labeled Q and $\overline{Q}$. The falling edge of a pulse in the clock input triggers an operation that depends on the setting of J and K. For the 7473 flip-flop Table 7-3 shows the response of the outputs Q and $\overline{Q}$.

If the clear input line is set low, the output Q is low, regardless of the J and K settings. When the clear line is set high, the output Q responds to the falling edge of the clock in a manner dependent upon the setting of the J and K inputs, as Table 7-3 shows. The symbol Q_o means the old value of Q and $\overline{Q_o}$ the old value of $\overline{Q}$.

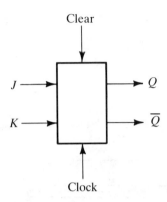

Fig. 7-30. A J-K flip-flop with clear.

Table 7-3. Functions of 7473 J-K Flip-Flop

L = low, H = high, X = don't care

Clear	Clock	J	K	Q	$\overline{Q}$
L	X	X	X	L	H
H	↓	L	L	Q_o	$\overline{Q_o}$
H	↓	H	L	H	L
H	↓	L	H	L	H
H	↓	H	H	$\overline{Q_o}$	Q_o

One of the principal uses of flip-flops is as a "divide-by-two" counter. Such counters will be discussed next.

7-24 Divide-By Counters

The 7473 J-K flip-flop whose characteristics are shown in Table 7-3 could be used as a divide-by-two counter by setting J and K high and supplying the clock pin with a series of pulses, as shown in Fig. 7-31. Each time the signal to the clock pin falls, Q changes position, resulting in the frequency of pulses that Q transmits to be one-half that of the input.

7-25. Schmitt Trigger   145

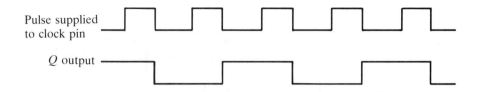

Fig. 7-31. J-K flip-flop as divide-by-two counter.

Several of these J-K flip-flops can be cascaded to form a counter of pulses that registers in binary numbers. The J and K pins are set high, and the Q output of one flip-flop is used as the clock input to the next chip. As Fig. 7-32 shows, if the start of the operation is assumed to occur when A, B, C, and D are all high, the next drop of the input pulse at the left will drop B, C, and D as well. If the states of A, B, C, and D are brought out as bits of a binary number, the number increases by one with each rise or fall of the input pulse.

Integrated circuits are available that provide other divide-by combinations, for example,

- 7490 divide-by-two, -five, and -ten

- 7492 divide-by-two, -six, and -twelve

The pin diagram for the 7492 chip is shown in Fig. 7-33. In order to activate the divide-by-twelve output the divide-by-two output must be connected to the B input at pin 1.

7-25 Schmitt Trigger

This chip contains the equivalent of a hysteresis, and in fact the symbol of the Schmitt trigger shows a hysteresis sign, as for example in the 7414 inverter Schmitt trigger shown in Fig. 7-34. The characteristic of the Schmitt trigger inverter is that the output drops to a logical zero when the input first rises past 1.7 V. The output does not go high again until the input drops below 0.9 V. The result of this characteristic is, as shown in Fig. 7-35, that a noisy input can be converted to a clear output. In addition to cleaning up a signal, the Schmitt trigger can be used as a threshold detector, as shown in Fig. 7-36.

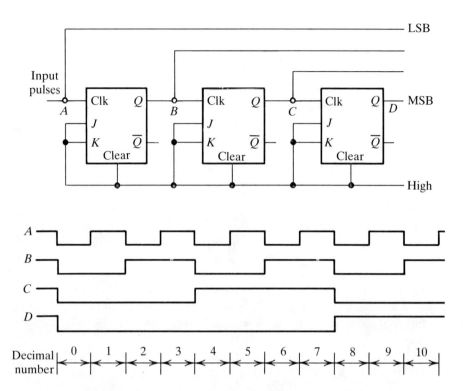

Fig. 7-32. Binary counter using a series of flip-flops.

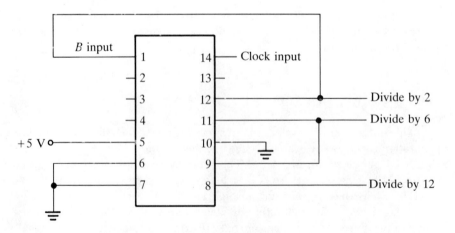

Fig. 7-33. Pin diagram of 7492 divide-by-two, -six, and -twelve counter.

7-25. Schmitt Trigger

Fig. 7-34. Schmitt trigger inverter.

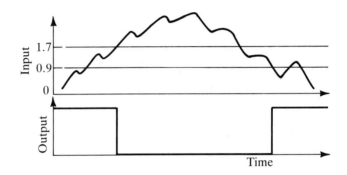

Fig. 7-35. Input and output of a Schmitt trigger inverter.

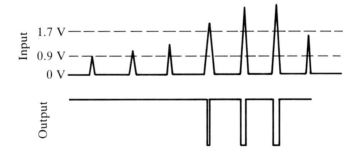

Fig. 7-36. Schmitt trigger as a threshold detector.

7-26 Monostable Multivibrator

Sometimes it is necessary to provide a pulse when an input changes from a low to a high signal. A chip available for this purpose is the 74121 monostable multivibrator for which the key connections are shown in Fig. 7-37. The input is applied to pin 5, and the output is received from pin 6. A resistance R and capacitance C are connected external to the chip, and the combination of magnitudes of R and C determines the pulse width, according to the approximate equation

$$t \text{ s} = 0.7(C, \text{ F})(R, \Omega)$$

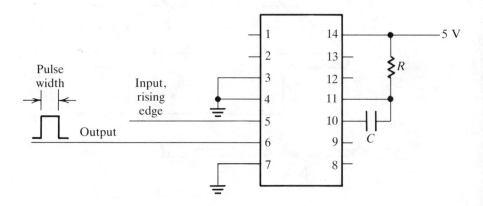

Fig. 7-37. Monostable multivibrator.

7-27 Low-Frequency Pulses

A frequent need in computer control systems is to obtain a pulse to prompt some action by the microcomputer every minute, or every hour, or every day, for example. A frequently used arrangement for achieving such low-frequency pulses is to start with a high-frequency pulse and feed this pulse through the required combination of cascaded divide-by counters. The output of this series of counters will be a falling (or rising) edge at the desired frequency, which can be converted to a pulse of desired width through a monostable multivibrator with a proper choice of the external resistance and capacitance.

7-28. Latches

The original source of high-frequency pulses might be from a circuit driven by a clock oscillator, or the source could be the 115 V ac power line, as illustrated in Fig. 7-38. The 60 Hz, 115 V line feeds a 6.3 V transformer, which provides a sine wave with an amplitude of from −9 to +9 V. A transistor limits the voltage between 0 to 5 V for protection of the Schmitt trigger. The Schmitt trigger converts the blunted positive part of the sine wave to a square signal for transmission to the desired combination of divide-by counters and finally to the monostable multivibrator.

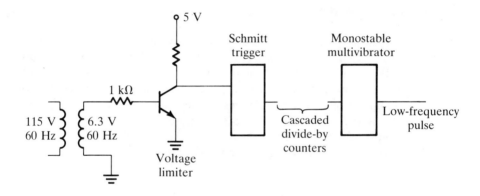

Fig. 7-38. Low-frequency pulse circuit activated by 60 Hz source.

7-28 Latches

The characteristic of microcomputers of accomplishing only one task at a time permeates operations external to the microcomputer. A frequent requirement, as illustrated in Fig. 7-39, is that at one instant the lines of a data bus (8 bit or 12 bit, for example) form a certain binary number that is to be held for a certain device until the cycle returns and a new binary number is to be provided to the device. The assignment can be accomplished by a latch between the bus and the device. The 74100 chip is a latch that transmits the input directly to the output when the enable line is high. When the enable voltage falls, the output remains the same as the last input, regardless of how the input is changed. One way to achieve a latching action is through use of the gate circuit shown in Prob. 7-3.

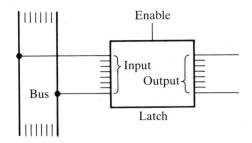

Fig. 7-39. Holding a digital number with a latch.

7-29 Comparators

Section 3-3 discussed the use of an op amp as a comparator to compare the voltage of two signals, for example A and B. If A is greater than B the output is 5 V, and if B is greater than A the output is 0 V. There are chips available (such as the LM139, LM239, and LM339) that are specifically designed for comparison operations. The advantages of the comparator over the op amp are that the comparator is usually faster, is more sensitive to differences of voltages, and draws lower current. Many comparators are also capable of receiving input voltage signals larger than the supply voltage without damaging the comparator.

7-30 Analog Switches—Field-Effect Transistors

The flow of current in the transistors studied in Chapter 4 was unidirectional. Another type of transistor, with the symbol shown in Fig. 7-40, is the field-effect transistor (FET). The feature of the FET is that it is bilateral and permits current to flow in either direction when supplied with gate current. With no gate current the transistor offers a very high resistance. The performance of the FET essentially duplicates a mechanical-contact switch.

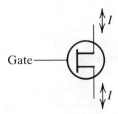

Fig. 7-40. A field-effect transistor that can be used as an analog switch.

7-31 Binary-Coded Decimal (BCD)

Computer users will generally find it more convenient to enter and receive numbers in decimal form, so there is often the need to perform a conversion between binary and decimal systems. One means of facilitating the conversion is through the use of the binary-coded decimal system in which each of the decimal digits is represented by a block of four binary bits. The conversion is shown in Table 7-4. In BCD the numbers 1010 and higher have no meaning. The decimal number 528_{10}, for example is represented in BCD by three 4-bit blocks, 0101 0010 1000.

Table 7-4. Conversion Between Decimal and BCD

BCD	Decimal	BCD	Decimal
0000	0	0101	5
0001	1	0110	6
0010	2	0111	7
0011	3	1000	8
0100	4	1001	9

7-32 Seven-Segment LEDs

A widely used means of displaying decimal numbers is the seven-segment LED in which lighting the proper combination of the segments displays numbers between 0 and 9. The seven segments are labeled on Fig. 7-41a, and the pin diagram of one of the commercial LEDs is shown in Fig. 7-41b. Only 11 pins of the 14 pin positions exist on the LED. The LED is supplied with 5 V, and a segment is turned on by grounding the appropriate pin through a resistor. The resistor in the range of 300 to 500 Ω limits the current through the diode.

Immediately one thinks in terms of circuitry for converting a BCD input directly to the combination of segments to represent the decimal number. That circuitry already exists in the 7447 BCD-to-seven-segment decoder/driver, whose pin diagram is shown in Fig. 7-42. The chip translates the BCD input (A, B, C, and D) to the decimal number by sinking current on the appropriate combination of output pins.

One 7447 combines with one LED, but if decimal numbers larger than one digit are to be represented, another chip is available to assist with that task— the 74185 binary-to-BCD decoder. A single 74185 with a 6-bit binary input feeding two decimal digits with blocks of BCD is shown in Fig. 7-43. The

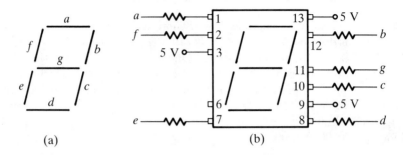

Fig. 7-41. (a) The seven-segment LED, and (b) the pins of the MAN 72A seven-segment LED.

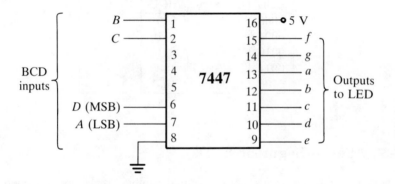

Fig. 7-42. The 7447 BCD-to-seven-segment decoder/driver.

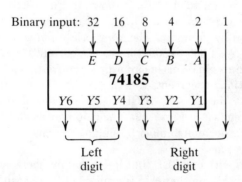

Fig. 7-43. The 74185 binary-to-BCD decoder.

Table 7-5. Typical Conversions Performed by the 74185 Binary-to-BCD Decoder

Binary Input						Base	Left Digit			Right Digit			
32	16	8	4	2	1	10	Y6	Y5	Y4	Y3	Y2	Y1	1
0	0	0	0	0	0	0	0	0	0	0	0	0	0
0	0	0	1	0	1	5	0	0	0	0	1	0	1
0	1	0	0	1	1	19	0	0	1	1	0	0	1
1	0	0	0	1	1	35	0	1	1	0	1	0	1
1	1	0	0	1	0	50	1	0	1	0	0	0	0
1	1	1	1	1	1	63	1	1	0	0	0	1	1

internal logic of the chip converts the binary input to BCD, several samples of which are shown in Table 7-5. Binary-to-BCD decoders can be ganged (see Prob. 7-7) to decode binary inputs with higher numbers of bits.

7-33 Summary

This chapter has introduced digital circuitry and thus provides the platform from which to move to principles and devices that convert between analog and digital signals. Digital signals are the natural ones for inputs to and outputs from microcomputers. Data books from major manufacturers of integrated circuits are sources of information about other special-purpose components.

General References

1. J. V. Becker, "Designing Safe Interlock Systems," *Chemical Engineering*, vol. 86, no. 22, pp. 103–110, Oct. 15, 1979.

2. E. M. Cohen and W. Fehervari, "Sequential Control," *Chemical Engineering*, pp. 61–66, April 29, 1985.

Problems

7-1. Fill out the truth table below for the three gate circuits in Figs. 7-44a, 7-44b, and 7-44c.

A	B	C	D in Circuit		
			a	b	c
0	0	0			
0	0	1			
0	1	0			
0	1	1			
1	0	0			
1	0	1			
1	1	0			
1	1	1			

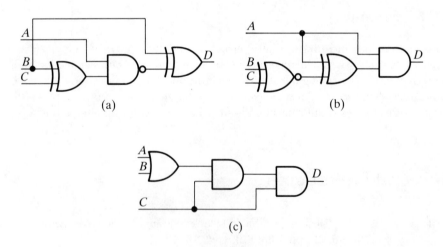

Fig. 7-44. Gate circuits in Prob. 7-1.

7-2. Devise a gate circuit with an output L and inputs of I, J, and K. The output L is to provide $I + J$ if K is high; otherwise L is low if K is low for any I and J.

7-3. For the gate diagram shown in Fig. 7-45 fill in the table below:

Assign Enable	Then Assign Data	C	P	Q
1	0			
0	0			
0	1			
1	1			
0	1			
0	0			

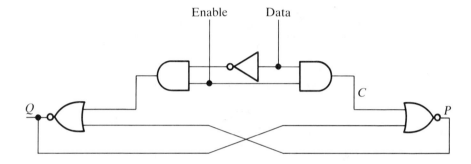

Fig. 7-45. Gate circuit in Prob. 7-3. (This circuit is used in the 74100 chip as described in Sec. 7-28.)

7-4. The counter shown in Fig. 7-32 is called an "up counter," because the binary number increases with each input clock pulse. Show that the circuit of J-K flip-flops in Fig. 7-46 is a down counter. Start with all Q's $= 0$ and all $\overline{Q}$'s $= 1$.

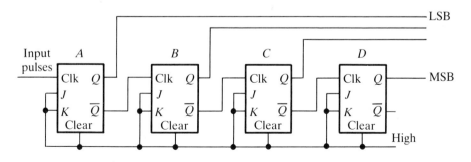

Fig. 7-46. Down counter in Prob. 7-4.

7-5. Devise a gate circuit with three inputs A, B, and C and one output D. The output D is high if two and only two inputs are high; otherwise D is low.

7-6. Draw a gate circuit diagram to achieve the function shown in Fig. 7-17 with the use of only one chip, an SN7402N.

7-7. Three binary-to-BCD decoders are combined to accept an 8-bit binary input in an arrangement shown in Fig. 7-47. Use the decoding of a single 74185 as the basis, and designate on Table 7-6 the status of the various pins if the decimal equivalent of the binary input is 125.

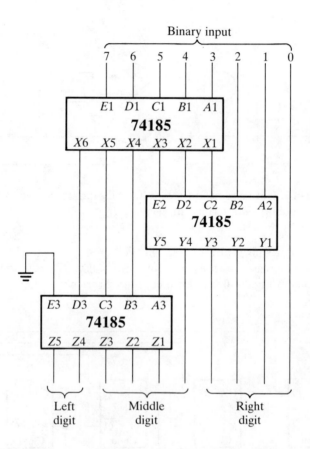

Fig. 7-47. An 8-bit binary-to-BCD decoder in Prob. 7-7.

Table 7-6. Table to Complete in Prob. 7-7.

			E1	D1	C1	B1	A1			
	X6	X5	X4	X3	X2	X1				
				E2	D2	C2	B2	A2		
				Y5	Y4	Y3	Y2	Y1		
E3	D3	C3	B3	A3						
Z5	Z4	Z3	Z2	Z1	Y4	Y3	Y2	Y1	Bit 0

Left digit   Middle digit   Right digit

7-8. The gate circuit in Fig. 7-48 provides the desired logic, but uses three different types of gates. Use De Morgan's laws to construct an equivalent circuit with just two types of gates.

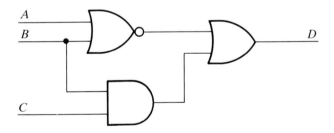

Fig. 7-48. Gate circuit in Prob. 7-8 that is to be replaced by an equivalent one with fewer gate types.

7-9. Refrigeration systems often are equipped with a low-pressure protector, because when the suction pressure drops too low the reduced flow of refrigerant over the compressor motor may cause the motor to overheat. A hot-gas valve,

158   Chapter 7. Binary Numbers and Digital Electronics

as in Fig. 7-49, can be opened to protect the motor, but it is inefficient, and it is advisable to inform the operator that the condition has occurred. Design a control system using gates that opens the hot-gas bypass when the pressure drops below 120 kPa, closes the valve when the pressure rises above 160 kPa, and also turns on an alarm light when the pressure drops below 120 kPa which stays on until reset by the operator.

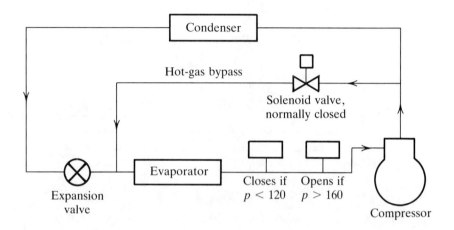

Fig. 7-49. Refrigeration system with low-pressure protection.

Chapter 8

Conversion Between Digital and Analog

8-1 Elements of a Microcomputer Controller

The overall structure of a computer controller often fits the pattern shown in Fig. 8-1. The transducers and status sensors provide voltages to the analog channels of a multiplexer (MUX). The MUX selects one of these channels at a time and sends that signal on to the analog-to-digital converter (ADC), which feeds a binary number of perhaps eight bits to the microcomputer. Acting in

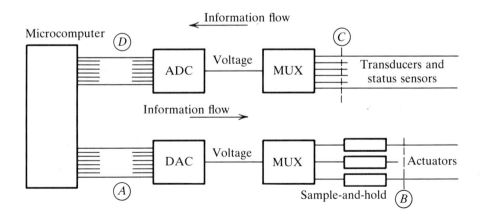

Fig. 8-1. Typical components of a computer control system.

accordance with the program stored in the microcomputer by the user, the microcomputer takes action based on this program and the signal received. This action may be to output an 8-bit binary number that can be converted to a voltage by a digital-to-analog converter (DAC). In other cases the output of a single bit may be all that is necessary to activate one of the actuator devices discussed in Chapter 8. If there is more than one actuator, a MUX transfers the voltage from the DAC to the appropriate channel. Since the MUX is connected momentarily to only one channel, the signal leaving the MUX is as shown in Fig. 8-2. The voltage appears as a series of pulses, but the actuator will need a continuous signal. A "sample-and-hold" device in the circuit following the MUX will provide a continuous voltage to the actuator at a magnitude of the last pulse delivered by the DAC.

This chapter concentrates on the chain of components in Fig. 8-1 from point A to point B and from C to D. Thus, four devices are explored: the DAC, MUX, ADC, and the sample-and-hold circuit.

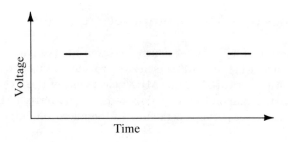

Fig. 8-2. Signal leaving multiplexer destined for an actuator.

8-2 A Simple DAC

A DAC is shown symbolically in Fig. 8-3a receiving an 8-bit binary number and converting it to a voltage. The DAC is to deliver a voltage proportional to the magnitude of the input binary number, as in Fig. 8-3b. A simple DAC could be constructed through the use of a summing amplifier circuit (Sec. 3-9), as shown in Fig. 8-4. A supply voltage V_s is connected to a group of lines each of which contains a switch and a resistance. Closure of the top switch is equivalent to activating the LSB and providing an output voltage $V_o = -V_s(R/256R) = -V_s/256$. In general, the output voltage is

8-2. A Simple DAC

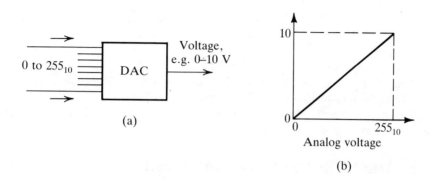

Fig. 8-3. (a) A DAC, and (b) the relationship of the output to the input.

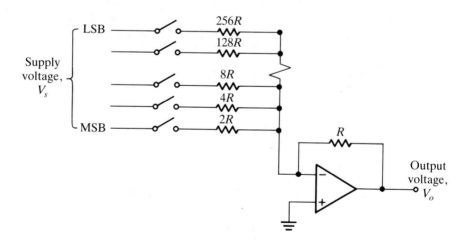

Fig. 8-4. A rudimentary DAC.

$$V_o = -V_s \left(\underbrace{\overbrace{\frac{1}{2}}^{\text{MSB}} + \frac{1}{4} + \frac{1}{8} + \cdots + \overbrace{\frac{1}{256}}^{\text{LSB}}}_{\text{include whichever bits are "on"}} \right)$$

so V_o varies between 0 and $-V_s(255/256)$.

This elementary DAC will work, but it poses fabrication problems in holding the resistance of these various resistors to the precise multiples required. Instead of the circuit shown in Fig. 8-4, the R-$2R$ ladder circuit explained next is commonly used.

8-3 DAC Using R-$2R$ Ladder Circuit

The customary circuit used in DACs is the R-$2R$, which is a combination of two groups of resistances—one group with identical resistances of R and the other group with twice that resistance, namely $2R$. The circuit for a 4-bit DAC is shown in Fig. 8-5. When a bit is 1, the switch is in its leftmost position such that 5 V is fed into the circuit; otherwise the bit is 0 and the switch is connected to the ground bus.

To illustrate the characteristics of the R-$2R$ circuit, a 2-bit DAC will be analyzed, as in Fig. 8-6a where only the LSB is on. The equivalent circuit is shown in Fig. 8-6b. Point X is held at 0 V by the op amp, so if the current passing through the op amp resistor can be determined, V_o can also be calculated. The equivalent resistance between the 5 V level and ground is

$$R_{eq} = 2R + \frac{1}{1/R + 1/(2R)} = \frac{8R}{3}$$

$$I_3 = \frac{2I_1}{3} = \frac{5}{4R}$$

so

$$V_o = V_x - RI_3 = 0 - RI_3 = -\frac{5}{4}V = -V_s\left(\frac{1}{4}\right)$$

If only the MSB is on, $I_1 = I_2 = 0$, and $I_4 = 5/(2R)$. Then,

$$V_o = -\frac{5}{2R}R = -\frac{5}{2}V = -V_s\left(\frac{1}{2}\right)$$

If both the MSB and LSB are on, the restructured circuit in Fig. 8-7 shows that the current passing through the op amp resistance is $I_3 + I_4$, where

$$I_4 = \frac{5}{2R} \quad \text{and} \quad I_3 = \frac{5}{4R}$$

8-4. The 1408 DAC

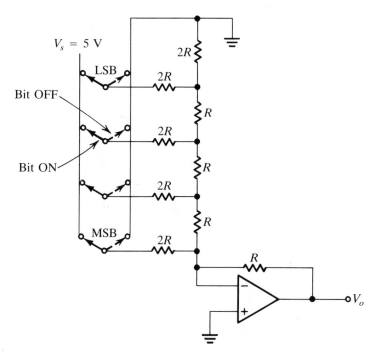

Fig. 8-5. A 4-bit DAC using an R-$2R$ circuit.

Thus

$$V_o = -5\left(\frac{3}{4}\right) = -V_s\left(\frac{3}{4}\right)$$

8-4 The 1408 DAC

One frequently used DAC chip is the MC1408 8-bit converter whose pin diagram is shown in Fig. 8-8. Expected connections to the DAC are power supplies, eight connections for the digital input, and a voltage output. The chip requires a $-15\,\text{V}$ power supply as well as $+5\,\text{V}$ supply and ground. Also the chip itself does not deliver an output voltage but instead regulates a current into pin 4 that varies from 0 to $2\,\text{mA}$ as the digital input varies from 0000 0000 to 1111 1111, as shown in Fig. 8-9. The magnitude of this current i_o must then be translated into a voltage through the use of an op amp.

164   Chapter 8. Conversion Between Digital and Analog

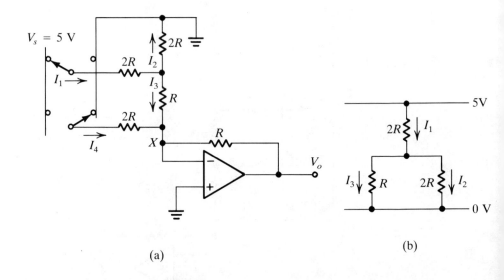

Fig. 8-6. (a) Two-bit DAC with only LSB on, and (b) the equivalent circuit.

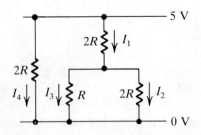

Fig. 8-7. Circuit when both LSB and MSB are ON in a 2-bit DAC.

8-5. Applying the 1408 DAC

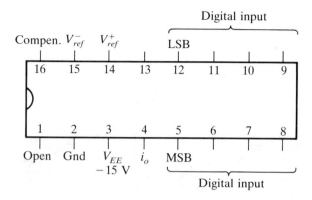

Fig. 8-8. Pin diagram of the MC1408 DAC.

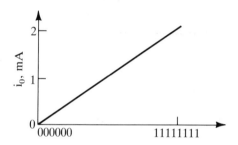

Fig. 8-9. Input current at pin 4 as a function of the digital input.

8-5 Applying the 1408 DAC

An operating circuit incorporating the 1408 DAC is shown in Fig. 8-10. The main chips are the 1408 DAC and a 741 op amp. For a 0 to 10 V output of the assembly, the feedback resistance of the op amp is 5 kΩ, which conforms to the 0 to 2 mA current drawn into pin 4. Figure 8-10 shows 0.1 μF capacitors to ground. The practice of installing such capacitors is a common one to protect the chip from voltage spikes.

The calibration procedure is as follows:

1. With all bits off, adjust the null pot on the op amp so that the analog output is zero.

2. With all bits on, adjust the R_{ref} pot that feeds pin 14 to achieve 9.961 V analog output. The magnitude of 9.961 derives from 10(255/256) V.

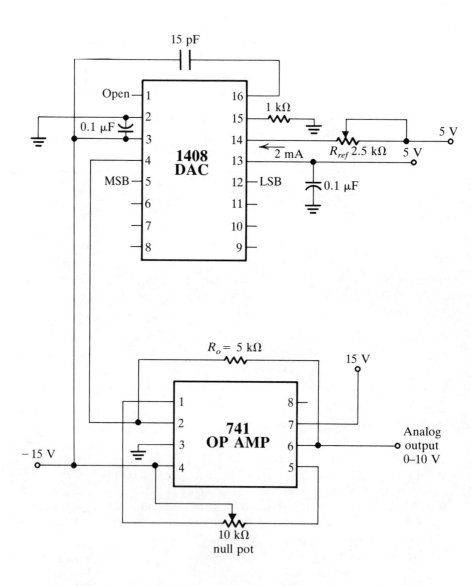

Fig. 8-10. The complete 1408 DAC with amplifier.

8-6 Multiplexers

A multiplexer (MUX) is the electronic equivalent of the rotary switch shown in Fig. 8-11. When in operation the common line is connected to one of the eight channels. The MUX can also be disabled, in which case the common is not connected to any of the channels. Rather than requiring a manual rotation of the wiper, as Fig. 8-11 implies, the selection should ideally be made by electric signals. For a four-channel MUX it should only require two binary lines to select any of the four different channels. Figure 8-12a shows a gate circuit that in conjunction with the four switches forms a rudimentary MUX. The channels selected by the four combinations of A and B are shown in Fig. 8-12b.

Some MUXs are unidirectional in that the signal can be transmitted in only one direction. On the other hand, many MUXs that use an analog switch of the type explained in Sec. 7-30 are designed to function as either multiplexers or demultiplexers. An example of this bilateral multiplexer is the 4051 (CD4051 of RCA, MC14051 of Motorola, etc.). The pin diagram of this eight-channel MUX is shown in Fig. 8-13. The control signals are lines A, B, and C. Channel 0 is selected by imposing 000 on CBA, channel 1 with 001 on CBA, etc. The inhibitor pin must be grounded for the MUX to be activated. The chip can pass a maximum current of 10 mA in either direction. There are some voltage limitations on the analog signals that can be switched by 0 and 5 V digital inputs at A, B, and C. If V_{DD} is 5 V and V_{EE} is -13.5 V, the range of voltage of analog signals can be between -13.5 and 5 V.

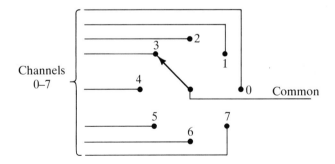

Fig. 8-11. A rotary switch that is duplicated by the MUX.

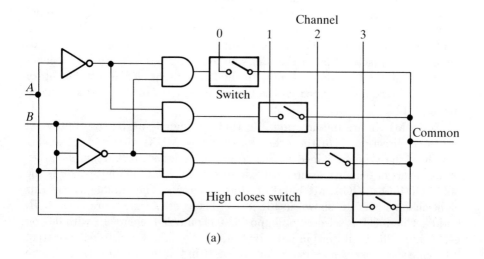

A	B	Channel
0	0	0
0	1	1
1	0	2
1	1	3

(b)

Fig. 8-12. (a) A gate circuit for an elementary four-channel MUX, and (b) channels selected by various combinations of control lines A and B.

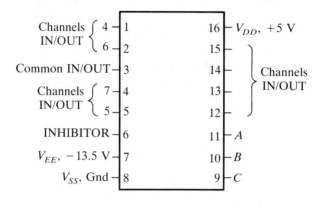

Fig. 8-13. Pin diagram of the 4051 MUX.

8-7 Fidelity of Voltage Transmission Through a MUX

When the configuration described in Fig. 8-1 is used, the magnitude of the voltages must be preserved when passing through the MUX. There is internal resistance in the MUX between the common and the selected channel that may be of the order of 100 Ω in a MUX such as the 4051. If the current passing through the MUX is 10 mA, the voltage drop would be approximately 1 V, which would probably be unacceptable. If the current passing through the MUX is restricted to a few microamps, the output voltage will essentially equal the input voltage—in either direction. Two techniques of preventing appreciable current flow through the MUX are illustrated in Fig. 8-14. If the device on the output side of the MUX has a very high impedance as in Fig. 8-14a, the voltage will pass through essentially undistorted. Most

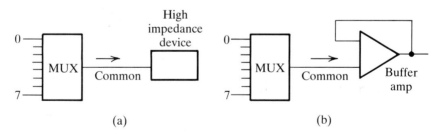

Fig. 8-14. Preventing current flow through the MUX to preserve the fidelity of voltage by (a) feeding output to a device with a high impedance, and (b) using a buffer amp.

analog-to-digital converters are high-impedance devices. When the device on the output does draw current, a buffer amp, as in Fig. 8-14b, is recommended.

Another concept for preserving the fidelity of the analog voltage is to use multiple DACs, each with its own latch, as illustrated in Fig. 8-15. In this arrangement the MUX is used to select the latch that will be enabled. When a given latch is not connected to the 5 V common, a pull-down resistor grounds the enable. One advantage of the configuration of Fig. 8-15 is that the MUX does not have to preserve an analog voltage but only needs to transmit a logical high or low.

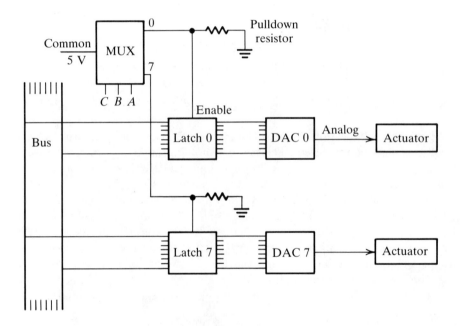

Fig. 8-15. Multiplexing on the digital side of multiple DACs.

8-8 Sample-and-Hold Circuits

A direct connection from the MUX to an actuator, as shown in Fig. 8-16a, would be unsatisfactory, because while the actuator would be supplied with the proper voltage the signal would appear in pulses as shown in Figs. 8-2 and 8-16b. The microcomputer has other channels to serve and other duties to perform as well, so the percentage of time that it can feed a given actuator

8-8. Sample-and-Hold Circuits

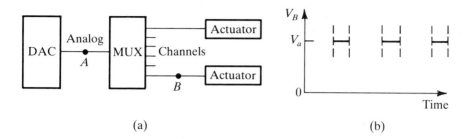

(a) (b)

Fig. 8-16. (a) This connection between the MUX and actuator would provide a signal to the actuator as shown in (b).

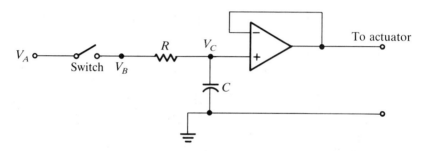

Fig. 8-17. A sample-and-hold circuit.

may be small. The actuator, however, needs a steady voltage that changes only gradually as the control signal from the DAC changes.

To translate the pulses into a constant or slowly changing voltage, a "sample-and-hold" circuit, as shown in Fig. 8-17, may be used. This circuit is an RC circuit with a buffer amp to prevent the actuator from draining the capacitor. When the MUX is disabled or serving another channel, the resistance that the sample-and-hold circuit sees is very high, giving the effect of a switch closing for a short time, then opening. When the switch closes (MUX connects the channel in question to the common line), current flows across R, bringing V_C closer to V_A. During the portion of the cycle that the switch is open (MUX disabled or serving another channel) some current drains through the op amp. A qualitative evaluation of the operating parameters of the sample-and-hold circuit is as follows:

1. A small value of R makes V_C more responsive to V_A but runs the risk of overloading the MUX or the op amp associated with the DAC (Fig. 8-10) if there is a large difference in voltage between V_C and V_A.

2. A large magnitude of capacitance C damps the rate of voltage decay caused by leakage through the op amp during the off cycle but makes V_C less responsive to changes in V_A.

3. A long pulse time during which the MUX serves a channel is desirable for matching V_C and V_A, but the length of time is limited by other duties required of the microcomputer.

Instead of using the 741 op amp discussed in Chapter 4, it is preferable to use the 3140 op amp for the sample-and-hold circuit, because the current drain of the 3140 is of the order of 1/10,000 of that of the 741.

8-9 Operating Sequence with Multichannel Control

Several characteristics of the DAC and the MUX should be kept in mind when programming the microcomputer for their use. Suppose that the assembly available for controlling several actuators is as shown in Fig. 8-18, where the microcomputer has two ports (port A and port B) available to control the MUX and feed a binary number to the DAC. The microcomputer is capable of only one operation at a time, but when an 8-bit number is placed on port A it remains latched, and similarly for port B. When changing channels on some MUX models, several of the output channels will overlap momentarily, so erroneous readings temporarily appear on the affected channels. Furthermore, when a given channel is turned on or off by the A-B-C control lines, there is a momentary conducting path from the channel to V_{EE}, which is approximately -13.5 V.

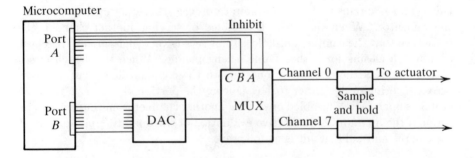

Fig. 8-18. Output system of a certain microcomputer for which the switching sequence is described.

8-10. Where Analog-to-Digital Conversion Is Needed

The MUX and the DAC do not operate instantaneously, and the microcomputer is so fast that some software delays are advisable. A sequence for switching from one channel to another in the circuit of Fig. 8-18 would be:

1. Inhibit MUX.
2. Send from port B to DAC the binary number intended for the next channel.
3. Allow a short delay for the DAC to settle.
4. Shift the A-B-C control lines to the next channel.
5. Provide a short delay for the MUX switching to occur.
6. Enable the MUX.
7. Allow time for the sample-and-hold circuit to digest the signal.
8. Return to Step 1 for the next signal.

The above sequence inhibits the MUX during switching operations. One deficiency that is not solved is that when the MUX is enabled by canceling the inhibit, there is a momentary dumping of the charge to V_{EE}. This dumping does not occur, incidentally, upon switching the inhibit on. There is no way to avoid this dumping, but the draining of the charge from the sample-and-hold circuit of Fig. 8-17 is lessened by using a large resistance R. It is clear, then, that an adequate pulse width in Fig. 8-16b should be maintained to provide a signal for an adequate period of time.

Some of the foregoing problems are bypassed by choosing the multiple DAC configuration of Fig. 8-15. In addition to avoiding the problem of input-to-output voltage distortion that was mentioned in Sec. 8-7, sample-and-hold circuits are not needed. The holding operation is performed on the digital, rather than on the analog, side. The sequence of controlling the multiple DAC configuration still needs to be planned properly (Prob. 8-5).

8-10 Where Analog-to-Digital Conversion Is Needed

We next concentrate on the sequence from point A to point B, as shown in Fig. 8-19. Analog voltages coming from transducers, perhaps through a multiplexer, must be converted into equivalent binary numbers to be received by the microcomputer, as shown in Fig. 8-19. The sequence of operations is that the microcomputer selects the MUX channel to be transmitted, the voltage of this channel is forwarded on to the analog-to-digital converter (ADC), and the ADC converts the voltage to a binary equivalent that is received by the microcomputer.

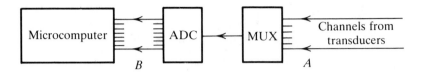

Fig. 8-19. Position of the analog-to-digital converter.

This chapter first presents a functional description of an ADC, describes some search routines commonly used in commercial ADCs, and then describes procedures for implementing ADCs.

8-11 Internal Functions of One Class of ADCs

The conceptual construction of the DAC-based ADC is shown in Fig. 8-20 as consisting of three major components: a DAC, a generator of digital numbers, and a comparator. Since the DAC is a component in this type of ADC, it is to be expected that the ADC will be more expensive and more complicated than a DAC. The operation is that the generator provides a digital number that serves as both the output of the ADC and the input of the DAC. If the voltage output of the DAC matches that of the analog input, then the conversion is complete and the output digital number is taken as the conversion. If the voltage supplied by the DAC is not equal to the analog input, the generator provides a different digital number whose value is influenced by whether the comparator senses the DAC output to be higher or lower than the analog input.

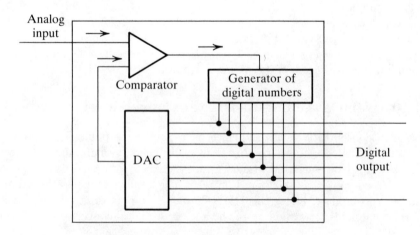

Fig. 8-20. Basic components of the ADC.

8-12. More Complete Description of an ADC   175

The above description implies that some sort of search must take place and that there will be a series of trials and checks. This requirement suggests the need of some binary number with which to begin the search, some means of telling the generator to move to another number, and an indication of when the ADC has been satisfied.

The component that sequences the ADC along in its operation is a clock, which was first discussed in Sec. 7-22.

8-12 More Complete Description of the Internal Functions of an ADC

The next stage of detail of the ADC is shown in Fig. 8-21, in which the digital number generator of Fig. 8-20 is broken into a sequence controller and a register. The register holds the digital number until it is changed by the sequence controller. The sequence controller performs several sophisticated operations that include sending the initial number to the register upon receipt of the start pulse, revising the number in response to the message from the comparator, and indicating when the conversion is complete.

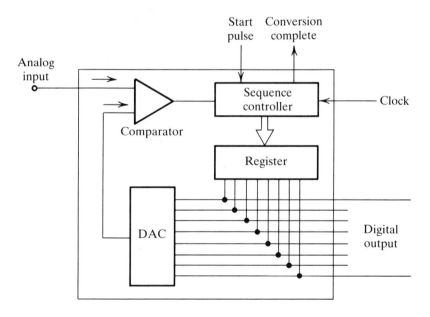

Fig. 8-21. Expanded diagram of internal components of an ADC.

8-13 Staircase and Successive Approximation Search Routines and Dual-Slope Integration

There are a number of different routines used by various commercial ADCs for adjusting the binary number, and three of them will be described here. The staircase is the simplest procedure, and the process starts with a digital number of zero, as shown in Fig. 8-22. The number then increments with each clock pulse until the voltage from the DAC exceeds the input voltage. The number of clock pulses is counted, and this total is the output digital number.

In the successive approximation technique the progression starts at the MSB and moves down toward the LSB one bit at a time. Initially the number in the register is assigned a zero value. Next the MSB is first assigned to 1, and if the DAC output is less than the input voltage the bit is left at 1; otherwise it is restored to zero. Then the test moves down to the next lower bit to repeat the process. The search thus determines successively which half, fourth, eighth, etc., of the range represents the correct conversion.

In comparison to the staircase routine, successive approximation will, in general, require fewer tests. In an 8-bit ADC, for example, the staircase may require anywhere from 1 to 255 tests while the successive approximation always requires 8.

The third ADC concept is dual-slope integration, which integrates the input voltage for a specified number of clock pulses, as shown in Fig. 8-23. Thereafter a negative reference voltage is switched to integrate down to zero voltage. The number of pulses required for the decay to zero is counted and is the digital output. An advantage of the dual-slope integration technique

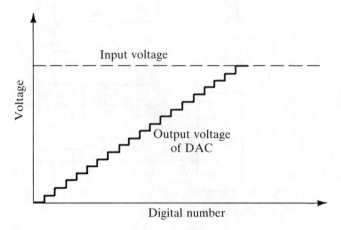

Fig. 8-22. Staircase routine.

8-14. Pin Diagram of an 8-Bit ADC

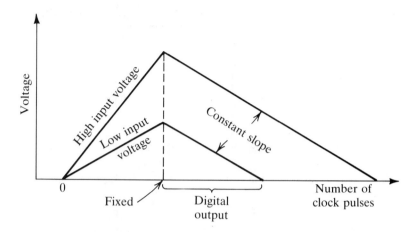

Fig. 8-23. Dual-slope integration.

is that the result is often less affected by noise in the input voltage signal. If a voltage spike occurs during the conversion by the staircase or successive approximation routine at an inopportune time, the digital output could show an appreciable error, while in dual-slope integration the error might be small if it occurs only during a small fraction of the integration interval.

8-14 Pin Diagram of an 8-Bit ADC

The choice of placement of the various connections varies from one manufacturer to another, but certain connections would be expected on all ADCs. In order to be specific, one ADC will be chosen for illustration, the National Semiconductor ADC 0800. The power connections include +5 V, −5 V, a reference voltage of −12 V, and a ground. The pin diagram is shown in Fig. 8-24. The input voltage for this ADC should range between −5 and +5 V and is connected to pin 12. The output bits starting with the MSB in descending order are pins 4, 3, 2, 1, 17, 16, 14, and 13. The clock is connected to pin 11, and the ADC 0800 can operate with frequencies between 5 kHz and 2 MHz. Pin 7 is an enable connection that requires +5 V in order to permit the chip to function.

To start the conversion a 5 V pulse is applied to the Start Conversion pin. At the end of conversion the voltage on pin 9 goes high. When the analog-to-digital conversion is a part of the computer controller, the end-of-conversion signal can inform the microcomputer to proceed. If time is not crucial, a delay can be built into the software to allow generous time for the conversion.

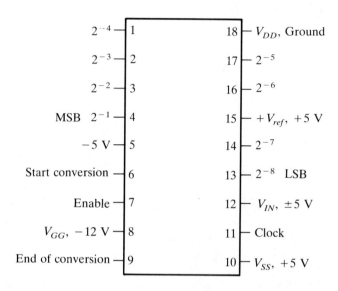

Fig. 8-24. Pin diagram of an ADC 0800 analog-to-digital converter.

The conversion on the 0800 chip requires approximately 40 μs, although the precise time required is a function of the clock frequency.

8-15 Characteristics of the ADC 0800

The expected input voltage range is from -5 to $+5$ V, which yields 255_{10} to 0_{10}, respectively, as shown in Fig. 8-25. If the actual input voltage range differs from this, the actual range should be converted into the -5 to $+5$ V range with a conditioning circuit.

The binary output lines can be in one of three conditions: (1) 0 V, (2) 5 V, or (3) providing an infinite resistance. Thus, the output is three-state, as described in Sec. 7-20. The utility of this output form is clear when the output of the ADC is connected to a bus as shown in Fig. 8-26. When the ADC is enabled, each bit delivers either 5 V or 0 V to the bus. When the ADC is not enabled, the bus sees a high resistance at the ADC, which is the equivalent of disconnecting it from the bus.

The ADC latches the digital output until a new conversion is ordered. It is possible to change the input voltage, but if no "start conversion" occurs there will be no change in the digital output.

8-15. Characteristics of the ADC 0800

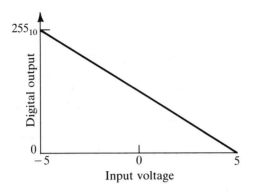

Fig. 8-25. Relationship of the digital output to the analog input.

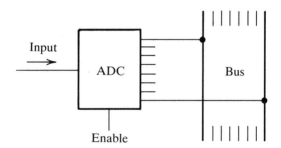

Fig. 8-26. Three-state options (5 V, 0 V, or high resistance) with the enable pin regulating whether the bus sees a voltage or a high resistance.

Figure 8-19 shows analog voltages passing through a multiplexer which selects one channel for passage to the ADC. Section 8-7 stressed the need of limiting the current passed through the MUX in order to preserve the fidelity of the voltage signal. Fortunately ADCs usually possess a high input impedance (of the order of 100 MΩ) so a buffer amp is not necessary.

8-16 Analog-to-Digital Conversion Using a DAC in Combination with Software

When the microcomputer is not heavily loaded with duties the analog-to-digital conversion can be accomplished with a low-cost DAC in combination with instructions from the computer program. With the circuit shown in Fig. 8-27 the flow diagram of Fig. 8-28 will perform the analog-to-digital conversion. The procedure is one of first sending 0000 0000 out from the computer on lines D0 to D7 into the DAC and comparing the output of the DAC with the analog signal to be converted. If the comparator shows the voltage M in Fig. 8-27 higher than voltage N, the computer increments (adds 1 to) the D0-D7 number and repeats the test. When the comparator switches from low to high, the binary number existing at the time is taken as the conversion of the analog voltage.

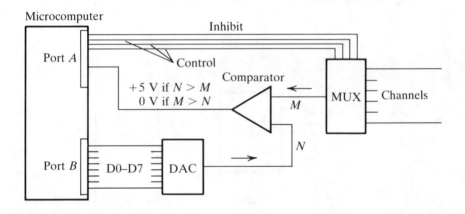

Fig. 8-27. Arrangement for using a DAC in combination with software to achieve an analog-to-digital conversion.

8-17 Choosing the ADC

Several types of ADCs have been explained in this chapter—two of them DAC-based and one the dual-slope integrator. Still other concepts are available, such as the use of transducers that provide a frequency output that is proportional to the variable being sensed. The computer can count the number of pulses during a fixed time interval and thereby obtain a digital equivalence of the magnitude of the variable.

8-17. Choosing the ADC

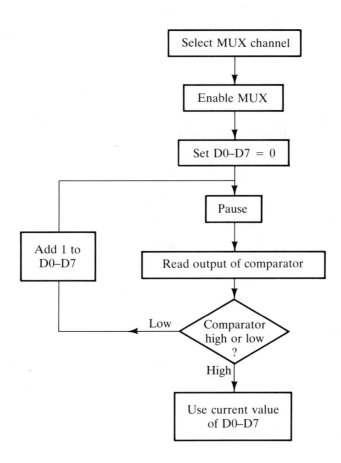

Fig. 8-28. Flow diagram for using a DAC as an ADC.

The illustrations chosen in this chapter were of 8-bit ADCs, but high-resolution ADCs with more bits are also available. Suppose, for example, that the range of interest of a sensed temperature is 100°C and that a resolution of 0.1°C is required. The ADC must therefore be capable of distinguishing 1000 different readings, which requires a 10-bit ADC. When the computer control system must scan variables at a high rate, the speed of the ADC may become an issue, and in general the faster converters are more expensive. Finally, the accuracy of the ADC should be chosen such that it is appropriate for the assignment.

Problems

8-1. For the R-$2R$ ladder circuit used in DACs, show by means of a current analysis for a three-bit ladder the output voltage expected for (a) 100_2 and (b) 101_2. **Ans.:** (a) $-V_s/2$; (b) $-V_s(5/8)$.

8-2. Two 4051 MUXs are available for multiplexing 16 channels to one common channel. Five control lines are available; four of them select any one of the 16 channels, and the other inhibits both MUXs when the line goes high. Show the gate circuit to achieve this control.

8-3. Devise a gate circuit using only NOR gates that controls the four switches in Fig. 8-12 with the two input lines.

8-4. Develop a gate circuit that uses two inputs to control three MUXs according to the following table:

Inputs		Inhibits		
I	J	MUX 1	MUX 2	MUX 3
0	0	0	1	1
0	1	1	0	1
1	0	1	1	0
1	1	1	1	1

8-5. Develop the sequence for sending first a signal to one latch/DAC and then to another latch/DAC in the arrangement of Fig. 8-15. Two 8-bit latched output ports are available, one for the binary number on the bus and the other for controls. The functions to control are (1) MUX inhibit, (2) A-B-C MUX control, and (3) the binary number on the bus.

8-6. A certain 8-bit ADC using the successive approximation search routine gives a binary output from 0_{10} to 255_{10} in a linear relation to the input voltage as it varies from 0 to 10 V, respectively. If the input voltage is 6.1 V, make a table of the actions that result in the complete conversion:

Binary Number Sent to DAC	DAC output higher or lower than input?
etc.	

8-7. The input signal to an ADC is subjected to noise during the conversion process as shown in Fig. 8-29. The ADC is 8-bit and is intended to provide an output of 0 to 255_{10} based linearly on an input variation from 0 to 10 V, respectively. (a) What would be the digital conversion of 4.5 V without noise? (b) If the ADC is DAC-based using successive approximation and the noise distortions occur during the fourth and seventh determinations, respectively, what would be the digital output? (c) If the ADC uses dual-slope integration with the input during the 0–8 time period, what would be the digital conversion?

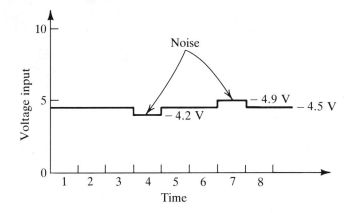

Fig. 8-29. Noise in the input signal to an ADC in Prob. 8-7.

Chapter 9

Memories

9-1 Function and Types of Memories

This chapter continues the progression toward the microcomputer by exploring memories. Memory chips store information that the computer is to process, the results of the processing, and the program that provides the instructions of what actions the computer should perform. Even though memories are rarely used apart from the computer, it is helpful to examine the types and functions of the various memories as an aid to a better understanding of the operation of the microcomputer.

Only zeros and ones can be stored, and each bit of memory can be visualized as a double-throw switch, as in Fig. 9-1, that is connected either to 5 V or to ground. When a bit of memory is being read, the memory either pulls the reader low by sinking current or supplies 5 V—in either case with very low current. The write operation assigns the bit to either 0 or 1, so the write operation is the equivalent of positioning the switch in Fig. 9-1.

Several types of memories that will be discussed in this chapter are the read-only memory (ROM), erasable programmable read-only memory (EPROM), random-access memory (RAM), and the electrically erasable programmable read-only memory (EEPROM).

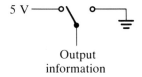

Fig. 9-1. A bit of memory symbolized by a double-throw switch.

9-2 ROMs

A ROM is a memory that is the most permanent of all and cannot be changed by the user. The ROM should contain, then, only that information that does not change as the user proceeds from one assignment to the next. Normally the information stored in a ROM is that which controls the routine operation of the microcomputer. ROMs are mask programmed, and a link is provided for each bit that is to be 1 and no link is provided when the bit is to be 0.

The manufacturing process is such that it is a factory operation and not one that can be accomplished by the user. The contents cannot be altered following manufacture.

Memories are connected to buses within the computer. The bus arrangement will be explained more fully in Chapter 11, but it is sufficient at this point to state that a bus is a series of parallel conductors. Two buses are standard—one a data bus and the other an address bus, as shown in Fig. 9-2, although in certain microcomputers the data and address buses are multiplexed. A memory, such as a ROM, is connected to each of these buses. In an 8-bit microcomputer the data bus has eight lines. The number of address lines varies, but even small microcomputers will likely have 16 or more. Not all of these address lines need be connected to the ROM, but the minimum number depends on the capacity, measured in bytes, of the memory. For example, nine lines would be needed to address 512 different memory locations, 10 lines for 1024 locations, etc. The setting of the address lines gives the data bus access to one particular byte within the memory.

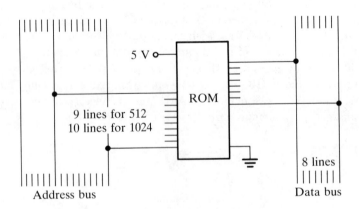

Fig. 9-2. Some necessary pins on a ROM, and their connections to the data and address buses.

9-3 EPROMs

The EPROM is one version of a programmable read-only memory (PROM), which is a memory that can be programmed by the user without extensive manufacturing facilities. The EPROM can be erased and reprogrammed by the user. Figure 9-3 is a drawing of an EPROM, which can be erased by shining ultraviolet light through the window. Special UV lamps are available for the erasing process, which may require a half hour. The EPROM is programmed through the application of high-voltage (perhaps 24 V) pulses. A popular EPROM is the 24-pin 2716 whose pin diagram is shown in Fig. 9-4. The EPROM must be erased before reprogramming, and this erasing sets all the bits to 1. The programming changes the appropriate bit locations from 1 to 0. The voltage to V_{PP} is normally 5 V, but it is +25 V during programming. To program a byte of memory, the address and data bit inputs are imposed on the EPROM whereupon a 5 V pulse is applied on E/Prog for a duration of several milliseconds. The programming operation then moves to the next memory location to be programmed. Special programming devices are available that can program one EPROM at a time and cost several hundred dollars. Some microcomputers are equipped with the required voltages to program an EPROM onboard the computer.

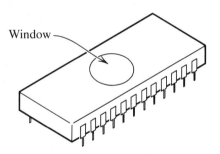

Fig. 9-3. An EPROM.

9-4 RAMs

The designation "random-access memory" is a poor choice, because other memories such as the ROM provide random access as well. It is not necessary to start at the first memory location and proceed through the locations in sequence to arrive at the one to be read or changed, such as occurs in a tape

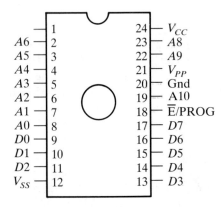

Fig. 9-4. Pin diagram of the 2716 EPROM.

drive. A better name for the RAM would be "read/write memory," because it has the capability of having its memory changed through the 0 and 5 V signals available in the microcomputer, but the contents of a memory location can also be read without altering the contents.

Two types of RAMs are "static" and "dynamic." The static RAM holds the contents of memory as long as power remains on the chip, while the dynamic RAM loses its memory within milliseconds unless it is refreshed. The signal for the dynamic RAM to refresh itself must be provided from an external source. The dynamic RAM requires more supporting logic than the static RAM but possesses a greater memory density. All RAMs are volatile in the sense that if power is withdrawn from the chip, the memory is lost. It is this characteristic of RAMs that makes it advisable to use battery backup on microcomputer controllers. Upon powering up after a power outage or a shutdown, the contents of the RAM are unpredictable. The discussion that follows concentrates on several different static RAMs.

The expected connections to be found on a RAM include:

1. Address pins to specify the memory location

2. Data pins

3. Some means of indicating whether to read (accessing what is already in the memory) or write (changing the contents of a memory location)

4. Chip selects that have the effect of closing or opening internal switches to the data pins

5. Power (5 V and ground)

9-5 The MCM6810 RAM

More insight about the characteristics of a RAM can be obtained by examining a few details. The RAM chosen is a small-capacity one, the 6810, whose pin diagram is shown in Fig. 9-5. The data pins are $D0$ through $D7$ with $D0$ being the LSB. The 6810 has a storage capacity of 128 bytes, so seven address lines are needed to access all these locations. Those address lines are labeled $A0$ through $A6$, with $A0$ being the LSB.

The read/write ($R/\overline{W}$) pin is number 16. A high input (5 V) to pin 16 equips the RAM location to be read; if the input to pin 16 is low, the memory location is prepared to be changed.

There are six "chip select" pins labeled $CS0$, $\overline{CS1}$, $\overline{CS2}$, $CS3$, $\overline{CS4}$, and $\overline{CS5}$. In order to enable the RAM, the pins with bars, namely $\overline{CS1}$, $\overline{CS2}$, $\overline{CS4}$, and $\overline{CS5}$, must be grounded while 5 V must be applied to $CS0$ and $CS3$. The availability of these chip selects permits one of a number of RAMs to be chosen to be read from or written into. Suppose, for example, that two 6810's combine to a total of 256 memory locations accessible to an 8-bit data bus and an 8-bit address bus, as shown in Fig. 9-6. The chip selects $CS0$ through $\overline{CS4}$ are tied to voltages such that the RAM is activated. Chip select $\overline{CS5}$ is connected to the MSB of the 8-bit address bus with the line that serves RAM 2 passing through an inverter. An alternative strategy for activating only one RAM at a time would be to connect the MSB of the address bus to $CS3$ of RAM 2. If the address bus feeds a binary number 0XXX XXXX, RAM 1 is activated and RAM 2 is disabled. On the other hand, the number 1XXX XXXX on the address bus selects a memory location in RAM 2.

```
         Gnd ──┤1        24├── V_CC, 5 V
          D0 ──┤2        23├── A0
          D1 ──┤3        22├── A1
          D2 ──┤4        21├── A2
          D3 ──┤5        20├── A3
          D4 ──┤6        19├── A4
          D5 ──┤7        18├── A5
          D6 ──┤8        17├── A6
          D7 ──┤9        16├── R/W̄
         CS0 ──┤10       15├── CS5̄
         CS1̄ ──┤11       14├── CS4̄
         CS2̄ ──┤12       13├── CS3
```

Fig. 9-5. Pin diagram of the 6810 RAM.

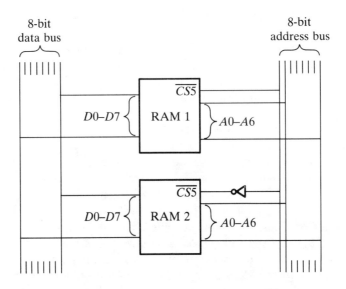

Fig. 9-6. Using chip select pins to address a memory location in one of two RAMs.

The data lines on the RAM have a three-state output, as discussed in Sec. 7-20. When all the chip selects are set to receive or send data, each data pin is at either 0 or 5 V. During the reading operation the 6810 is capable of supplying approximately 0.2 mA for a 1 and can sink about 1.6 mA for a 0. When the chip selects are such that the RAM is disabled, the data pins are seen from the data bus as having an extremely high resistance. This condition is desirable so that data transmitted on the data bus to and from other components is not distorted by the contents of the memory of an idle RAM. The sequence of operations during the reading process would normally be to first set the desired address, set the read/write pin high, activate the chip selects, and then read the information on the data bus. Following a reading or writing operation, the RAM is first disabled by the chip selects.

9-6 Four-Bit RAMs—the MCM2114

The quest for large memory capacity on small chips is relentless, and an example of an improvement of the capacity/size ratio over the 6810 memory chip is to use two 4-bit 2114 memories. The pin diagram of this chip is shown in Fig. 9-7a and the connections to the data and address buses in Fig. 9-7b. The 10 address lines are associated with the 1024 4-bit memory locations on

9-6. Four-Bit RAMs—the MCM2114

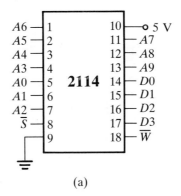

(a)

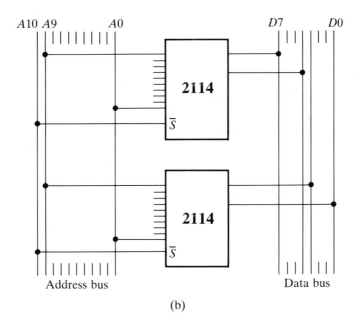

(b)

Fig. 9-7. (a) Pin diagram of the 4-bit 2114 memory chip, and (b) the connections to the data and address buses.

each chip. The data bus is divided with the upper four bits connected to one chip and the lower four bits to the other chip. The $\overline{W}$ pin equips the chip to write when low and read when high. Only one chip select is used on the 2114.

9-7 Dynamic RAMs

As mentioned earlier, dynamic RAMs have greater memory capacity than static RAMs for a given size chip. Static RAMs incorporate arrays of flip-flops, which retain their status as long as they continue to be powered. The memory of a single bit in the dynamic RAM, on the other hand, is a charged capacitor subject to continuous leakage. Most dynamic RAMs are one-bit devices, which requires them to be used in gangs of eight for an 8-bit data bus. On the MCM4027 dynamic RAM, for example, a signal to one of the pins activates the refresh cycle, which must be performed at intervals of 2 ms or less.

9-8 EEPROMs

The process of reprogramming the EPROM described in Sec. 9-3 was to remove the chip from the circuit, place it under an ultraviolet lamp, and then reprogram it either in a dedicated programming device or in a microcomputer suitably equipped for the purpose. The feature of the electrically erasable programmable read-only memory (EEPROM), such as the Intel 2817, is that the memory chip can remain on the board during reprogramming. Anywhere from one byte to the entire memory capacity of 2 kbytes can be reprogrammed, although one byte is altered at a time and this process requires between 10 and 50 ms. The 2817 EEPROM requires 21 V to one pin during the write operation. The chip has a long life, because it is capable of being reprogrammed 10,000 times, in contrast to the 2716 EPROM, which can be erased only 30 to 40 times.

9-9 Memories on the Microcomputer

Essentially all microcomputers incorporate RAMs and one or more ROMs. Some of the functions served by the RAM are to store nonpermanent programs and to store data that are part of the current control functions. The ROM is programmed at the factory and contains utility programs applicable to the unique hardware of the microcomputer. EPROMs are devoted to programs and other storage that are probably intended to be permanent, but since they can be reprogrammed by a user with modest equipment, the contents can be revised if necessary. Memories are not peripheral equipment but are integral components to the microcomputer. They were studied as individual elements in this chapter only to isolate the characteristics and capabilities of the various types of memories found in a microcomputer.

General References

1. *Motorola Memory Data Manual*, Motorola Semiconductor Products, Inc., Austin, TX, 1982.

2. L. Wheeler, "The Practical EEPROM," *Byte*, pp. 460–468, July 1983.

3. *Memory Components Handbook*, Intel Corporation, Santa Clara, CA, 1986.

Problems

9-1. An 18-pin RAM is being designed with the intent that the chips will be used in pairs. Allowing a pin for each of the following functions: 5 V, ground, R/W, and a chip select, what is the maximum total capacity in bytes of the two chips, if (a) the chips store eight bits at each memory location, and (b) the chips store four bits at each location, as done by the MCM2114?

9-2. In a certain microcomputer that incorporates a 6810 RAM, $\overline{CS1}$ through $\overline{CS5}$ are tied permanently to activate the RAM, and $CS0$ is connected to the eighth bit in the address bus. List the sequence of operations the microcomputer should use to write the number 1001 0110 to memory location 57_{10}.

9-3. Four pairs of 2114 RAMs providing 4096 bytes of memory are addressed by 12 address lines. The ten lowest address lines connect to the address pins of the RAMs, and the highest two lines are fed through a gate circuit to the chip select to activate the desired pair. Develop a gate circuit to perform this assignment.

Chapter 10

Binary Arithmetic

10-1 The Eight-Bit Microcomputer

A "bit" is a binary digit expressing either the number 0 or 1. In digital electronics the 0 and 1 are expressed by the nominal 0 V or 5 V, respectively, although there is a range of voltages recognized as 0 V and a range recognized as 5 V, as explained in Chapter 7. The microcomputer in which we are interested in these chapters is the 8-bit microcomputer typified by such microprocessor families as the Motorola 6800, Intel 8080/8085, TI 990, and Rockwell 6502.

Sixteen-bit microcomputers such as the Motorola 68000, Intel 8086, and TI 9900 are in common use, particularly in desktop microcomputers. It is possible for the 8-bit microcomputer to perform 16-bit arithmetic, as will be studied later in this chapter, but the direct use of the 8-bit microprocessor limits us to expressing 256 different values. Thus, a maximum of 256 different gradations of the sensed variable can be identified, and a modulating actuator may be controlled to 256 different positions. In most mechanical engineering systems the 256 gradations are adequate for both the sensed and controlled variable. Using 8-bit arithmetic, however, imposes the limit of 256 different numbers, so 8-bit arithmetic does not lend itself to number crunching. For dedicated controllers of small machines, the 8-bit microprocessor dominates, and in fact 85% of all microprocessors sold in the United States[1] are 8-bit.

10-2 Two's Complement Arithmetic—Subtraction

The subtraction process could be performed manually by proceeding systematically from right to left using the appropriate choice of one of the four operations:

$$\begin{array}{cccc} 0 & 1 & 1 & 10 \\ -0 & -0 & -1 & -1 \\ \hline 0 & 1 & 0 & 1 \end{array}$$

The computer is able to utilize the same circuitry to subtract as it uses to add by using the "two's complement" representation of numbers. One number (the subtrahend) is subtracted from another (the minuend) by adding the two's complement of the subtrahend to the minuend. The digit that appears in the ninth position following the operation is ignored. The "one's complement" of a binary number is formed by replacing all 0's with 1's and all 1's with 0's. The two's complement is formed by adding 1 to the one's complement. Thus the two's complement of 1010 1110 is 0101 0001 + 1, or 0101 0010.

Example 10-1. Subtract the binary equivalent of the decimal number 13 from the binary equivalent of decimal 22.

Solutions. The binary equivalent of decimal 13 is 0000 1101 and its two's complement is 1111 0011. The binary equivalent of decimal 22 is 0001 0110, so the results of the subtraction is

$$\begin{array}{l} 0001\ 0110 \\ +\ \underline{1111\ 0011} \\ 1\ 0000\ 1001 \quad = \text{decimal } 9 \\ | \\ \text{ignore} \end{array}$$

10-3 Multiplication

The basic rules of multiplication are

$$\begin{aligned} 0 \times 0 &= 0 \\ 0 \times 1 &= 0 \\ 1 \times 0 &= 0 \\ 1 \times 1 &= 1 \end{aligned}$$

An algorithm for multiplying multi-bit numbers is similar to that of decimal multiplication; for example,

```
decimal 11:      1011
× decimal 9:     1001
                 1011
                 0000
                 0000
                 1011
                 ‾‾‾‾‾‾‾
                 1100011 = 64 + 32 + 2 + 1 = decimal 99
```

In computer control it may be necessary to perform a multiplication in order to scale the output of a transducer. When two one-byte numbers are multiplied, provision must be made for a two-byte product.

10-4 Hexadecimal System

Writing out binary numbers is tedious, and conversion at a glance between binary and decimal numbers requires much practice. Another system closely related to the binary one is the hexadecimal (hex) system, which has a base of decimal 16. The equivalence of the decimal, binary, and hex systems is shown in Table 10-1. The hex system uses the letters A through F for decimal numbers 10 through 15. Conversion between binary and hex is achieved by working with four-bit blocks of the binary number. Thus, for example,

$$\underbrace{1011}_{B} \; \underbrace{0111}_{7}$$

A one-byte number is represented by two hex characters. Conversion into and out of decimal numbers is still cumbersome, so a conversion table between hex and decimal numbers, such as Table 10-2, may be useful.

Example 10-2. What is the binary equivalent of decimal 138?

Solution. From Table 10-2 the hex equivalent of 138 is 8A, and in binary, 8 is 1000 and A is 1010, so the binary number is 1000 1010.

10-5 Labeling Conventions

When there is some question as to which system a number is being represented in, the system should be designated. A frequently used convention that will be adopted here is illustrated by the examples:

Table 10-1. Equivalence of Decimal, Binary, and Hexadecimal Systems

Decimal	Binary	Hexadecimal
0	0000	0
1	0001	1
2	0010	2
3	0011	3
4	0100	4
5	0101	5
6	0110	6
7	0111	7
8	1000	8
9	1001	9
10	1010	A
11	1011	B
12	1100	C
13	1101	D
14	1110	E
15	1111	F

Binary 1101   : 1101_2
Decimal 53   : 53_{10}
Hexadecimal 23F: $23F_{16}$ or \$23F

Later on when we are studying the microcomputer, the hex number \$F2 may represent a memory location, and the contents of \$F2 will be designated as #\$F2.

10-6 Signed and Unsigned Numbers

The microprocessor operates simultaneously in two different modes—with signed and with unsigned numbers. The microprocessor further provides information following each addition and subtraction that permits the user to properly interpret the result in accordance with whatever mode the user has chosen. The unsigned number mode constitutes the numbers 0 to 255, while the signed mode comprises both positive and negative numbers that range from -128_{10} through zero to 127_{10}. The two modes may be pictured in a circular arrangement, as shown in Fig. 10-1, with Fig. 10-1a showing the unsigned and Fig. 10-1b the signed mode. In the unsigned mode, which contains only positive numbers and zero, a straightforward addition is represented by

10-6. Signed and Unsigned Numbers

Table 10-2. Conversion between Decimal and Hexadecimal Numbers

	0	1	2	3	4	5	6	7	8	9	A	B	C	D	E	F
00	000	001	002	003	004	005	006	007	008	009	010	011	012	013	014	015
01	016	017	018	019	020	021	022	023	024	025	026	027	028	029	030	031
02	032	033	034	035	036	037	038	039	040	041	042	043	044	045	046	047
03	048	049	050	051	052	053	054	055	056	057	058	059	060	061	062	063
04	064	065	066	067	068	069	070	071	072	073	074	075	076	077	078	079
05	080	081	082	083	084	085	086	087	088	089	090	091	092	093	094	095
06	096	097	098	099	100	101	102	103	104	105	106	107	108	109	110	111
07	112	113	114	115	116	117	118	119	120	121	122	123	124	125	126	127
08	128	129	130	131	132	133	134	135	136	137	138	139	140	141	142	143
09	144	145	146	147	148	149	150	151	152	153	154	155	156	157	158	159
0A	160	161	162	163	164	165	166	167	168	169	170	171	172	173	174	175
0B	176	177	178	179	180	181	182	183	184	185	186	187	188	189	190	191
0C	192	193	194	195	196	197	198	199	200	201	202	203	204	205	206	207
0D	208	209	210	211	212	213	214	215	216	217	218	219	220	221	222	223
0E	224	225	226	227	228	229	230	231	232	233	234	235	236	237	238	239
0F	240	241	242	243	244	245	246	247	248	249	250	251	252	253	254	255

	0	1	2	3	4	5	6	7	8	9	A	B	C	D	E	F
10	256	257	258	259	260	261	262	263	264	265	266	267	268	269	270	271
11	272	273	274	275	276	277	278	279	280	281	282	283	284	285	286	287
12	288	289	290	291	292	293	294	295	296	297	298	299	300	301	302	303
13	304	305	306	307	308	309	310	311	312	313	314	315	316	317	318	319
14	320	321	322	323	324	325	326	327	328	329	330	331	332	333	334	335
15	336	337	338	339	340	341	342	343	344	345	346	347	348	349	350	351
16	352	353	354	355	356	357	358	359	360	361	362	363	364	365	366	367
17	368	369	370	371	372	373	374	375	376	377	378	379	380	381	382	383
18	384	385	386	387	388	389	390	391	392	393	394	395	396	397	398	399
19	400	401	402	403	404	405	406	407	408	409	410	411	412	413	414	415
1A	416	417	418	419	420	421	422	423	424	425	426	427	428	429	430	431
1B	432	433	434	435	436	437	438	439	440	441	442	443	444	445	446	447
1C	448	449	450	451	452	453	454	455	456	457	458	459	460	461	462	463
1D	464	465	466	467	468	469	470	471	472	473	474	475	476	477	478	479
1E	480	481	482	483	484	485	486	487	488	489	490	491	492	493	494	495
1F	496	497	498	499	500	501	502	503	504	505	506	507	508	509	510	511

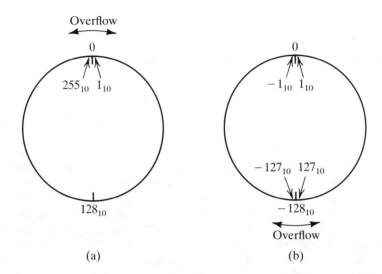

Fig. 10-1. (a) The unsigned and (b) the signed number mode.

a clockwise motion around Fig. 10-1a and a subtraction by a counterclockwise motion.

In the signed mode of Fig. 10-1b the 256 numbers span the range from -128_{10} to 127_{10}. The sequences for decimal, binary, and hex are shown in Table 10-3.

The negative binary numbers are the two's complements of the corresponding positive numbers. Thus, the two's complement of 2_{10} is $1111\ 1110_2$ or FE_{16}. In the signed number mode the number is negative when it contains 1 in the MSB.

When the addition or subtraction operation passes in either direction through the 0 position in the unsigned mode or through the -128_{10} position in the signed mode, the result is invalid.

10-7 Unsigned Numbers—The Carry Flag

The microprocessor performs additions in accordance with the rules listed in Sec. 7-4 and subtractions according to procedures in Sec. 10-2. Thus, if 95_{10} is added to 43_{10}, the binary operation and its visualization on the number circle would be as shown in Fig. 10-2a. The result is $\$8A = 138_{10}$.

In the second example shown in Fig. 10-2b, 87_{10} is added to 184_{10}, which should yield 271_{10}. The binary operation shows a result of $1\ 0000\ 1111 = 271_{10}$, but the microprocessor is limited to eight bits and can only show $0000\ 1111$

10-7. Unsigned Numbers—The Carry Flag

Table 10-3. Computer Interpretation of its 256_{10} Numbers

Decimal	Binary		Hexadecimal
−128	1000 0000	Two's	80
−127	1000 0001	complement	81
—	— —	of	—
−2	1111 1110	positive	FE
−1	1111 1111	number	FF
0	0000 0000		00
1	0000 0001		01
2	0000 0010		02
—	— —		—
126	0111 1110		7E
127	0111 1111		7F

MSB indicates sign

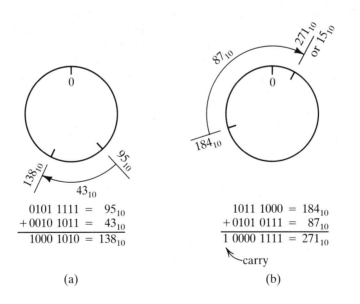

Fig. 10-2. Additions of unsigned numbers: (a) without carry, and (b) with carry.

or 15_{10}. Microprocessors are equipped to provide a flag when 1 is pushed out into the ninth position during an addition. This flag is one bit of storage that is assigned a zero if there was no carry, or a 1 if there was a carry. The function of this carry bit is to serve one of two purposes: (1) indicate an invalid result in unsigned 8-bit arithmetic, or (2) to add 1 to the addition of high-order bytes in 16-bit arithmetic.

In 16-bit addition the two low-order bytes are first added, then the high-order bytes. Microprocessors have available two different instructions to add—"add-without-carry" and "add-with-carry." The add-with-carry instruction performs the addition of the two bytes and adds 1 to the result if the carry bit was 1 before the operation. In the 16-bit addition:

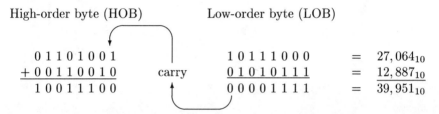

The add-without-carry is used for the LOB, because even if the carry bit were 1 from a previous operation it must not be allowed to affect the LOB addition. The addition of the HOB immediately follows the addition of the LOB, and for this operation the add-with-carry instruction is used. If the carry bit is 1 after the LOB addition, as will be the case in the above example, the add-with-carry instruction adds the two HOBs and then adds a 1 to the result.

The subtraction process moves counterclockwise around the number circle, as shown in Fig. 10-3. The subtraction in Fig. 10-3a gives a valid result in the unsigned number mode:

$$
\begin{array}{r}
1\,0\,1\,0\,0\,0\,1\,1 \\
-0\,1\,0\,0\,1\,1\,0\,0
\end{array}
\rightarrow
\begin{array}{r}
1\,0\,1\,0\,0\,0\,1\,1 \\
+1\,0\,1\,1\,0\,1\,0\,0 \\ \hline
1\,0\,1\,0\,1\,0\,1\,1\,1
\end{array}
= 87_{10}
$$

ignore

In the subtraction of Fig. 10-3b,

$$
\begin{array}{r}
0\,0\,1\,0\,1\,0\,1\,1 \\
-0\,1\,1\,0\,1\,0\,0\,0
\end{array}
\rightarrow
\begin{array}{r}
0\,0\,1\,0\,1\,0\,1\,1 \\
+1\,0\,0\,1\,1\,0\,0\,0 \\ \hline
1\,1\,0\,0\,0\,0\,1\,1
\end{array}
= 195_{10}
$$

no bit

10-7. Unsigned Numbers—The Carry Flag

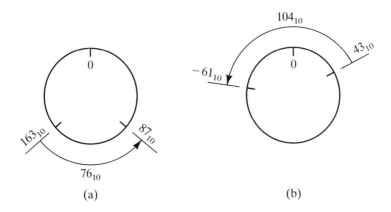

Fig. 10-3. (a) Subtraction of unsigned numbers without carry, and (b) with carry.

the result is 195_{10} in the unsigned mode. This result is incorrect, because the result should have been -61_{10}, which is a number that cannot be expressed in the unsigned mode. A further discrepancy is that the bit normally pushed out into the ninth position and ignored did not appear. If 256_{10} were subtracted from the result, 195_{10}, the difference is -61_{10}, which is correct. These several observations are all unified by recognizing a special carry-in operation or a borrow. Microprocessors assign the carry bit to 1 if in a subtraction the absolute value of the minuend, the number from which the other is subtracted, is less than the subtrahend.

Just as there are two forms of addition available, there are also two forms of subtractions—subtract-without-carry and subtract-with-carry. In the subtract-with-carry operation the contents of the carry bit from the previous operation is subtracted from the result. In 16-bit arithmetic the subtract-without-carry is used for the LOB and subtract-with carry for the HOB.

Microprocessors determine whether a 0 or a 1 is placed in the carry bit, not by detecting the actual carry, but by performing a Boolean algebra operation.[1] In both the add-with-carry and add-without-carry the formula for the carry bit C is

$$C = X_7 \cdot Y_7 + Y_7 \cdot \overline{S_7} + \overline{S_7} \cdot X_7 \qquad (10\text{-}1)$$

where the operation is adding the numbers X and Y in order to obtain a sum S. The subscript 7 indicates the seventh bit, the dot represents an AND, the + an OR, and the bar an inverse. The test is made using the MSB of X, Y, and S. In the addition shown in Fig. 10-2a,

$$C = 0 \cdot 0 + 0 \cdot 0 + 0 \cdot 0 = 0$$

In the operation shown in Fig. 10-2b, however,

$$C = 1 \cdot 0 + 0 \cdot 1 + 1 \cdot 1 = 1$$

The Boolean formula to determine the carry bit in both subtract-with-carry and subtract-without-carry is

$$C = \overline{M_7} \cdot S_7 + S_7 \cdot D_7 + D_7 \cdot \overline{M_7} \qquad (10\text{-}2)$$

where the operation is

$$\text{Minuend } M - \text{subtrahend } S = \text{difference } D$$

In the subtraction of Fig. 10-3a,

$$C = 0 \cdot 0 + 0 \cdot 0 + 0 \cdot 0 = 0$$

while for Fig. 10-3b,

$$C = 1 \cdot 0 + 0 \cdot 1 + 1 \cdot 1 = 1$$

10-8 Signed Numbers—Two's Complement Overflow

Figure 10-1b shows the number circle where the 256 numbers are interpreted as the series from -128 to 127. The results of addition and subtraction are valid when the operations stay in this range of numbers. For example, if 106_{10} is added to -46_{10}, the result is 60_{10},

$$\begin{array}{rcr}
1\,1\,0\,1\,0\,0\,1\,0 & = & -46_{10} \\
+\,0\,1\,1\,0\,1\,0\,1\,0 & = & 106_{10} \\
\hline
1\,0\,0\,1\,1\,1\,1\,0\,0 & & \\
0\,0\,1\,1\,1\,1\,0\,0 & = & 60_{10}
\end{array}$$

The operation is shown on the number circle of Fig. 10-4a and gives a valid result. If 51_{10} is added to 93_{10} as in Fig. 10-4b,

$$\begin{array}{rcr}
0\,1\,0\,1\,1\,1\,0\,1 & = & 93_{10} \\
0\,0\,1\,1\,0\,0\,1\,1 & = & 51_{10} \\
\hline
1\,0\,0\,1\,0\,0\,0\,0 & = & -112_{10}
\end{array}$$

The result in the signed number mode is -112_{10}, which is invalid. Passage in either direction through the $-128/127$ boundary, called two's complement overflow, will cause an invalid result.

10-8. Signed Numbers—Two's Complement Overflow

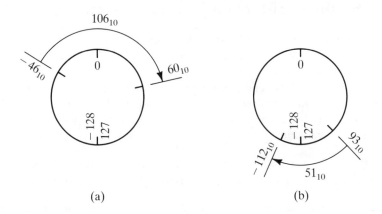

Fig. 10-4. Addition of signed numbers (a) with a valid result, and (b) with two's complement overflow.

Subtraction of two numbers in the signed number mode may result in a valid positive number, a valid negative result, or an invalid result if two's complement overflow occurs. The Boolean equation to test for a two's complement overflow (TCO) is

$$\text{Addition TCO} = X_7 \cdot Y_7 \cdot \overline{S_7} + \overline{X_7} \cdot \overline{Y_7} \cdot S_7 \tag{10-3}$$

and

$$\text{Subtraction TCO} = M_7 \cdot \overline{S_7} \cdot \overline{D_7} + \overline{M_7} \cdot S_7 \cdot D_7 \tag{10-4}$$

In a subtraction that gives a valid result, $43_{10} - 79_{10}$,

```
  0 0 1 0 1 0 1 1    →      0 0 1 0 1 0 1 1
- 0 1 0 0 1 1 1 1    →    + 1 0 1 1 0 0 0 1
                           1 1 0 1 1 1 0 0   = -36₁₀
```

and

$$\text{TCO} = 0 \cdot 1 \cdot 0 + 1 \cdot 0 \cdot 1 = 0$$

A subtraction causing a two's complement overflow is $-56_{10} - 92_{10}$,

```
  1 1 0 0 1 0 0 0    →      1 1 0 0 1 0 0 0
- 0 1 0 1 1 1 0 0    →    + 1 0 1 0 0 1 0 0
                         1 0 1 1 0 1 1 0 0   = 108₁₀
```

and

$$\text{TCO} = 1 \cdot 1 \cdot 1 + 0 \cdot 0 \cdot 0 = 1$$

The TCO flag indicates an invalid result, since $-56_{10} - 92_{10}$ should be -148_{10} rather than 108_{10}.

10-9 Status Registers on Microprocessors

Following most operations (add, subtract, compare, etc.) microprocessors indicate the status of some conditions resulting from the operation. These status bits are placed in a one-byte memory, called a register, that is incorporated in the microprocessor chip. The flag indicators contained in the status registers of three different microprocessors[2-4] are shown in Table 10-4. In addition to the carry and two's complement overflow flags, the following additional indicators are available:

- Sign bit, S = MSB, so if $S = 0$ the result of the previous operation is positive, and if $S = 1$ the result is negative in the signed number mode.

- Zero bit, $Z = 1$ if the result of the previous operation was zero, otherwise $Z = 0$.

- Half carry, $H = 1$ if a bit was carried from the third to the fourth position—information that is useful in BCD arithmetic.

Table 10-4. Status Registers on Several Microprocessors

Status Bit	Motorola 6800	Intel 8080/8085	Rockwell 6502
S - sign	•	•	•
Z - zero	•	•	•
C - carry	•	•	•
TCO - two's complement overflow	•		•
H - half carry	•	•	•
Other flags	•	•	•

Example 10-3. What will be the result in binary and the status of the S, Z, C, and TCO bits after the following operations? (a) Add 120_{10} to -16_{10}; (b) add 53_{10} to 98_{10}; (c) subtract 84_{10} from 37_{10}; (d) subtract 89_{10} from -52_{10}.

Solutions. (a) 0111 1000 is added to 1111 0000 to get 1 0110 1000. $S = 0 =$ MSB of the result, $Z = 0$ because the result does not equal 0, $C = 1$ because the operation passes through the zero boundary, and TCO $= 0$ because it does not pass through the $-128/127$ boundary.

(b) The result of adding 0011 0101 to 0110 0010 is 1001 0111 or 151_{10}, which is the correct sum in the unsigned mode. In the signed mode the sum is -105_{10}, which is invalid. $S = 1$, which indicates a negative number, $Z = 0$, $C = 0$, but TCO = 1 because of passage through the $-128/127$ boundary.

(c) When the two's complement of 84_{10}, 1010 1100, is added to 0010 0101 the result is 1101 0001, which is -47_{10} in the signed mode. $S = 1$, $Z = 0$, $C = 1$ because the absolute magnitude of the subtrahend is greater than that of the minuend, and TCO=0.

(d) In binary, -52_{10} is 1100 1100, and when 0101 1001 is subtracted the result is 1 0111 0011.

$$S = 0$$
$$Z = 0$$
$$C = 0 \cdot 0 + 0 \cdot 0 + 0 \cdot 0 = 0$$
$$TCO = 1 \cdot 1 \cdot 1 + 0 \cdot 0 \cdot 0 = 1$$

The operation passed in a counterclockwise direction through the $-128/127$ boundary.

References

1. U.S. Market Report—Semiconductors, *Electronics*, p. 55, Jan. 6, 1986.

2. *M6800 Programming Reference Manual*, Motorola Semiconductor Products, Phoenix, AZ, 1976.

3. *MCS-80/85^{TM} Family User's Manual*, Intel Corporation, Santa Clara, CA, 1985.

4. *R6500 Microcomputer System Programming Manual*, Rockwell International, Anaheim, CA, 1979.

Problems

10-1. For the following additions, compute the resulting binary number, sketch the operation on the number circle, using Eq. (10-1) determine the carry bit C, and using Eq. (10-3) determine the two's complement overflow bit TCO: (a) $2B + $51, (b) $98 + $49, (c) $56 + $3F, and (d) $A1 + $73.

10-2. For the following subtractions, compute the resulting binary number, sketch the operation on the number circle, using Eq. (10-2) determine the carry bit C, and using Eq. (10-4) determine the two's complement overflow bit TCO: (a) $6D - $44, (b) $D4 - $38, (c) $4A - $7B, and (d) $91 - $37.

10-3. In the signed number mode, subtract a negative number, for example, $26_{10} - (-73_{10})$, using the same binary arithmetic as the computer uses to determine whether the result given by the computer is valid.

10-4. In 16-bit signed-number arithmetic, (a) what is the range of numbers in base 10 and in binary? (b) In the following operation:

$$0101\ 1100\ 1010\ 1001$$
$$+0011\ 0100\ 0011\ 0111$$

what instructions are to be used to add the LOBs and the HOBs, respectively? (c) What is the indicator of an integer overflow?

10-5. The Boolean algebra equation to determine the carry bit in an addition, Eq. (10-1), shows three combinations, any of which will result in $C = 1$. Show and explain the combinations of X_7, Y_7, and S_7 that cause each of the three AND operations to equal 1.

Chapter 11

Programming a Microprocessor

11-1 A Generic Microprocessor

The goal of this chapter is to investigate two commercial microprocessors and to learn to program them. There are many microprocessors on the market, and studying just two of them may seem limiting, but this is not the case. The microprocessors selected are the Motorola 6800 and the Intel 8080/8085 families, which are widely used in their own right. In addition, the Rockwell 6502 has many similarities to the Motorola 6800 and the Zilog Z-80 is related to the Intel 8080/8085.

Rather than initially approaching either of these microprocessors in their commercial forms, the strategy of this chapter will be to develop a generic microprocessor. This generic microprocessor is hypothetical but serves the purpose of leading step by step to actual microprocessors. This procedure provides some insight into how the microprocessor performs its tasks and builds expectations for the capabilities and programming techniques of commercial microprocessors.

11-2 Data and Address Buses in a Generic Microcomputer

Chapter 13 will describe several elementary microcomputers in more detail, but it is sufficient for this chapter on programming to show the connections between a microprocessor and a RAM as in Fig. 11-1. The two buses are an address bus and a data bus, which were first introduced in Chapter 9 in the

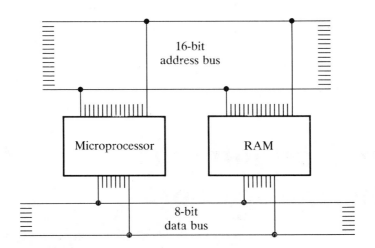

Fig. 11-1. A microprocessor and a RAM connected by a data bus and an address bus.

study of memories. The microcomputers with which we deal use an 8-bit data bus and a 16-bit address bus. Other components that are also connected to these buses, such as ROMs and I/O to the outside world, are not shown.

The program on which the microprocessor operates will be stored in memory—either RAM or ROM. A non-permanent program developed by the user will be placed in RAM. If the purposes of the program are to store, access, and perform arithmetic operations on these values, the data must be stored in a different section of RAM from that where the program is located to keep the program section intact. The direction of flow on the address bus is from the microprocessor to the RAM, while data may flow from the microprocessor to memory or at other times in the opposite direction.

11-3 The Accumulator with its Arithmetic, Logic, and Transfer Operations

Within the microprocessor is an accumulator that can hold one byte of data, just as a memory location, but the accumulator can do much more. It has the capability of participating in arithmetic and logic operations by calling on another resource of the microprocessor—the arithmetic and logic unit (ALU). The operations to be introduced now with their explanations are:

Arithmetic and logic operations

ADD Add the contents of a specified memory location to the contents of the accumulator, and store the result in the accumulator.

SUB Subtract the contents of a specified memory location from the contents of the accumulator, and store the result in the accumulator.

AND Perform a logical AND operation using the contents of a specified memory location and the contents of the accumulator. Store the results in the accumulator.

OR Perform a logical OR operation using the contents of a specified memory location and the contents of the accumulator. Store the results in the accumulator.

Transfer operations

LDA Load the contents of a specified memory location into the accumulator, displacing previous contents of the accumulator.

STA Store the contents of the accumulator into a specified memory location, displacing the original contents of the memory location. The contents of the accumulator remain unchanged.

These operations are only the start of what will become an extensive set of operations of which the microprocessor is capable. Before introducing more operations, the sequence used by the microprocessor to execute operations will be explained.

11-4 The Fetch-Decode-Execute Sequence

In the three-step sequence used by the microprocessor the fetch operation means that the microprocessor goes to the memory (RAM or ROM) where the program is stored and extracts one byte. The understanding of the microprocessor is that this byte is an instruction to perform some function. Within its own circuitry the microprocessor decodes this instruction, which might be, for example, to perform an ADD, SUB, etc. Having decided what to do, the microprocessor then executes that step, which may involve sending addresses and writing or reading data to or from the data bus. After this fetch-decode-execute sequence has been completed, the microprocessor automatically goes to the next position in the memory where its program is stored and withdraws the next instruction.

Figure 11-2 visualizes the fetch-decode-execute sequence for an ADD instruction. Two sections of RAM are indicated—the section in which the program is located and the section from which the byte of data is extracted to add to the accumulator. The instruction that indicates to the microprocessor to perform an ADD is one byte, but as soon as the microprocessor decodes that instruction it recognizes that it must ask the program where in memory the byte is located that is to be added to the accumulator. The memory address, which requires two bytes, is located in the two bytes of the program following the add instruction. The microprocessor reads those two bytes and then goes to that memory location (which is in another section of RAM memory) to read the byte. The final step is to add the contents of that memory location to what is already in the accumulator and store the contents back in the accumulator.

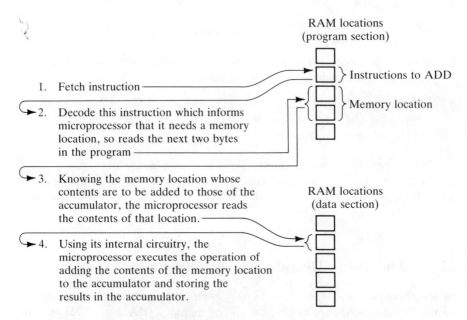

Fig. 11-2. The fetch-decode-execute sequence for an ADD instruction.

11-5 Preliminary Instruction Set

A one-byte instruction designates to the microprocessor the operation that is to be performed. The entire group of instructions is called an instruction set and is often displayed in a form similar to Fig. 11-3. The arithmetic, logical, and transfer commands listed in Sec. 11-3 are shown in the figure.

11-5. Preliminary Instruction Set

The subtract instruction SUB, for example, is $02 and the AND instruction is $08. Numerous blanks appear in the chart of Fig. 11-3, which suggests that some or all of the blanks will be filled in as the instruction set grows.

Left	Right Hex Digit															
	0	1	2	3	4	5	6	7	8	9	A	B	C	D	E	F
0	ADD		SUB		LDA		STA		AND		OR						
1																
2																
3																

Fig. 11-3. Machine code for instruction set with arithmetic, logical, and transfer operations.

Example 11-1. Write the machine-language program to add the contents of memory location $0070 to the contents of memory location $0071 and store the results in memory location $007A.

Solution. A convention will first be introduced that interprets the enclosure of a memory location in parentheses to mean the contents of that location. There are three separate steps in accomplishing the assignment: (1) bring ($0070) into the accumulator, (2) add ($0071) to the accumulator, and (3) store the sum that is now in the accumulator into memory location $007A. The machine-language program is:

```
                Comments
   04       LDA from $0070
   00            "
   70            "
   00       ADD ($0071)
   00            "
   71            "
   06       STA to $007A
   00            "
   7A            "
```

Following each of the operational codes, or op codes for short, is the pertinent 16-bit memory location. Here we use the high-order byte (HOB) first and the low-order byte (LOB) next, which is the convention for the Motorola 6800 microprocessor. The Intel 8080/8085, on the other hand, arranges the two bytes of the address in the opposite sequence.

11-6 Program Counter

The microprocessor must have some means of knowing the current position within the program. That indication is kept by a register called the program counter. In the program of Example 11-1 the microcomputer needed to be informed of the memory location where the first byte of the program, the $04, was located. If that location happens to be $E100, for example, the program counter is originally set to $E100 whereupon the LDA instruction is executed. The microprocessor knows that the complete LDA instruction (op code plus memory location) requires three bytes, so the program counter automatically advances three locations after the execution of the LDA and directs the microprocessor to $E103 for the next instruction.

11-7 Status Register and Jumps

The need and use of several registers in the microprocessor (the accumulator and program counter) have already been encountered. It is appropriate at this point to reveal that there are several additional registers that will be incorporated in the generic microprocessor. These registers are shown in Fig. 11-4, and the ones not described yet will be explained in the next several sections. Accumulator A is an 8-bit register, while the program counter is 16-bit in order to accommodate memory addresses of that size. The next register to identify is the status register, which has already been introduced in Sec. 10-9. The status register is an 8-bit register in which individual bits indicate the sign, zero, carry, half carry, two's complement overflow, and perhaps several other flags.

The existence of the status register opens the possibility of the next class of operations: jumps. The four jump instructions to be selected for the generic microprocessor are

JUN Unconditional jump
JEQ Jump if equal
JGT Jump if greater than
JLT Jump if less than

11-7. Status Register and Jumps

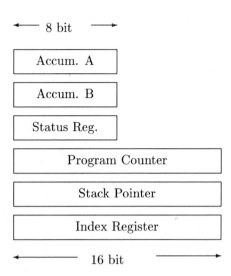

Fig. 11-4. Registers in the generic microprocessor.

The op codes for these jump operations are included in the chart of Fig. 11-5. A jump operation alters the normal incremental advance of the program counter and instead assigns the program counter a value that shifts the execution to a new location within the program.

Left	Right Hex Digit															
	0	1	2	3	4	5	6	7	8	9	A	B	C	D	E	F
0	ADD		SUB		LDA		STA		AND		OR		JUN			
1													JEQ			
2													JGT			
3													JLT			

Fig. 11-5. Machine code for instruction set that includes jump operations.

Example 11-2. A gap is to be left in a program between $002A and $0037 to allow for possible later insertion of instructions. Show the pertinent sections of the program.

Solution.

Location	Op Code	Comment
002A		Last byte before jump
002B	0C	Unconditional jump
002C	00 ⎫	to location $0037
002D	37 ⎭	
⋮		Skip $002E through $0036
0037		Continuation of program

The conditional jumps—JEQ, JGT, and JLT—need to be interpreted. In the case of JEQ, for example, what two variables are being checked for equality? In the case of JGT, what variable must be greater than what other one in order for the jump to take place? The answers are that the jumps are triggered by the bits in the status register, as indicated in Table 10-4. The status register is assigned values based on the nature of the operation, so the jump is activated or bypassed as dictated by the results of the previous operation. A jump typically follows a subtraction of a number from the contents of an accumulator, a subtraction of the contents of a memory location from the accumulator, or decrementing or incrementing an accumulator. Thus, for the three conditional jumps:

If (ACCUM) = (M) JEQ causes a jump
If (ACCUM) > (M) JGT causes a jump
If (ACCUM) < (M) JLT causes a jump

Example 11-3. Write the section of a program for a controller that continually compares a reading stored in $0048 (continually updated by another segment of the program) with a setpoint stored in $006D. If the reading is greater than the setpoint, jump to location $0039 in the control program.

Solution.

Location	Op Code	Comment
000D	04 00 48	LDA A ($0048), loading reading in A
0010	02 00 6D	SUB A ($006D), subtracting setpoint
0013	2C 00 39	JGT if reading exceeds setpoint
0016	0C 00 0D	JUN returns to start of loop

11-8 Another Accumulator—Incrementing and Decrementing

The accumulator is a key element in the operation of the microprocessor, and the power of the microprocessor can be enhanced by equipping the processor with an additional accumulator. Call one of these accumulators A and the other B. All of the arithmetic, logic, and transfer operations shown in the instruction set of Fig. 11-3 are now duplicated in Fig. 11-6 so that they apply to either ACCUM A or ACCUM B. Two more operations to be added to the elementary microprocessor at this stage are the capability of decrementing and incrementing the accumulators. The mnemonic for decrementing, the reduction of the value by 1, is DEC, while incrementing, which increases the value by 1, is designated by INC.

Left	Right Hex Digit															
	0	1	2	3	4	5	6	7	8	9	A	B	C	D	E	F
0	ADD A	ADD B	SUB A	SUB B	LDA A	LDA B	STA A	STA B	AND A	AND B	OR A	OR B	JUN	DEC A	DEC B	
1													JEQ	INC A	INC B	
2													JGT			
3													JLT			

Fig. 11-6. Machine code for the instruction set that incorporates two accumulators as well as incrementing and decrementing.

Example 11-4. Write a program to multiply ($005D) by ($005E). Assume that the product does not exceed $FF.

Solution. The strategy is that of adding ($005D) the number of times indicated by ($005E). ACCUM A will store the sums and will at termination contain the product. ACCUM B will count the number of additions.

Location	Op Code	Comments
0000	05 00 5E	LDA B the multiplier
0003	04 00 5D	LDA A initial contents of A
0006	0E	DEC B
0007	1C 00 10	JEQ when B = 0, jump to end
000A	00 00 5D	ADD A adding ($005D)
000D	0C 00 06	JUN to 0006 for another add
0010		(End)

11-9 Additional Addressing Modes

While it has not been pointed out yet, a certain mode of addressing has been used in the mathematical, logical, and transfer instructions. This mode of addressing, used in the instruction sets included in machine codes between 00 and 0B, specifies a memory location. In the case of the ADD, SUB, AND, and OR operations the contents of a memory location were processed together with the contents of the accumulator. In the LDA instructions, the contents of a memory location specified in the instruction are loaded into the accumulator. This mode of addressing, designated by *adr* in the expanded instruction set of Fig. 11-7 is extremely useful but needs to be supplemented by other modes. A question may arise as in Example 11-4 of how to specify various choices of the values in memory locations $005D and $005E. The immediate addressing mode offers a means of providing numbers that can be directly loaded into the accumulator or added or subtracted from the accumulator. The immediate addressing mode consists of the op code followed by one byte of data—thus the entire instruction consists of two bytes. Several examples of immediate addressing, using op codes shown in Fig. 11-7, are:

Adding $13 to A	10 13
Subtracting $2C from B	13 2C
Loading B with $C4	15 C4
AND A with $08	18 08

Storing a number directly into a memory location with a STA *imm* is not an available option. Storing a number into a memory location requires that it be passed through one of the accumulators. The final mode of addressing that will be provided for the generic microprocessor is called register addressing. This mode initially seems cumbersome but will prove to be convenient to program. Figure 11-4 showed a register in the microprocessor called an index register. The contents of this 16-bit register designate the memory location where the contents are found for the instruction (the ADD, SUB, LDA, etc.).

Left	Right hex digit															
	0	1	2	3	4	5	6	7	8	9	A	B	C	D	E	F
0	ADD A adr	ADD B adr	SUB A adr	SUB B adr	LDA A adr	LDA B adr	STA A adr	STA B adr	AND A adr	AND B adr	OR A adr	OR B adr	JUN	DEC A	DEC B	
1	ADD A imm	ADD B imm	SUB A imm	SUB B imm	LDA A imm	LDA B imm			AND A imm	AND B imm	OR A imm	OR B imm	JEQ	INC A	INC B	
2	ADD A reg	ADD B reg	SUB A reg	SUB B reg	LDA A reg	LDA B reg	STA A reg	STA B reg	AND A reg	AND B reg	OR A reg	OR B reg	JGT			
3													JLT			

Fig. 11-7. Machine code for the instruction set that includes memory, immediate, and register modes of addressing.

Suppose, for example, that the current contents of the index register are $E07F. The action taken by several different instructions are:

21   Adds ($E07F) to B and stores the result in B
24   Loads ($E07F) into A
26   Stores the contents of A into location $E07F

The instructions using register addressing are one byte, which is adequate for the generic microprocessor. Commercial microprocessors expand the capability of register addressing by using two- or three-byte instructions, as will be seen later in this chapter.

11-10 The Index Register and the Use of Register Addressing

In the previous section the contents of the index register was $E07F, and the first point to explain is how the index register could be loaded with this or any other desired value. The instruction available for this purpose is LDX, which appears in two different addressing forms in Fig. 11-8.

Suppose that the index register is to be loaded with $E008. One option is to use LDX *imm*, which would be 31 E0 08. Another possibility would be to first load E0 and 08 into two consecutive memory locations, for example, $0072 and $0073, respectively. Thereafter the memory addressing instruction, LDX *adr*, and specifically 30 00 72 would store ($0072) as the HOB and ($0073) into the LOB of the index register.

In certain situations register addressing in combination with DEX or INX to decrement or increment, respectively, the index register will offer a convenient programming strategy. Refer to Sec. 11-24 for index addressing using offset.

Left	Right Hex Digit															
	0	1	2	3	4	5	6	7	8	9	A	B	C	D	E	F
0	ADD A adr	ADD B adr	SUB A adr	SUB B adr	LDA A adr	LDA B adr	STA A adr	STA B adr	AND A adr	AND B adr	OR A adr	OR B adr	JUN	DEC A	DEC B	DEX
1	ADD A imm	ADD B imm	SUB A imm	SUB B imm	LDA A imm	LDA B imm			AND A imm	AND B imm	OR A imm	OR B imm	JEQ	INC A	INC B	INX
2	ADD A reg	ADD B reg	SUB A reg	SUB B reg	LDA A reg	LDA B reg	STA A reg	STA B reg	AND A reg	AND B reg	OR A reg	OR B reg	JGT	JTS adr	JTS reg	RFS
3	LDX adr	LDX imm	STX adr	STX imm	LDS adr	LDS imm							JLT			

Fig. 11-8. Machine code for complete instruction set for the generic microprocessor.

Example 11-5. Using register addressing, shift to one lower memory location the contents of the block of memory currently in $0061 through $0072.

Solution.

Location	Op Code	Comments Applicable to First Loop
0000	15 12	LDA B $12, use B to count shifts
0002	31 00 61	LDX *imm* $0061, load index register
0005	24	LDA A *reg*, load ($0061) into A
0006	0F	DEX, decrement index register
0007	26	STA A *reg*, store into $0060
0008	1F	INX get ready for next shift
0009	1F	INX
000A	0E	DEC B, decrement B
000B	2C 00 05	JGT, loop if another shift called for (End)

11-11 Subroutines and the Stack

Even the generic microprocessor should have the capability of using subroutines because of their usefulness in programming. Jump instructions have already been explained, but a jump to a subroutine has the special characteristic that it can be called from various locations in the main program and always returns to the next instruction following the subroutine call.

The jump-to-subroutine call, JTS, is available through either memory or index addressing, as shown in Fig. 11-8. The final statement of the subroutine is the return-from-subroutine, RFS.

Example 11-6. Write a subroutine that receives a memory location from the main program and compares the contents of this memory location with the contents of the next memory location in sequence. If the contents of the two locations are equal, the subroutine returns 0 in ACCUM B, otherwise it returns a 1 in ACCUM B.

Solution. An example of a subroutine call and the subroutine itself (this one comparing the contents of $E056 and $E057) are:

Location	Op Code	Comments
xxxx	31 E0 56	LDX $E056, load index register
xxxx+3	2D 00 6A	JTS at $006A, jump to subroutine
xxxx+6		(take action based on contents of B)

The subroutine, starting at $006A is:

Location	Op Code	Comments
006A	15 00	LDA B $00, default value into B
006C	24	LDA A ($E056), register addressed
006D	1F	INX, increment the index register
006E	22	SUB A *reg*, subtract ($E057) from A
006F	1C 00 73	JEQ, jump to $0073 if numbers equal
0072	1E	INC B, values not equal so B = 1
0073	2F	RFS, return from subroutine

In jumping to a subroutine the microprocessor must store the contents of the program counter so that operation can return to that position upon completion of the subroutine. The program counter at the time of the subroutine call is stored on what is called a stack. The stack is a section of memory, shown symbolically in Fig. 11-9, with its position indicated by the stack pointer. The stack pointer is a 16-bit register in the microprocessor (see

Fig. 11-4). When the JTS command is executed, the contents of the program counter are loaded onto the stack by storing the two bytes of the program counter. Suppose the stack pointer contains $E71A when the subroutine is called. If the program counter is loaded onto the stack it is placed in positions $E71A and $E719. Simultaneously the stack pointer is decremented by 2 so that if another byte must later be loaded onto the stack it would be loaded into position $E718.

Upon jumping to the subroutine, the program counter takes on the value indicated by the JTS command and begins stepping through the subroutine. When the RFS is reached, the original contents of the program counter are withdrawn from the stack, and execution proceeds in sequence from that memory location further through the main program.

Upon powerup, the microcomputer is assigned a default value for the stack, and since the stack is located within RAM, care must be taken to avoid using that section of memory for other purposes. If the programmer would like the stack to be located elsewhere from the default position, the option is available to assign the stack position through the LDS (load stack pointer) command, which is a three-byte instruction.

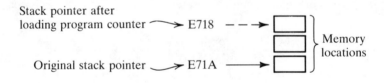

Fig. 11-9. The stack and stack pointer.

11-12 The Intel 8080/8085 Microprocessor

The preceding description of the generic microprocessor prepares for the introduction of two families of commercial microprocessors, the Intel 8080/8085 and the Motorola 6800. The sequence to be followed in explaining more about each of these microprocessors will begin by describing the registers and their capabilities, which makes the listing of the instruction set more meaningful. The machine code corresponding to the instruction set will then be presented. Thereafter will follow some of the programming features, such as addressing, branches and loops, index registers, and stack pointers. The commercial microprocessors incorporate all the features of the hypothetical generic microprocessor and many more features as well, but a few of the fundamental operations will now be familiar.

11-12. The Intel 8080/8085 Microprocessor

The Intel Corporation introduced the first microprocessor in the early 1970s and followed in 1974 with the 8080 microprocessor serving an 8-bit data bus and a 16-bit address bus.[1] A further Intel development was the 8085 microprocessor, which has all the programming instructions of the 8080 plus two additional ones. The 8085 also has some additional hardware facilities.

There are eight 8-bit and two 16-bit registers on the 8085, as shown in Fig. 11-10. The most-active register is the accumulator that participates in most operations. The B, C, D, E, H, and L registers can function as 8-bit registers, but some instructions that will be explained later expect the B-C, D-E, or H-L pairs to work as 16-bit combinations. The stack pointer and program counter perform in the same manner as those in the generic microprocessor. An outline of the instruction set for the 8080/8085 is shown in Table 11-1. A brief description of each of the mnemonics demonstrates that the capabilities of this commercial microprocessor are an order of magnitude greater than the those of the generic microprocessor. Many of the instructions introduced with the generic machine appear, but the 8080/8085 offers several versions of the instructions. In Table 11-1, v indicates one byte of immediate data, vv represents two bytes of data, and aa signifies a two-byte address. A register pair surrounded by parentheses, $(H\&L)$ for example, denotes the memory location whose address is contained in the register pair.

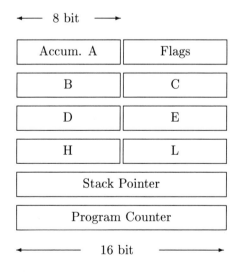

Fig. 11-10. Registers on the 8085 microprocessor.

Table 11-1. Outline of 8080/8085 Instruction Set

Mnemonic	Op-code (hex)	Description	Mnemonic	Op-code (hex)	Description
ADD A	87	Add A to A	CC aa	DD	Call if carry
ADD B	80	Add B to A	CNC aa	D4	Call if no carry
ADD C	81	Add C to A	CPE aa	EC	Call if parity even
ADD D	82	Add D to A	CPO aa	E4	Call if parity odd
ADD E	83	Add E to A	CMA	2F	Complement A
ADD H	84	Add H to A	CMC	3F	Complement carry
ADD L	85	Add L to A	CMP A	BF	Compare A with A
ADD M	86	Add $(H\&L)$ to A	CMP B	B8	Compare B with A
ADI v	C6	Add v to A	CMP C	B9	Compare C with A
ADC A	8F	Add w/carry, A to A	CMP D	BA	Compare D with A
ADC B	88	Add w/carry, B to A	CMP E	BB	Compare E with A
ADC C	89	Add w/carry, C to A	CMP H	BC	Compare H with A
ADC D	8A	Add w/carry, D to A	CMP L	BD	Compare L with A
ADC E	8B	Add w/carry, E to A	CMP M	BE	Compare A with $(H\&L)$
ADC H	8C	Add w/carry, H to A			
ADC L	8D	Add w/carry, L to A	CPI v	FE	Compare v with A
ADC M	8E	Add w/carry $(H\&L)$ to A	DAA	27	Decimal Adjust A
			DAD B	09	Double add $B\&C$ to $H\&L$
ACI v	CE	Add w/carry v to A			
ANA A	A7	AND A with A	DAD D	19	Double add $D\&E$ to $H\&L$
ANA B	A0	AND B with A			
ANA C	A1	AND C with A	DAD H	29	Double add $D\&E$ to $H\&L$
ANA D	A2	AND D with A			
ANA E	A3	AND E with A	DAD SP	39	Double add SP to $H\&L$
ANA H	A4	AND H with A			
ANA L	A5	AND L with A	DCR A	3D	Decrement A
ANI v	E6	AND v with A	DCR B	05	Decrement B
CALL aa	CD	Call a subroutine at address aa	DCR C	0D	Decrement C
			DCR D	15	Decrement D
CZ aa	CC	Call if zero	DCR E	1D	Decrement E
CNZ aa	C4	Call if not zero	DCR H	25	Decrement H
CP aa	F4	Call if plus	DCR L	2D	Decrement L
CM aa	FC	Call if minus	DCR M	35	Decrement memory $(H\&L)$

All mnemonics ©Intel Corporation 1976.

11-12. The Intel 8080/8085 Microprocessor

Table 11-1. Outline of 8080/8085 Instruction Set (continued)

Mnemonic	Op-code (hex)	Description	Mnemonic	Op-code (hex)	Description
DCX B	0B	Decrement $B\&C$	**LXI** B,vv	01	Load $B\&C$ with vv
DCX D	1B	Decrement $D\&E$	**LXI** D,vv	11	Load $D\&E$ with vv
DCX H	2B	Decrement $H\&L$	**LXI** H,vv	21	Load $H\&L$ with vv
DCX SP	3B	Decrement SP	**LXI** SP,vv	31	Load SP with vv
DI	F3	Disable interrupts	**MOV** A,B	78	Move B to A
EI	FB	Enable interrupts	**MOV** A,C	79	Move C to A
HLT	76	Halt until interrupt	**MOV** A,D	7A	Move D to A
IN v	DB	Input from port v	**MOV** A,E	7B	Move E to A
INR A	3C	Increment A	**MOV** A,H	7C	Move H to A
INR B	04	Increment B	**MOV** A,L	7D	Move L to A
INR C	0C	Increment C	**MOV** A,M	7E	Move $(H\&L)$ to A
INR D	14	Increment D	**MOV** B,A	47	Move A to B
INR E	1C	Increment E	**MOV** B,C	41	Move C to B
INR H	24	Increment H	**MOV** B,D	42	Move D to B
INR L	2C	Increment L	**MOV** B,E	43	Move E to B
INR M	34	Increment $(H\&L)$	**MOV** B,H	44	Move H to B
INX B	03	Inc. extended $B\&C$	**MOV** B,L	45	Move L to B
INX D	13	Inc. extended $D\&E$	**MOV** B,M	46	Move $(H\&L)$ to B
INX H	23	Inc. extended $H\&L$	**MOV** C,A	4F	Move A to C
INX SP	33	Inc. extended SP	**MOV** C,B	48	Move B to C
JMP aa	C3	Jump to aa	**MOV** C,D	4A	Move D to C
JZ aa	CA	Jump if zero	**MOV** C,E	4B	Move E to C
JNZ aa	C2	Jump if not zero	**MOV** C,H	4C	Move H to C
JP aa	F2	Jump if plus	**MOV** C,L	4D	Move L to C
JM aa	FA	Jump if minus	**MOV** C,M	4E	Move $(H\&L)$ to C
JC aa	DA	Jump if carry	**MOV** D,A	57	Move A to D
JNC aa	D2	Jump if no carry	**MOV** D,B	50	Move B to D
JPE aa	EA	Jump if parity even	**MOV** D,C	51	Move C to D
JPO aa	E2	Jump if parity odd	**MOV** D,E	53	Move E to D
LDA aa	3A	Load A from aa	**MOV** D,H	54	Move H to D
LDAX B	0A	Load A from aa	**MOV** D,L	55	Move L to D
LDAX D	1A	Load A from aa	**MOV** D,M	56	Move $(H\&L)$ to D
LHLD aa	2A	Load A from aa			

All mnemonics ©Intel Corporation 1976.

Table 11-1. Outline of 8080/8085 Instruction Set (continued)

Mnemonic	Op-code (hex)	Description	Mnemonic	Op-code (hex)	Description
MOV E,A	5F	Move A to E	NOP	00	No operation
MOV E,B	58	Move B to E	ORA A	B7	Test A
MOV E,C	59	Move C to E	ORA B	B0	OR B with A
MOV E,E	5A	Move E to E	ORA C	B1	OR C with A
MOV E,H	5C	Move H to E	ORA D	B2	OR D with A
MOV E,L	5D	Move L to E	ORA E	B3	OR E with A
MOV E,M	5E	Move $(H\&L)$ to E	ORA H	B4	OR H with A
MOV H,A	67	Move A to H	ORA L	B5	OR L with A
MOV H,B	60	Move B to H	ORA M	B6	OR memory with A
MOV H,C	61	Move C to H	ORI v	F6	OR v with A
MOV H,D	62	Move D to H	OUT v	D3	Output A to device v
MOV H,E	63	Move E to H	PCHL	E9	Jump indirect via $(H\&L)$
MOV H,L	65	Move L to H			
MOV H,M	66	Move $(H\&L)$ to H	POP B	C1	Pop $B\&C$ from stack
MOV L,A	6F	Move A to L	POP D	D1	Pop $D\&E$ from stack
MOV L,B	68	Move B to L	POP H	E1	Pop $H\&L$ from stack
MOV L,C	69	Move C to L	POP PSW	F5	Pop PSW from stack
MOV L,D	6A	Move D to L	PUSH B	C5	Push $B\&C$ on stack
MOV L,E	6B	Move E to L	PUSH D	D5	Push $D\&E$ on stack
MOV L,H	6C	Move H to L	PUSH H	E5	Push $H\&L$ on stack
MOV L,M	6E	Move $(H\&L)$ to L	PUSH PSW	F5	Push PSW on stack
MOV M,A	77	Move A to memory	RAL	17	Rotate $CY+A$ left
MOV M,B	70	Move B to memory	RAR	1F	Rotate $CY+A$ right
MOV M,C	71	Move C to memory	RLC	07	Rotate A left and into CY
MOV M,D	72	Move D to memory			
MOV M,E	73	Move E to memory	RRC	0F	Rotate A right and into CY
MOV M,H	74	Move H to memory			
MOV M,L	75	Move L to memory	RIM*	20	Read interrupt mask
MVI A,v	3E	Move v to A	RET	C9	Return
MVI B,v	06	Move v to B	RZ	C8	Return if zero
MVI C,v	0E	Move v to C	RNZ	C0	Return if not zero
MVI D,v	16	Move v to D	RP	F0	Return if plus
MVI E,v	1E	Move v to E	RM	F8	Return if minus
MVI H,v	26	Move v to H	RC	D8	Return if carry
MVI L,v	2E	Move v to L	RNC	D0	Return if no carry
MVI M,v	36	Move v to memory	RPE	E8	Return if parity even
			RPO	E0	Return if parity odd

All mnemonics ©Intel Corporation 1976.

11-12. The Intel 8080/8085 Microprocessor

Table 11-1. Outline of 8080/8085 Instruction Set (continued)

Mnemonic	Op-code (hex)	Description	Mnemonic	Op-code (hex)	Description
RST *0*	C7	Restart subroutine at address $00	**SBB** *A*	9F	Sub. w/borrow A from A
RST *1*	CF	Restart at $08	**SBB** *B*	98	Sub. w/borrow B from A
RST *2*	D7	Restart at $10			
RST *3*	DF	Restart at $18	**SBB** *C*	99	Sub. w/borrow C from A
RST *4*	E7	Restart at $20			
RST *5*	EF	Restart at $28	**SBB** *D*	9A	Sub. w/borrow D from A
RST *6*	F7	Restart at $30			
RST *7*	FF	Restart at $38	**SBB** *E*	9B	Sub. w/borrow E from A
SIM*	30	Set interrupt mask	**SBB** *H*	9C	Sub. w/borrow H from A
SPHL	F9	Load *SP* from *H&L*			
SHLD *aa*	22	Store *H&L* at memory *aa*	**SBB** *L*	9D	Sub. w/borrow L from A
STA *aa*	32	Store *A* at memory *aa*	**SBB** *M*	9E	Sub. w/borrow $(H\&L)$ from A
STAX *B*	02	Store *A* at (*B&C*)	**SBI** *v*	DE	Sub. w/borrow v from A
STAX *D*	12	Store *A* at (*D&E*)			
STC	37	Set *CY* flag	**XCHG**	E3	Exchange *D&E* with *H&L*
SUB *A*	97	Clear *A*			
SUB *B*	90	Subtract *B* from *B*	**XRA** *A*	AF	Clear *A*
SUB *C*	91	Subtract *B* from *C*	**XRA** *B*	A8	XOR *B* with *A*
SUB *D*	92	Subtract *B* from *D*	**XRA** *C*	A9	XOR *C* with *A*
SUB *E*	93	Subtract *B* from *E*	**XRA** *D*	AA	XOR *D* with *A*
SUB *H*	94	Subtract *B* from *H*	**XRA** *E*	AB	XOR *E* with *A*
SUB *L*	95	Subtract *B* from *L*	**XRA** *H*	AC	XOR *H* with *A*
SUB *M*	96	Subtract *B* from (*H&L*)	**XRA** *L*	AD	XOR *L* with *A*
			XRA *M*	AE	XOR (*H&L*) with *A*
SUI *v*	D6	Subtract *v* from *A*	**XRI** *v*	EE	XOR *v* with *A*

All mnemonics ©Intel Corporation 1976.

*8085 only

11-13 Loading Into and Storing From the Accumulator

Several means of inserting contents into the accumulator are available using the LDA aa, LDAX r, and appropriate move operations. The three move operations are MOV r_i, r_j, MOV r, M, and MVI r, v. The first of these move operations transfers the contents of register j to register i, leaving the contents of register j unaffected. The MOV r, M operation, specifically MOV A,M, transfers the contents of a memory location into the accumulator. The MVI A,v transfers immediate data into A. Examples of the op code for loading the accumulator are:

Op Code	Mnemonic	Comment
7A	MOV A,D	Loads contents of register D into accumulator
3A 00 E1	LDA $E100	Loads contents of location $E100 into accumulator (note LOB followed by HOB)
3E B2	MVI A,$B2	Loads $B2 into accumulator
0A	LDAX B	If registers B&C contain $005F, for example, loads ($005F) into accumulator
7E	MOV A,M	If registers H&L contain $007A, for example, loads ($007A) into accumulator

The commands that transfer the contents of the accumulator to the following destinations are: MOV r_i,A into another register; MOV M,A into the memory location designated by H&L; STA aa into memory location aa; STAX B into the memory location designated by B&C; and STAX D into the memory location designated by D&E. Examples of the foregoing instructions are:

Op Code	Mnemonic	Comments
4F	MOV C,A	Contents of accumulator to register C
32 3D 00	STA aa	Contents of accumulator to location $003D
02	STAX B	If B&C contain $004A, for example, the contents of accumulator to location $004A
77	MOV M,A	If H&L contain $E056, for example, the contents of accumulator to location $E056

11-14 Forms of Addressing on the 8080/8085

Section 11-9 summarized three different forms of addressing for the generic microprocessor, namely memory, immediate, and register addressing. The instruction set for the 8080/8085 given in Table 11-1 and even the examples for

11-15. Flag Register

loading into and storing from the accumulator given in Sec. 11-13 have already illustrated the forms of addressing available on this microprocessor. Those forms might be called inherent, memory, immediate, and register addressing, and they are illustrated in that order by the loading into and storing from accumulator operations in Sec. 11-13, with one exception. The one instruction not available is a store of a value directly into a memory. It is necessary to pass data through one of the working registers to a memory location.

Inherent addressing signifies that the specific op code determines the source and destination of the data. In memory addressing the memory location is a part of the instruction, in contrast to register addressing where the memory address is stored in the B&C, D&E, or H&L register combinations.

The illustrations of the forms of addressing given in Sec. 11-13 applied to loading into and storing data from the accumulator, but these forms of addressing are available to other classes of operations, such as adds, subtracts, ANDs, ORs, etc.

Example 11-7. Perform the following operation and place the results in location $003B: $63 − 2[($003A) − (B)].

Solution. The term in the brackets will be computed first, doubled by adding to itself, then moved to a temporary location.

Location	Op Code	Mnemonic	Comments
0000	3A 3A 00	LDA $003A	Load ($003A) into A
0003	90	SUB B	Subtract (B) from (A)
0004	87	ADD A	Add (A) to itself
0005	4F	MOV C,A	Move (A) to C
0006	3E 63	MVI A, $63	Load A with $63
0008	91	SUB C	Subtract (C) from (A)
0009	32 3B 00	STA $003B	Store in $003B
000C	C7	RST 0	Terminate

11-15 Flag Register

The flag register, as shown in Fig. 11-11, contains five flags that are assigned values in the process of performing some, but not all, instructions. The sign flag S, following an arithmetic or logical operation, is assigned the same as the MSB of the result in the accumulator. Thus, if the result is negative in the signed number system, S is set.

The Z bit is set if the result of certain operations is zero and cleared if the result is not zero.

7	6	5	4	3	2	1	0
S	Z		AC		P		C

S = sign flag
Z = zero flag
AC = auxiliary carry
P = parity
C = carry

Fig. 11-11. The flag register in the 8080/8085.

The auxiliary carry bit AC is set if there is an overflow or carry from bit 3. This flag is used in BCD arithmetic.

The parity flag P counts the number of 1 bits in the accumulator. If the accumulator has an even number of 1s, even parity exists and P is set.

The final flag is the carry bit C, which is set or cleared in arithmetic and certain other operations. During an addition C is set if there is an overflow from the eighth bit. In subtraction C is set if the minuend is less than the subtrahend (the number being subtracted).

Certain operations affect the status of the flags, and other operations do not. Arithmetic operations (additions and subtractions), logical operations, comparisons, and the decrementing or incrementing of a one-byte register or memory location all act on the flags. Operations that do not change the status of the flags are calls, complements, jumps, loading of registers, moves, returns, and stores from the accumulator. Additional operations that do not affect the status of the flags are decrements or increments of 16-bit registers.

There is no operation that can transfer the current condition of the status flags into a register or a memory location, but conditional jumps or calls can determine the status of a particular flag.

Example 11-8. Sixteen numbers are located in memory locations $0050 to $005F. Write a program that determines how many times the number $1A appears in this block of memory. Place that count in register B.

Solution. The plan will be to place $1A in A, use B as the counter of appearances, and H&L as the indicator of which number within the memory block is currently being evaluated. Register addressing will be used to perform the repetitive comparison process.

11-16. Subroutines

Location	Op Code	Mnemonic	Comments
0000	06 00	MVI B,v	Clear B
0002	21 50 00	LXI H,vv	Load H&L
0005	0E 0F	MVI C,v	Number of locs. examined
0007	3E 1A	MVI A,v	$1A into A
0009	96	SUB M	Subtract (H&L) from A
000A	C2 0E 00	JNZ aa	Jump to $000E if not zero
000D	04	INR B	Increment B
000E	23	INX H	Increment (H&L)
000F	0D	DCR C	Decrement C
0010	C2 07 00	JNZ aa	Loop for another number
0013	C7	RST 0	Stop microprocessor

11-16 Subroutines

Subroutines are often convenient programming tools, and large programs in high-level languages often consist of a main program and subroutines with the principal function of the main program to call subroutines. Subroutines are also useful when the same subprogram is needed at several locations throughout the program. The 8080/8085 microprocessor activates subroutines through the CALL statements, which are available in both conditional and unconditional form. Subroutines end with the return (RET) statement, which directs the operation back to the instruction following the CALL statement.

Example 11-9. Use as much as possible of the program in Example 11-8 as a subroutine that can be called to count the appearances of several different numbers in the given memory block.

Solution. Two different calls will be shown, one to count the appearances of $1A and the other $1B. The program in Example 11-8 is revised to delete the instruction at location $0000 and insert:

0014	C9		RET	Return from subroutine

The two CALL statements for $1A and $1B are, respectively,

nnn0	3E 1A	MVI,A,v	$1A into A
nnn2	CD 02 00	CALL aa	Call subroutine
nnn5	...		(store or use contents of B)

and

mmm0	3E 1B	MVI,A,v	$1B into A
mmm2	CD 02 00	CALL aa	Call subroutine
mmm5	...		(store or use contents of B)

11-17 The 8080/8085 Programming Guide

Each manufacturer of a microprocessor publishes a reference manual containing information on the microprocessor and its programming. For the 8080/8085, this booklet is *8080/8085 Assembly Language Programming*.[2] Each instruction or class of instruction is described with pertinent information about each. Figure 11-12 shows a reproduction of a page describing two different ADD instructions, one using register and the other memory addressing. The 8-bit number in the box is the op code, which for the ADD M is clearly $86. The last three bits of the ADD *reg* class depend upon which register is added to the accumulator. Table 11-1 shows that the SSS bits advance in sequence from 111 to 000, 001, etc., corresponding to registers A through L.

The number of states required for each instruction is also shown, which permits the programmer to compute the approximate time required for an operation. The time required for each state is a function of the clock period and may range from 320 ns to 2 μs on the 8085, so the minimum total time required for the operation is the product of the number of states and the cycle time. The total time may be somewhat longer than the calculated time if the microprocessor must wait for a memory to respond.

Figure 11-12 also shows whether the flags are affected by the operation, and for the ADD instructions all are adjusted according to their respective rules.

11-18 The Motorola 6800 Family

Another group of widely used 8-bit microprocessors are those in the Motorola 6800 family.[3] The original chip that appeared early in the history of microprocessors was the 6800. Later the 6802 was developed; it has the same instruction set as the 6800 but is somewhat faster and has some RAM built into the chip. All of the explanations on the 6800 family that follow apply to both the 6800 and 6802 chips.

11-19 Registers in the 6800 Microprocessor

Incorporated in the 6800 microprocessor are six registers, as shown in Fig. 11-13, three of which are 8-bit and three 16-bit. All of the load and store, logic, and arithmetic operations can be performed with either accumulator. The only difference in function between the accumulators is that when the contents of the two accumulators are added or subtracted, the result is placed in accumulator A.

ADD   **ADD**

> The ADD instruction adds one byte of data to the contents of the accumulator. The result is stored in the accumulator. Notice that the ADD instruction excludes the carry flag from the addition but sets the flag to indicate the outcome of the operation.
>
> *Add Register to Register*
>
>   Opcode   Operand
>
>   ADD   reg
>
> The operand must specify one of the registers A through E, H or L. The instruction adds the contents of the specified register to the contents of the accumulator and stores the result in the accumulator.
>
> | 1 | 0 | 0 | 0 | 0 | S | S | S |
> |---|---|---|---|---|---|---|---|
>
> Cycles: 1
> States: 4
> Addressing: register
> Flags: Z,S,P,CY,AC
>
> *Add From Memory*
>
>   Opcode   Operand
>
>   ADD   M
>
> This instruction adds the contents of the memory location addressed by the H and L registers to the contents of the accumulator and stores the result in the accumulator. M is a symbolic reference to the H and L registers.
>
> | 1 | 0 | 0 | 0 | 0 | 1 | 1 | 0 |
> |---|---|---|---|---|---|---|---|
>
> Cycles: 2
> States: 7
> Addressing: register indirect
> Flags: Z,S,P,CY,AC

Fig. 11-12. Excerpt from a page of the programming reference manual for the 8080/8085 microprocessor (From Ref. 2).

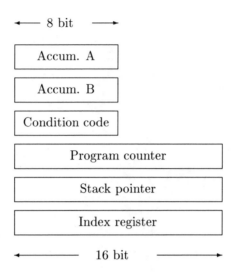

Fig. 11-13. Registers on the 6800 microprocessor.

The program counter specifies the position in the memory where the program is currently executing and performs in the same manner as explained for the generic and 8080/8085 microprocessors. The condition code, index register, and stack pointer will be explained in Secs. 11-21, 11-24, and 11-26, respectively.

11-20 The Instruction Set of the 6800

The instruction set, shown in Table 11-2, presents 72 instructions that are identified by three-letter mnemonics. The instruction set includes arithmetic and logical operations, shifts and rotations, branches and jumps, loads, stores, and interrupt controls.

Table 11-3 presents the op codes for all the programming commands available on the 6800. The names of the instructions that were introduced in Table 11-2 all appear in Table 11-3, and since there are approximately 200 entries in Table 11-3 and only 72 instructions in Table 11-2, some of the Table 11-2 instructions are repeated with variations in Table 11-3. Those variations are particularly the modes of addressing and the applicable accumulator. There are, for example, eight LDAs (load accumulators) given in Table 11-3 that are distinguished by the contents of the box. The B6 op code shown in Fig. 11-14 is an LDA applicable to accumulator A (ACCA) using

11-20. The Instruction Set of the 6800

Table 11-2. Instruction Set for the 6800 Microprocessor

ABA	Add Accumulators	**INC**	Increment
ADC	Add with Carry	**INS**	Increment Stack Pointer
ADD	Add	**INX**	Increment Index Pointer
AND	Logical And	**JMP**	Jump
ASL	Arithmetic Shift Left	**JSR**	Jump to Subroutine
ASR	Arithmetic Shift Right	**LDA**	Load Accumulator
BCC	Branch if Carry Clear	**LDS**	Load Stack Pointer
BCS	Branch if Carry Set	**LDX**	Load Index Register
BEQ	Branch if Equal to Zero	**LSR**	Logical Shift Right
BGE	Branch if Greater or Equal Zero	**NEG**	Negate
		NOP	No Operation
BGT	Branch if Greater than Zero	**ORA**	Inclusive OR Accumulator
BHI	Branch if Higher	**PSH**	Push Data
BIT	Bit Test	**PUL**	Pull Data
BLE	Branch if Less or Equal	**ROL**	Rotate Left
BLS	Branch if Lower or Same	**ROR**	Rotate Right
BLT	Branch if Less than Zero	**RTI**	Return from Interrupt
BMI	Branch if Minus	**RTS**	Return from Subroutine
BNE	Branch if Not Equal to Zero	**SBA**	Subtract Accumulators
BPL	Branch if Plus	**SBC**	Subtract with Carry
BRA	Branch Always	**SEC**	Set Carry
BSR	Branch to Subroutine	**SEI**	Set Interrupt Mask
BVC	Branch if Overflow Clear	**SEV**	Set Overflow
BVS	Branch if Overflow Set	**STA**	Store Accumulator
CBA	Compare Accumulators	**STS**	Store Stack Pointer
CLC	Clear Carry	**STX**	Store Index Register
CLI	Clear Interrupt Mask	**SUB**	Subtract
CLR	Clear	**SWI**	Software Interrupt
CLV	Clear Overflow	**TAB**	Transfer Accumulators
CMP	Compare	**TAP**	Transfer Accumulators to Condition Code Register
COM	Complement	**TBA**	Transfer Accumulators
CPX	Compare Index Register	**TPA**	Transfer Condition Code Register to Acc. A
DAA	Decimal Adjust		
DEC	Decrement	**TST**	Test
DES	Decrement Stack Pointer	**TSX**	Transfer Stack Pointer to Index Register
DEX	Decrement Index Register		
EOR	Exclusive OR	**WAI**	Wait for Interrupt

Mnemonics ©Motorola, Inc., 1976

Table 11-3. Machine Code for the 6800 Operations

MSB\LSB	0	1	2	3	4	5	6	7	8	9	A	B	C	D	E	F
0	•	NOP (INH)	•	•	•	•	TAP (INH)	TPA (INH)	INX (INH)	DEX (INH)	CLV (INH)	SEV (INH)	CLC (INH)	SEC (INH)	CLI (INH)	SEI (INH)
1	SBA	CBA	•	•	•	•	TAB (INH)	TBA (INH)	•	DAA (INH)	•	ABA (INH)	•	•	•	•
2	BRA (REL)	•	BHI (REL)	BLS (REL)	BCC (REL)	BCS (REL)	BNE (REL)	BEQ (REL)	BVC (REL)	BVS (REL)	BPL (REL)	BMI (REL)	BGE (REL)	BLT (REL)	BGT (REL)	BLE (REL)
3	TSX (INH)	INS (INH)	PUL (A)	PUL (B)	DES (INH)	TXS (INH)	PSH (A)	PSH (B)	•	RTS (INH)	•	RTI (INH)	•	•	WAI (INH)	SWI (INH)
4	NEG (A)	•	•	COM (A)	LSR (A)	•	ROR (A)	ASR (A)	ASL (A)	ROL (A)	DEC (A)	•	•	INC (A)	TST (A)	CLR (A)
5	NEG (B)	•	•	COM (B)	LSR (B)	•	ROR (B)	ASR (B)	ASL (B)	ROL (B)	DEC (B)	•	•	INC (B)	TST (B)	CLR (B)
6	NEG (IND)	•	•	COM (IND)	LSR (IND)	•	ROR (IND)	ASR (IND)	ASL (IND)	ROL (IND)	DEC (IND)	•	•	INC (IND)	TST (IND)	CLR (IND)
7	NEG (EXT)	•	•	COM (EXT)	LSR (EXT)	•	ROR (EXT)	ASR (EXT)	ASL (EXT)	ROL (EXT)	DEC (EXT)	•	•	INC (EXT)	TST (EXT)	CLR (EXT)
8	SUB (A) (IMM)	CMP (A) (IMM)	SBC (A) (IMM)	•	AND (A) (IMM)	BIT (A) (IMM)	LDA (A) (IMM)	•	EOR (A) (IMM)	ADC (A) (IMM)	ORA (A) (IMM)	ADD (A) (IMM)	CPX (A) (IMM)	BSR (REL)	LDS (IMM)	•
9	SUB (A) (DIR)	CMP (A) (DIR)	SBC (A) (DIR)	•	AND (A) (DIR)	BIT (A) (DIR)	LDA (A) (DIR)	STA (A) (DIR)	EOR (A) (DIR)	ADC (A) (DIR)	ORA (A) (DIR)	ADD (A) (DIR)	CPX (A) (DIR)	•	LDS (DIR)	STS (DIR)
A	SUB (A) (IND)	CMP (A) (IND)	SBC (A) (IND)	•	AND (A) (IND)	BIT (A) (IND)	LDA (A) (IND)	STA (A) (IND)	EOR (A) (IND)	ADC (A) (IND)	ORA (A) (IND)	ADD (A) (IND)	CPX (A) (IND)	JSR (IND)	LDS (IND)	STS (IND)
B	SUB (A) (EXT)	CMP (A) (EXT)	SBC (A) (EXT)	•	AND (A) (EXT)	BIT (A) (EXT)	LDA (A) (EXT)	STA (A) (EXT)	EOR (A) (EXT)	ADC (A) (EXT)	ORA (A) (EXT)	ADD (A) (EXT)	CPX (A) (EXT)	JSR (EXT)	LDS (EXT)	STS (EXT)
C	SUB (B) (IMM)	CMP (B) (IMM)	SBC (B) (IMM)	•	AND (B) (IMM)	BIT (B) (IMM)	LDA (B) (IMM)	•	EOR (B) (IMM)	ADC (B) (IMM)	ORA (B) (IMM)	ADD (B) (IMM)	•	•	LDX (IMM)	•
D	SUB (B) (DIR)	CMP (B) (DIR)	SBC (B) (DIR)	•	AND (B) (DIR)	BIT (B) (DIR)	LDA (B) (DIR)	STA (B) (DIR)	EOR (B) (DIR)	ADC (B) (DIR)	ORA (B) (DIR)	ADD (B) (DIR)	•	•	LDX (DIR)	STX (DIR)
E	SUB (B) (IND)	CMP (B) (IND)	SBC (B) (IND)	•	AND (B) (IND)	BIT (B) (IND)	LDA (B) (IND)	STA (B) (IND)	EOR (B) (IND)	ADC (B) (IND)	ORA (B) (IND)	ADD (B) (IND)	•	•	LDX (IND)	STX (IND)
F	SUB (B) (EXT)	CMP (B) (EXT)	SBC (B) (EXT)	•	AND (B) (EXT)	BIT (B) (EXT)	LDA (B) (EXT)	STA (B) (EXT)	EOR (B) (EXT)	ADC (B) (EXT)	ORA (B) (EXT)	ADD (B) (EXT)	•	•	LDX (EXT)	STX (EXT)

DIR = Direct Addressing Mode
EXT = Extended Addressing Mode
IMM = Immediate Addressing Mode
IND = Index Addressing Mode
INH = Inherent Addressing Mode
REL = Relative Addressing Mode

A = Accumulator A
B = Accumulator B

• = Unimplemented Op Code

Courtesy Motorola, Inc.

11-21. Condition Codes

what is called extended addressing (see Sec. 11-22), while the D6 op code loads ACCB using a different form of addressing.

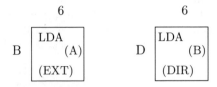

Fig. 11-14. Two LDA op codes.

11-21 Condition Codes

One of the 8-bit registers in the microprocessor is the condition code that is structured as shown in Fig. 11-15. Ones are automatically inserted in bit

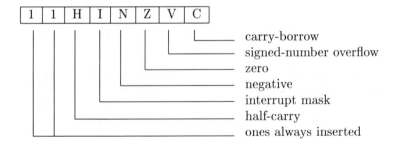

Fig. 11-15. Condition code register in the 6800 microprocessor.

locations 6 and 7, and bits 0 through 5 indicate the status of the carry-borrow, signed number overflow, zero, negative, interrupt mask (to be discussed in Chapter 14), and the half-carry, which indicates a carry from bit 3 to bit 4 and is used in BCD arithmetic. The interrupt mask is changed only by the CLI and SEI commands. In general the H, N, Z, V, and C bits are affected by arithmetic operations, and the N, Z, and V bits by logic as well as load and store operations.

11-22 Forms of Addressing

There are eight modes of addressing available to the 6800 family: (1) extended, (2) direct, (3) immediate, (4) inherent, (5) relative, (6) indexed, (7) accumulator, and (8) dual. The first six modes are of interest to us.

Extended addressing is the same as what was called memory addressing on the generic and the 8080/8085 microprocessors and entails a complete command:

Op code   16 bit address

Direct addressing can be used only if the memory location is in the block between $00 and $FF. The form of the instruction is

Op code   One-byte address

Immediate addressing has the same meaning on the 6800 as on the 8080/8085 and the generic microprocessor in that a one-byte number that is part of the instruction is placed directly in a register,

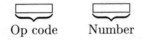

Op code   Number

Inherent addressing applies to some operations where only the registers of the microprocessor are involved and the operation leaves no doubt which those registers are. For example, the TPA instruction transfers the condition code to ACCA, and no further specification of registers or memory locations is needed.

Relative addressing is used in branch operations that will be explained in the next section, and indexed addressing will be presented in Sec. 11-24.

11-23 Branches—Relative Addressing

There are 17 branch instructions, and all use relative addressing. The branch instructions are all designated by B⎵⎵. In the instruction box in Table 11-3 the relative addressing is designated by (REL). The branch command is a two-byte instruction that shifts the program counter forward or backward. The branch-if-equal instruction BEQ, for example, has an op code of 27 ⎵⎵, where the last byte of the code indicates the displacement to be made in the program. The instruction 27 04 would advance four memory locations in the

11-23. Branches—Relative Addressing

program, while the instruction 27 FC moves backward four memory locations, utilizing two's complement arithmetic. Suppose a section of a program were

Location	Op code	Mnemonic	Comment
0041	BB 00 31	ADD A	Add ($0031) to ACCA
0044	27 04	BEQ	Branch if equal

The question arises, "From what location does the branch advance 04 memory locations?" The answer is, "From location $0046," which is the location following the complete BEQ instruction. Thus if the branch occurs, the next instruction executed will be the one located at $004A.

Another question is, "What is being checked as equal to zero in the BEQ instruction?" Clearly we want ACCA to be compared to zero, and the way the decision is described is

Operation: $PC \leftarrow (PC) + 0002 + \text{Rel if } (Z) = 1$.

The operation changes the program counter, PC, from the value it had before execution of the BEQ operation by two plus the amount of the relative address. This activation of the relative-address shift transpires if (Z), which is the zero condition code, is set. The previous operation sets or clears Z.

Another branch operation is BGT, branch if greater than zero, whose operation is

$PC \leftarrow (PC) + 0002 + \text{Rel if } (Z) \vee [(N) \oplus (V)] = 0$

where $\vee$ represents a logical OR and $\oplus$ is an exclusive OR. So the branch occurs if $Z = 0$ and both N and V are the same. Those two situations are:

Case 1
- Z = 0 when previous result not zero
- N = 0 previous result not negative
- V = 0 previous result not a two's-complement overflow

Case 2
- Z = 0 previous result not zero
- N = 1 previous result negative
- V = 1 previous result a two's-complement overflow

Case 1 handles results between $01 and $7F, while Case 2 covers numbers between $80 and $FF, as long as the result arrives there by wrapping around into the usual two's complement negative numbers.

11-24 Index Register—Indexed Addressing

The index register is one of the registers shown in Fig. 11-13, and its most frequent use is in connection with indexed addressing. The instructions that use the index register end in X and include, among others,

LDX   Load the index register

STX   Store from index register to two memory locations

INX   Increment the index register

DEX   Decrement the index register

CPX   Compare the contents of the index register with the contents of two consecutive memory locations

TSX   Load the stack pointer plus 1 into the index register

Our use of the index register will be for indexed addressing, which is abbreviated in Table 11-3 as (IND). The second byte in the op code that uses indexed addressing specifies the offset from the memory location stored in the index register. Suppose that ($E080) is to be added to (ACCA). The number $E080 could be loaded into the index register, and then executing the instruction AB 00 would perform the operation as indicated by the instruction block shown in Fig. 11-16.

The 00 byte specifies the offset from the number stored in the index register. If it were desired to add ($E081) to (ACCA), the instruction would be AB 01. In order to offset negatively from the number stored in the index register, the two's complement of the desired offset is used in the instruction.

Fig. 11-16. An add operation using indexed addressing.

11-25 Loops

The instructions described so far equip a user to construct loops in a program, as Example 11-10 demonstrates.

11-25. Loops

Example 11-10. Write a program to compute the summation

$$\sum_{i=1}^{15_{10}} i$$

Solution. The strategy will be to use ACCA as the running total of the summation and ACCB as the loop counter. A possible flow diagram is shown in Fig. 11-17. The program that executes the flow diagram of Fig. 11-17 is shown in Fig. 11-18 with the program stored in memory beginning at $0005. The subtraction of 1 from ACCB is achieved by decrementing ACCB. The branch instruction BGT continues to displace back four memory locations until (ACCB) is 0. The backward displacement is achieved by adding the two's complement of 4, which is $FC.

The solution to the summation is extractable from the microprocessor after executing the program by reading ACCA, which gives the result $78, or 120_{10}.

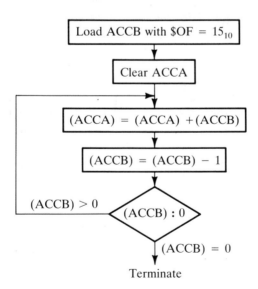

Fig. 11-17. Flow diagram for summation in Example 11-10.

Location	Op Code	Mnemonic	Comments
0005	C6 0F	LDA B $0F	Loading ACCB with $0F
0007	4F	CLR A	Clear ACCA
0008	1B	ABA	Add (ACCB) to (ACCA)
0009	5A	DEC B	Decrement (ACCB)
000A	2E FC	BGT to $0008	Skip back 4 bytes until (ACCB) = 0
000C	3F	SWI	Terminate

Fig. 11-18. Program for evaluating the summation in Example 11-10.

11-26 Stack Pointer

The stack pointer is a 16-bit register in the microprocessor that stores a memory location. An important use of the stack and stack pointer is to store the status of the microprocessor when there is a software interrupt (SWI) or an interrupt from outside the microcomputer. Another frequent role of the stack pointer is to remember where in the program the execution had reached when a subroutine was called. Following execution of the subroutine, the computer knows where to return in the main program to continue execution.

The stack consists of a number of locations of RAM memory. When a byte of information is stored in the stack, it is stored at the address contained in the stack pointer. The stack pointer is decremented immediately following the storage in the stack of each byte of information. Immediately before retrieving a byte of information from the stack, the stack pointer is incremented. Figure 11-19 illustrates the storage in the stack of the program counter when a subroutine is called.   Suppose that the stack pointer holds a value of $E418 and is thus at position A. Suppose also that the position in the program to which operation is to return following the subroutine is $005F. When the subroutine is called, $5F is stored in $E418 and the stack pointer decrements to $E417 to position B. To complete the storing of the program

Memory Location	Contents	Stack Pointer
E416		← C
E417	00	← B
E418	5F	← A

Fig. 11-19. Storing the program counter (PC) on the stack when a subroutine is called.

11-27. Subroutines 243

counter, $00 is stored in $E417 and the stack pointer decrements to $E416 at position C. When the return instruction is executed in the subroutine, the reverse process takes place, the contents of the stack pointer register revert to $E418, and the program counter becomes $005F.

The immediate concern about the stack and the use of the stack pointer—and perhaps the only concern we may encounter—is the knowledge that each microcomputer will normally have a default location for the stack pointer. If, for example, that default location is $E418 as in Fig. 11-19, the RAM memory locations in the neighborhood of $E418 would be used in stack operations. A simple precaution is to stay away from this region of RAM with the operating program, particularly if the stack is to be used by interrupts and subroutines. It is also possible to change the default location of the stack pointer by assigning it a desired value.

11-27 Subroutines

When the same series of instructions is to be used repeatedly in a program, this series may be set into a subroutine and called by a main program whenever it is to be executed. The essential commands are the JSR (Jump to Subroutine) that appears in the main program and the RTS (Return from Subroutine) that is included in the subroutine that sends the operation back to the main program.

Example 11-11. Determine the values of the two summations

$$\sum_{i=1}^{12_{10}} i \quad \text{and} \quad \sum_{i=1}^{15_{10}} i$$

by calling a subroutine.

Solution. The program shown previously in Fig. 11-18 for evaluating a summation can be used as the basis of the subroutine. The main program will load ACCB with the limit of the summation, call the subroutine, and then store the result in designated memory locations, specifically in $0070 and $0071. The program is shown in Fig. 11-20.

11-28 The 6800 Microprocessor Programming Guide

A sample page from the *M6800 Programming Reference Manual*[4] is shown in Fig. 11-21. The first line shows the operation symbolically, and the second line provides a brief word description. The left-pointing arrow in the first

Label	Location	Op Code	Mnemonic	Comments
Main	0001	C6 0C	LDA B $0C	Load ACCB with 12_{10}
	0003	BD 00 20	JSR $0020	Jump to subroutine
	0006	97 70	STA A $0070	Store in $0070
	0008	C6 0F	LDA B $0F	Load ACCB with 15_{10}
	000A	BD 00 20	JSR $0020	Jump to subroutine
	000D	97 71	STA A $0071	Store in $0071
	000F	3F	SWI	Software interrupt
Subr	0020	4F	CLR A	Clear ACCA
Loop	0021	1B	ABA B	Add (ACCB) to (ACCA)
	0022	5A	DEC B	Decrement (ACCB)
	0023	2E FC	BGT	Loop back 4 bytes
	0025	39	RTS	Return from subroutine

Fig. 11-20. Program for Example 11-11.

line indicates "replace by," so the operation is one of replacing the contents of ACCA or ACCB with the sum of the original contents of the accumulator plus the contents of memory location M.

The next group in Fig. 11-21 indicates under what situations the condition codes are set or cleared. For this ADD operation all of the codes are affected, but for some operations certain or all of the condition codes are not affected.

The table at the bottom lists the eight ADD instructions—including the combination of four modes of addressing and two accumulators. The execution time is presented in number of clock cycles required for the operation. This information may be useful in determining the length of time required for a program to execute. The next column is the number of bytes of machine code, which is two for all these operations except when extended addressing is used, which requires 2-byte addresses. Finally, the first byte of the machine code is presented, and this byte corresponds to that indicated in Table 11-3.

11-29 Summary

All commercial microprocessors are somewhat different from one another, but because they all must perform many of the same functions there is considerable similarity. The chapter is based on the assumption that in implementing computer control there will be times when it is useful to know how a microprocessor works. It is accepted that a user would prefer to work with the highest level programming available, but it is also true that the microprocessor operates fastest when using machine language. Some exposure to machine

11-29. Summary

Add Without Carry   **ADD**

Operation: ACCX ← (ACCX) + (M)

Description: Adds the contents of ACCX and the contents of M and places the result in ACCX.

Condition Codes: H: Set if there was a carry from bit 3; cleared otherwise.
                        I: Not affected.
                        N: Set if most significant bit of the result is set; cleared otherwise.
                        Z: Set if all bits of the result are cleared; cleared otherwise.
                        V: Set if there was two's complement overflow as a result of the operation; cleared otherwise.
                        C: Set if there was a carry from the most significant bit of the result; cleared otherwise.

Boolean Formulae for Condition Codes:

$H = X_3 \cdot M_3 + M_3 \cdot \bar{R}_3 + \bar{R}_3 \cdot X_3$

$N = R_7$

$Z = \bar{R}_7 \cdot \bar{R}_6 \cdot \bar{R}_5 \cdot \bar{R}_4 \cdot \bar{R}_3 \cdot \bar{R}_2 \cdot \bar{R}_1 \cdot \bar{R}_0$

$V = X_7 \cdot M_7 \cdot \bar{R}_7 + \bar{X}_7 \cdot \bar{M}_7 \cdot R_7$

$C = X_7 \cdot M_7 + M_7 \cdot \bar{R}_7 + \bar{R}_7 \cdot X_7$

Addressing Formats:

See Table A-1

Addressing Modes, Execution Time, and Machine Code (hexadecimal/octal/decimal):

(DUAL OPERAND)

Addressing Modes	Execution Time (No. of cycles)	Number of bytes of machine code	Coding of First (or only) byte of machine code		
			HEX.	OCT.	DEC.
A IMM	2	2	8B	213	139
A DIR	3	2	9B	233	155
A EXT	4	3	BB	273	187
A IND	5	2	AB	253	171
B IMM	2	2	CB	313	203
B DIR	3	2	DB	333	219
B EXT	4	3	FB	373	251
B IND	5	2	EB	353	235

Fig. 11-21. ADD instruction page from *M6800 Programming Reference Manual.*[4]

language can be advantageous to the engineer applying computer control to systems. The next higher level language, assembly language, is explored in the next chapter.

Having studied the functioning of a microprocessor, Chapter 13 expands the scope to that of a microcomputer that integrates the microprocessor with memories and I/O's, still continuing at an elementary level of computer hardware.

References

1. R. L. Tokheim, *Theory and Problems of Microprocessor Fundamentals*, Schaum's Outline Series, McGraw-Hill Book Co., New York, 1983.

2. *8080/8085 Assembly Language Programming*, Intel Corporation, Santa Clara, CA, 1979.

3. R. Bishop, *Basic Microprocessors and the 6800*, Hayden Book Company, Inc., Rochelle Park, NJ, 1979.

4. *M6800 Programming Reference Manual*, Motorola Semiconductor Products, Inc., Phoenix, AZ, 1976.

Problems

For all programs indicate the memory location, list the op code and mnemonic, and provide comments when appropriate.

11-1. Write a program for the generic microprocessor that multiplies ($0053) by 4. Perform the multiplication using a loop that adds the number to itself the required number of times. Store the result in the accumulator.

11-2. Write a program for the generic microprocessor that provides a time delay by counting down from 127_{10}.

11-3. Write a program for the generic microprocessor that compares ($0037) with ($0038) and if they are equal jumps the program operation to $0020.

11-4. Write a program for the generic microprocessor that executes the ladder diagram branch shown in Fig. 11-22 based on the status of push buttons A and B and cutout C. The memory locations $002A, $002B, and $002C indicate the status of A, B, and C, respectively, with 1 if the switch is closed, and 0 if the switch is open. The program loops continuously through the scan

Problems

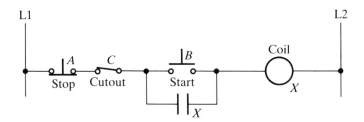

Fig. 11-22. Ladder diagram in Prob. 11-4.

of the A, B, and C memory locations, makes the decision about the status of the coil, and stores 1 in location $0040 if the coil is energized and a 0 if it is not.

11-5. Write a program for the 8080/8085 microprocessor that uses register addressing to decrement the contents of all memory locations in the block $0060 to $006F.

11-6. Write a program for the 8080/8085 microprocessor that adds two 2-byte numbers that are located in the $0070/$0071 and $0072/$0073 pairs, where the first location of the pair contains the high-order byte. Store the result in locations $0074/$0075.

11-7. Write a program for the 8080/8085 microprocessor that operates an on/off controller with a dead band. The set value is available in $0060 and the sensed value in $0061. The dead band extends 4 on either side of the setpoint. If the sensed value is less than ($0060 − 4, a 1 is placed in memory location $0040. If the sensed value is greater than ($0060)+4, a 0 is placed in $0040. If the sensed value is between the two boundaries, no change is made in ($0040).

11-8. Write a program for the 6800 microprocessor that provides a time delay by counting down from the two-byte number $5000 to zero.

11-9. Write a program for the 6800 microprocessor that calls a subroutine 16 times. The complete subroutine starts at $005A. The subroutine processes the number in ACCB and returns its result in ACCB. The 16 numbers to be processed are located in sequence in memory locations $0000 to $000F, and the results returned in ACCB are placed sequentially in memory locations $0010 to $001F. Use indexed addressing.

11-10. Write a program for the 6800 microprocessor that multiplies two 8-bit numbers with the expectation that the product will not be larger than eight bits. One number to be multiplied is in $0066 and the other in $0067. Place the product in ACCB.

Chapter 12

Assembly Language Programming

12-1 Machine Language and Assembly Language

In the previous chapter, we have been writing small programs by looking up the hexadecimal values that represent the instructions the microcomputer is to execute. These values are known as machine language, since they are interpreted directly by the machine. It has been convenient to annotate the machine language programs with mnemonic descriptions of each instruction and with explanatory comments for the benefit of the human programmer. This chapter will introduce a computer program called an assembler, which translates the mnemonic instructions into their machine language equivalents.

12-2 An Overview of the Assembly Process

The details of the assembly process vary from one system to another, but in all cases the goal is the same: to convert mnemonic instructions into an executable form, as shown in Fig. 12-1. The assembly language instructions reside in a file called the source program, and the output is either loaded directly into memory by the assembler or placed in an output file called an object module for later loading into memory. This subsequent loading may be simply typing the assembled code into an evaluation board, or a larger system may have facilities for combining the output of several separate assembler runs and loading the merged result into memory.

A useful product of the assembler, in addition to the object module, is a program listing. This printout will show the input source program, as well

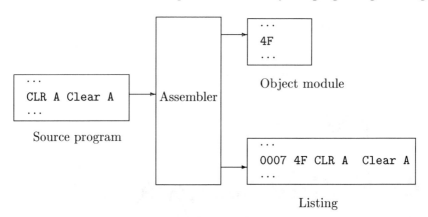

Fig. 12-1. The assembly process.

as the machine language produced by the assembler. Ready access to such a listing aids debugging and documents the program.

The assembler may be run on the machine that is to eventually execute the program, or the resources of a larger machine may be used to do the assembling. If the machine running the assembler is of a different type than the target machine, the assembler is called a cross assembler. The process of transmitting the results of the assembly to the target machine is called downloading, as depicted in Fig. 12-2.

One characteristic of assemblers is that one line of the source program corresponds to one machine instruction, whereas a single line of a higher level language such as Fortran or BASIC will produce many machine instructions. In contrast to machine language programming, the assembler frees the programmer from the tedium of looking up the op code corresponding to each instruction and addressing mode and calculating offsets for branches. This aid becomes especially important in the typical debug-edit cycle of program development, because inserting or deleting instructions is likely to change branch offsets throughout the program. Assembly language listings also provide more readable documentation for the programmer and program maintainer.

12-3 Major Components of the Program

The experience with the 8080/8085 and the 6800 microprocessors in Chapter 11 demonstrated that each has its unique instruction set as well as its own set of mnemonics. An assembler, therefore, is unique to each microprocessor. References 1, 2, and 3 are valuable in developing assembler programs for these microprocessors. While no one description could apply to all assemblers,

12-4. Assembly Language Statements

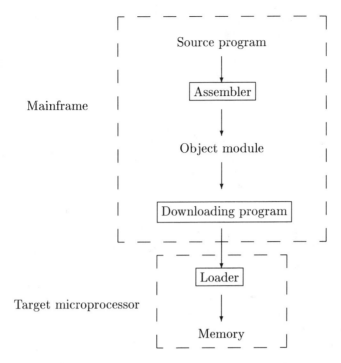

Fig. 12-2. Cross assembly and downloading.

the ultimate objective of various assembler programs is similar. In particular there are two important groups of statements in most programs: (1) assembly language statements, and (2) assembler directives.

The assembly language statements are what eventually result in machine language in the assembled program. The assembler directives define data locations that will be accessed by the program, specify the location in memory where the program will reside, and also provide format controls for a listing of the program if such a printout is called for.

12-4 Assembly Language Statements

Each line of an assembly language program is divided into the following four fields as illustrated in Fig. 12-3:

1. An optional label, for use in naming data storage locations or in identifying the target instruction of a branch or jump

2. The op-code, or mnemonic instruction

3. One or more operands that signify the source and destination of the operation and the addressing mode to be used

4. An optional comment that is ignored by the assembler but may be used by the programmer to indicate the purpose of this instruction or portion of the program

Fig. 12-3. Four fields of an assembly language statement.

Assemblers differ in how they expect the fields to be separated, but the most common methods are to assign special symbols as separators or to require some of the fields to appear in specific columns. For example, a colon (:) is frequently used to terminate a label, and another symbol such as * can indicate the beginning of a comment. Alternatively, the assembler may require that a label, if it appears, must begin in column 1 of the input. In this scheme, the op-code must appear in column 2 or later. The assembler can determine how many operands should appear, based on which op-code is specified, and any characters on the line after the operands are treated as a comment. Figure 12-4 shows typical assembly language statements with the fields identified. The first statement clears accumulator A, and the second performs an immediate add of 2 to accumulator A.

```
LOOP        CLR A   Comments begin after last operand
       ADD A  #$2   Opcode starts anywhere after column 1
↑
Col. 1
```

(a)

```
LOOP:       CLR A   *Comments begin after special symbol
ADD A #$2           *Opcode begins anywhere, if no : appears
```

(b)

Fig. 12-4. Examples of two styles of assembly language statements: (a) fixed input format, and (b) free format with separators.

12-5. Assembler Directives

The microprocessor manufacturer specifies the mnemonic corresponding to each instruction, and how the operands (registers or memory locations) and addressing modes are to be written. In addition to decimal notation, the assembler may allow binary, octal, or hexadecimal constants to be entered, so some convention is needed to distinguish the number base. For example, we have been preceding hexadecimal constants with a "$".

12-5 Assembler Directives

The second major component in an assembler program includes assembler directives that are intended for the assembler itself, not the program that is being assembled. For example, to convey the organization of a program, it is desirable to start each significant subroutine on a new page of the listing, and an assembler instruction such as "PAGE" will be provided to do this. Instructions such as these are called assembler directives or pseudo-ops because the operations do not apply to the microprocessor for which the code is written.

In addition to the program instructions in a program, it is usually necessary to reserve portions of memory for data storage or to load pre-determined values for use as constants or tables. Example assembler directives such as those listed below for the 6800 microprocessor handle the data allocation duties:

BYTE 5,10 Generate two bytes whose values are given. The list can be one or more byte-sized numbers. This directive can be used to generate constants and tables needed by the program.

WORD 4096,8192 Generate a sequence of 16-bit words whose values are specified in the list.

SPACE 16 Skip 16 bytes, making that memory area available for storing data. The contents of this memory area are undefined when the program begins execution but may be used by the program for data storage and retrieval.

ORG $E100 Set the memory address at which subsequent instructions are placed. The "origin" is set at the beginning of the program to define where in memory the program is to reside, and may be used multiple times within one program to position sub-units of the program, such as interrupt service routines and data tables.

END Indicates the end of the program and data and must be the last statement of the assembler program.

Corresponding directives for the 8080/8085 are DB, DW, DS, ORG, and END.

Example 12-1. Show the directives needed to position a Motorola 6800 assembly language program at memory location $E000 with an interrupt service routine at $8150. Allocate a byte to contain a temperature reading, and define an alarm limit of 150_{10}.

Solution.

```
            ORG $E000
MAIN:       CLR A           * start of program
            :
LOOP:       LDA A PORT      * take a temp. reading
            STA A TEMP      * save current temperature
            CMP A MAXTEMP   * compare with alarm limit
            BLT LOOP        * loop back if safe
            :               * process high temperature
            SWI             * stop program execution
TEMP:       SPACE 1         * temperature variable
MAXTEMP:    BYTE 150        * alarm set at 150 deg F

            PAGE
            ORG $8150       * ISR loaded at $8150
CLOCKISR:   :               * beginning of ISR
            END
```

12-6 The Location Counter

Just as the microprocessor maintains a program counter to point to the next instruction to be executed, the assembler maintains a counter that points to the next instruction to be assembled. This counter is available to the assembly language programmer as a dot symbol (.) and is useful for specifying the target of a short branch without having to invent a new label for the destination, and for defining some address-related constants.

```
LDA A PORT1   * loop here waiting for non-zero port reading
BEQ .-3       * branch to start of previous 3-byte-long
              * instruction
```

A discussion of the dot symbol is included here as an introduction to the assembler's location counter, although explicit symbols are preferable. Explicit symbols eliminate the need to count the number of bytes in the nearby instructions or to modify the branch statement when intervening instructions are inserted or deleted. For example, in the previous program fragment, if the programmer inserted a masking instruction without altering the branch statement, the branch instruction would skip over the new 2-byte instruction and try to execute the last byte of the address portion of the LDA—chaos would ensue.

```
LOOP:   LDA A PORT1    * watch for high bit---a better approach
        AND A #$80     * inserted instruction
        BEQ LOOP       *
```

12-7 Using Assembler Labels and Symbols

The assembler maintains a record of label names and their corresponding location within the program. Being able to refer to memory locations by label names rather than absolute memory address frees the programmer from having to assign unused locations and adjust all of the references to any location that changes. For instance, in Example 12-1, the storage location of MAXTEMP is assigned by the assembler to occupy the first byte following the main program, and the variable is used as the operand in storing and loading from the accumulator simply by specifying its name. If new portions are added to the program, the assembler will calculate a new address for the MAXTEMP variable, but the LDA and STA instructions need not change (even though the machine language output of the assembler will have different offsets or absolute addresses in those instructions).

Suppose that some I/O port or hardware register is located at a fixed address. How does one notify the assembler to assign that fixed address to some symbol? One could use ORG followed by a labeled SPACE or use an equivalent pseudo-op called EQU to "equate" a symbol with a particular value. For example, suppose that a register we wish to read is located at address $A020. First assign this value to a name by writing:

```
PORT1:   EQU $A020    * Location of Port 1 data register
```

Then the data from port 1 can be fetched by writing

```
        LDA A PORT1    * Read Port 1 data
```

If later the location of port 1 changes, only the EQU statements needs to change.

The value associated with a symbol can be used in other ways as well. Various constants can be defined by "EQUating" a symbol with the desired constant value and then using the symbol within the program. For example, after writing

```
        MAXPRES:   EQU 55    * Maximum pressure
```

a subsequent comparison for a pressure measurement being within range would appear:

```
CMP A #MAXPRES    * Is measurement in Acc.  A > MAXPRES?
BGT BADPRES       * Yes---handle bad pressure reading
       :          * No---process the reading
```

The values associated with symbols can also be used in expressions. Symbolic expressions can appear on the right side of an EQU statement or as an operand of a machine instruction. To define a constant whose value is the size of a table, write

```
TOP:     BYTE 16           * First value in table
         :                 * Other values in the table
BOT:     BYTE 64           * Last value in table
TSIZE:   EQU BOT-TOP+1     * Number of bytes in the table
```

Subsequent instructions can use the calculated value for the table size just as any other symbolic constant, e.g., LDA A #TSIZE. The +1 in the expression for TSIZE may at first appear curious, but Fig. 12-5 illustrates why this is necessary.

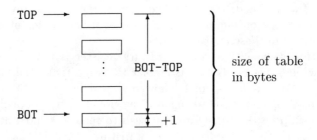

Fig. 12-5. Calculating the size of a table.

Some precautions apply to symbolic expressions. Consider the circular reasoning produced by a program fragment like this:

```
TBLSTART:  SPACE TBLSIZE           * Allocate some space
NEXTDATA:  BYTE 64                 * Last value in the table
TBLSIZE:   EQU NEXTDATA-TBLSTART   * Number of bytes in table
```

The value of TBLSIZE is dependent upon the size of the table allocated by the SPACE instruction, which in turn is dependent on the value of TBLSIZE. To avoid conflicts such as this the assembler requires that the operand in a SPACE directive be a constant or previously defined symbol (a pass 1 constant in the terminology of Sec. 12-9).

12-8 Relocating Assemblers and Loaders

The examples so far have assumed that the programmer explicitly sets the eventual memory location of all program segments with an ORG statement. It is convenient to defer this determination until the time when the assembler output is actually loaded into memory, at which time the location and size of other concurrently loaded modules are known. A program that does this loading into memory is known as a relocating loader, and provisions must be made by the assembler for it to do its job. In particular, references to addresses within the program must be adjusted by the position in memory at which it is loaded. For example, assume that the program is assembled as if it started at address 0, and the loader chooses to load the assembled output at location $500 in memory. Then a jump instruction with a target address of $25 (an offset of $25 from the start of the program) should be changed to have a target of $525 (the loading start address is added to the assembled address). In order for the loader to adjust all such addresses, the assembler includes in its object module output a list of locations that must be patched. This list, although part of the file that the assembler produces, is not part of the program itself and is not loaded into memory.

In order for the assembler to know which locations must be patched during loading, it associates with each symbol a designation of either "absolute" or "relocatable." Examples of absolute symbols are explicit constants and symbols that are EQUated to a constant; symbols that appear as labels are relocatable, since their value will change when the starting address of the program changes. Absolute and relocatable symbols can be combined in expressions, yielding attributes as shown in Table 12-1.

Table 12-1. Operations on Symbolic Expressions and their Resulting Types

Operand1	Operator	Operand2	Result	Example
absolute	+, −	absolute	absolute	512-128
relative	−	relative	absolute	TBLEND-TBLSTART
absolute	+	relative	relative	34+TBLSTART
relative	+	absolute	relative	TBLSTART+34
relative	−	absolute	relative	TBLEND-16
other			illegal	

12-9 The Operation of an Assembler

An outline of the processing steps needed for each line of assembly language would include the following operations:

1. Examine the input line, and extract the various fields (label, op-code, operands, and comment).

2. Search a table in the assembler to find an entry that matches the op-code on the current line.

3. Write the machine language equivalent of the op-code found in the table to the assembler output file.

These steps ignore all of the complexities of performing these tasks quickly and efficiently, but even more fundamentally, they ignore the dilemma presented the assembler by a program fragment such as this:

```
        BGT ERROR   * Process error if > 0
            :       * Normal processing
ERROR:              * Handle error cases here
```

The problem arises that when the `BGT ERROR` statement is encountered, the op-code `BGT` can be looked up in the assembler's table, but the label `ERROR` has not been encountered yet, so it is impossible to compute the branch offset needed for the second byte of this instruction. Such a use of an as-yet unknown symbol is called a forward reference. Several approaches are possible for surmounting this difficulty: either the machine language output is generated as much as possible and later corrected when the symbol is located, or the problem is broken into two stages. These alternatives are called one-pass and two-pass assemblers, respectively. The two-pass method is much more common and will be described in more detail here.

A "pass" of the assembler refers to reading the assembly language source program, so the passes of a two-pass assembler comprise the following steps.

1. Read the source program and determine how large each instruction is. For each instruction advance the location counter by the size of the instruction. When a label is encountered, record that label name and the current value of the location counter in a table called the symbol table. Do not generate any output during this pass.

2. Read the source program from the beginning again. Reset the location counter to the starting value. Repeat the calculation of instruction size and advance the location counter as before, but this time the location

of all labels in the program is known, so operands that contain forward references can be evaluated. Write the machine language instruction code and evaluated operand values to the output file.

During the processing of the assembly language program, numerous errors can be detected by the assembler and flagged as errors or warnings. Among these errors are a misspelled op-code, the wrong number of operands, a reference to a label not present in the program, or the same label appearing more than once in a program. Naturally, flaws in program logic can still lurk in the program even after it assembles without error.

References

1. *8080/8085 Assembly Language Programming*, Intel Corporation, Santa Clara, CA, 1979

2. *M6800 Programming Reference Manual*, Motorola Semiconductor Products, Inc., Phoenix, AZ, 1976

3. *M6800 Resident Assembler Reference Manual*, Motorola Semiconductor Products, Inc., Phoenix, AZ, 1979

Problems

12-1. Write an assembly language program for the Motorola 6800 or Intel 8080 that counts the occurrences of a zero byte in memory locations $1000 to $2000.

12-2. Some machines provide two forms of branch addressing: a short form, which uses an 8-bit offset, and a long form, which specifies the full 16-bit address. Suppose you are to design an assembler that will automatically choose the short form if possible, or the long form otherwise. Describe the difficulty this poses for a two-pass assembler, and propose a compromise.

12-3. What does accumulator A contain when this program terminates?

```
        CLR A
        LDA B N
K:      BGT L
        INC A
L:      ASL B
        BNE K
        SWI
N:      BYTE $C0
        END
```

12-4. Match the following errors with the assembly language statement that produced them.

1. Illegal op-code

2. Duplicate symbol

3. Too many operands

4. Branch target out of range

5. Byte overflow

6. Undefined symbol

```
        ORG 0
CNST0:  BYT 0
CNST1:  BYTE 128
CNST2:  BYTE 512
DATA:   SPACE 1
TEMP:   SPACE 1
MAIN:   CLR A
        LDA A CNST1    * Load 128 into A
        LDX #CNST2     * Load index reg with address of CNST2
        ASL A #2       * Multiply by 4
        CMP A 0,X      * Compare with CNST2
        BNE ERROR
        JMP LOOP
        ORG $1000
DATA:   SPACE 16
ERROR:  STA A DATA     * Save the accumulator
        SWI            * Then stop
        END
```

Chapter 13

The Structure of an Elementary Microcomputer

13-1 Definition of an Elementary Microcomputer

We define an elementary microcomputer as the assembly incorporating a microprocessor, ROM(s), RAM(s), and means to accomplish inputs and outputs. These components are connected through data and address buses. The elementary microcomputer can perform useful calculations, make decisions, and control small systems. The elementary microcomputer is representative of a small unit dedicated to controlling an individual machine or a small number of processes. Such a dedicated microcomputer would be assembled, programmed, and even mounted on the machine at the factory. If the user is expected to make frequent changes or adjustments to the program, the elementary microcomputer may be inadequate, and a higher-level computer should be considered.

The elementary microcomputer that will be studied in this chapter and can execute the programs introduced in Chapter 11 is not an imaginary piece of equipment but is modeled after two commercial products—the Motorola Evaluation Kit and the Intel System Design Kit. These two microcomputers, the MEK6802D5 Evaluation Kit and the SDK-85 System Design Kit, respectively, are products of two companies sold to acquaint users with the characteristics and capabilities of the Motorola 6800 and Intel 8080/8085 families of microprocessors.

Clearly a person can be a competent user of a personal or higher-level computer without knowing the elements presented in this chapter. But it is also true that in applying computers for control purposes, interfacing them to transducers and actuators, and communicating between computers, knowledge of the internal functions of the computer is valuable.

This chapter first extends the capability of the bus structure that was first introduced in Sec. 11-2 to incorporate a monitor ROM and inputs/outputs. Then, for each of the two microprocessors covered in this chapter, it provides a description of the pin diagrams, the structure of these elementary microcomputers, and their memory maps. This explanation is followed by a description of how they operate.

13-2 The Bus Structure

A bus is an electrical conductor that connects three or more components. In microcomputers these buses are usually two sets of multiple lines, one an address bus and the other a data bus. The overall arrangement is shown in Fig. 13-1. The data bus consists of eight lines corresponding to the eight bits of the microprocessor, and the address bus is made up of 16 lines, thus 16 bits, which permits addressing \$FFFF or $65,536_{10}$ different memory locations. The number $65,536_{10}$ is usually referred to as 64K.

The directions of flow of the data and address information are shown by the arrows on the two buses in Fig. 13-1. All addresses flow from the microprocessor to the other components, because the microprocessor controls all operations. Data, on the other hand, may flow in and out of the microprocessor and in or out of other components as well, with the exception of the ROM and PROM, which can be read only. When data are to be read from the user

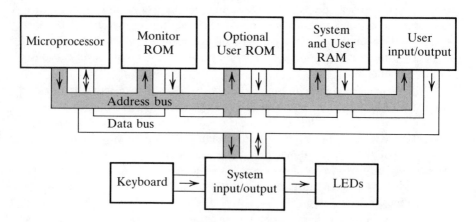

Fig. 13-1. The bus structure of an elementary microcomputer.

13-3. Flow of Information on the Buses

RAM, for example, it is essential that other components are not feeding data on the line. This requirement is met by the three-state capability, discussed in Sec. 7-20, in which all the idle components appear to the bus as extremely high resistances.

13-3 Flow of Information on the Buses During Execution of a Program

The next several paragraphs present an overview of where information is stored and how it flows on the data and address buses during the execution of a program. The capability of a microprocessor was described in Sec. 11-4 as a three-operation sequence of "fetch-decode-execute." One basic realization is that the program on which the microprocessor operates is stored in one section of memory, and other locations of memory are used for storing results used in calculation and control. Higher-level computers keep these compartments of memory separate, but in the elementary computer the programmer must be sure that none of the program is overwritten by the results of operations. If the program is of a nonpermanent nature it will normally be stored in RAM. Programs handling routine utilities (such as reading the keyboard) are stored in the monitor ROM provided by the manufacturer.

If it can be accepted that the microprocessor can move from one step in a program to the next, a preliminary function must also be envisioned to direct it to the starting point of the program. Elementary microcomputers require the operator to indicate where the program is to begin and then require the signal to start. Even before starting the user's program the microprocessor is operating within a higher-level program of the monitor ROM that equips it to start a program, read a key depression, or act on one of the other instructions that might be controlled by a dedicated key on the keyboard.

The reset operation can be called by pressing the RESET key on the board, and it is automatically triggered upon a powerup. Some provision must be made to get the microprocessor started in a program in the monitor ROM that prepares the microcomputer for further instructions. The microprocessors with which we deal are constructed to execute the operations illustrated by Fig. 13-2. The microprocessor is constructed so that whenever it receives a RESET it goes to a certain memory location, for example the two locations $FFFE and $FFFF. The monitor ROM has stored in those locations the memory location (also within the monitor ROM) that is the starting point of the monitor program. Thus, while every manufacturer of an elementary computer might have a different monitor ROM, all must store in specified locations ($FFFE and $FFFF in Fig. 13-2) the address where they choose to start their monitor program.

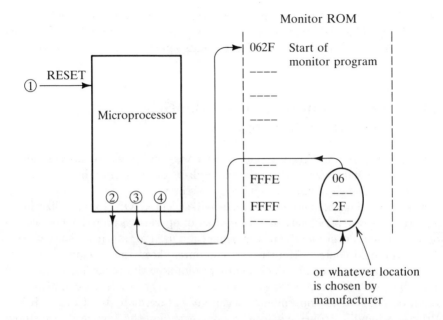

Fig. 13-2. A reset operation.

13-4 The Intel 8080 Microprocessor

The Intel Corporation introduced the first microprocessor in the early 1970s and followed in 1974 with the 8080 microprocessor serving an 8-bit data bus and a 16-bit address bus.[1] A further Intel development was the 8085 microprocessor, which has all the programming instructions of the 8080 plus two additional ones. The 8085 also has some additional hardware facilities. The pin diagram of the 40-pin chip is shown in Fig. 13-3 with certain pins readily identifiable, such as the +5 V power supply to pin 40 and the V_{SS} ground at pin 20. Sixteen address pins and eight data pins would be expected, but to conserve pin locations the 8085 uses pins 12 through 19 for combined data and address functions. This concept is called multiplexed address/data and works in this manner. A 16-bit address is first sent out on the combined AD0 to AD7 (the least significant bits) and the A8 to A15 lines. All of the memories and I/O's served by the microprocessor are signaled by a line from the address latch enable (ALE) pin whether the AD0 through AD7 lines are intended for an address or for data.

The pace of operations performed by a microprocessor is controlled by a clock, such as the one described in Sec. 7-22. Some microprocessors are served

13-4. The Intel 8080 Microprocessor

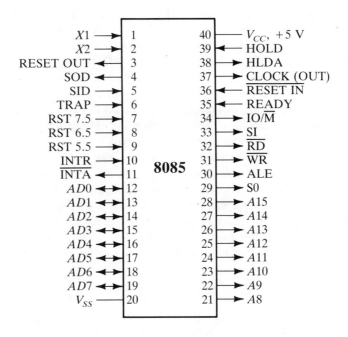

Fig. 13-3. Pin diagram of the 8085 microprocessor.

by an external clock, but the 8085 chip contains the clock circuit and needs only a crystal connected between X1 and X2 to establish the timing. The crystal may have a frequency up to 6.25 MHz. Pin 37 is designated as a clock connection, but it is an output operating at half the frequency of the internal clock and is available for any peripheral that has need of a clock pulse.

The $\overline{RD}$ and $\overline{WR}$ pins are the $\overline{READ}$ and $\overline{WRITE}$, respectively, by which the microprocessor notifies memories and I/O devices whether to send contents to or receive contents from the data bus.

Pin 36 is $\overline{RESET\ IN}$ and is activated by a special-purpose key on the microcomputer assembly but is also automatically activated by an external circuit when the microcomputer is first powered up. The booting operation described in Fig. 13-2 is one operation triggered by the $\overline{RESET\ IN}$. In addition, the registers (special memory locations within the microprocessor) are cleared (or in certain cases assigned a specified nonzero value). The RESET OUT pin sends a signal to the peripherals to inform them when the microprocessor is being reset.

13-5 Structure of the SDK-85 System Design Kit

Some of the major components that make up the SDK-85 elementary computer[2] and their ties to the several buses are shown in Fig. 13-4. The three buses are for data, address, and control. The major components shown are the microprocessor, memories (ROM, PROM, and RAM), chips to facilitate I/O, and a decoder. The data bus is shown separate from the address bus, but indeed it is multiplexed with eight of the address bus lines as described in the previous section. The 8155 chip contains 256 bytes of RAM memory and in addition provides the user with two 8-bit parallel I/O ports plus another port that can be used for control lines. The parallel I/O function will be discussed further in Chapter 14. The 8355 monitor ROM has a capacity of 2048 bytes, and space is available on the board for an 8755 user-programmed EPROM. The 8279 keyboard/display controller appears to the microprocessor as several memory locations, but its function is to translate the signals coming from the keyboard or flowing to the LEDs. The 8205 decoder receives addresses from the microprocessor and converts them to the appropriate enable instructions to the memory chips and the 8279.

13-6 Memory Map of the SDK-85

The 16 address lines provide the capability of $FFFF or $65,536_{10}$ different memory locations, but elementary microcomputers use only a fraction of this potential number. Different areas of the available memory are assigned to the various functions and chips, and this assignment is determined by the hardwire connections on the board. The memory map of the SDK-85 is shown in Fig. 13-5. If no expansion chips are being used, there are three blocks of memory active:

Address Range	Device	Enabled by
$0000 to $07FF	Monitor ROM	CS0
$1800 to $1FFF	Keyboard/display	CS3
$2000 to $20FF	RAM and parallel I/O	CS4

Each of those blocks is enabled by a line from the 8205 decoder that translates an address within the block to a single signal to the chip enable. The RAM foldback areas are address spaces that are unused, but unavailable for expansion, because these locations are multiple mappings of the basic locations. The user of an elementary microcomputer must be aware of the memory map, particularly because the operating program must be stored in the basic RAM location, and this same block must accommodate locations used in calculations, storage of control setpoints, and I/O data for the system being controlled.

13-6. Memory Map of the SDK-85

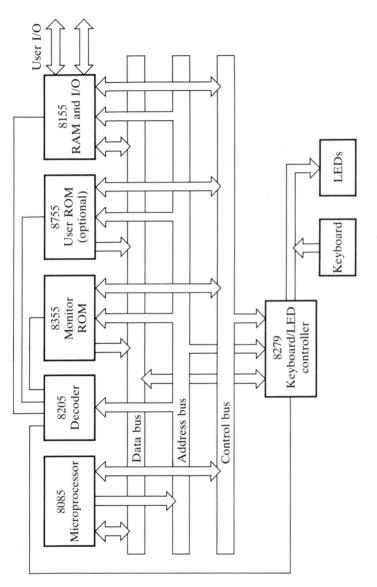

Fig. 13-4. Structure of the SDK-85.

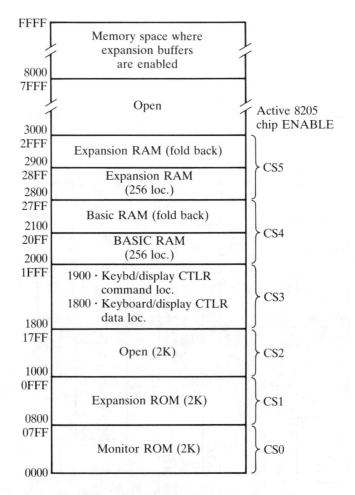

Fig. 13-5. Memory map of the SDK-85.

Not all of the 256 memory locations in the $2000 to $20FF block are always available to the user, because the monitor ROM claims locations $20C2 through $20FF for its own use. If an additional 8155 RAM is mounted on the SDK-85 board (locations $2800 to $28FF), all of these added locations are available to the user.

13-7 The Motorola 6802 Microprocessor

The 6802 microprocessor is a refinement of the original M6800 and incorporates some additional features but retains the same programming

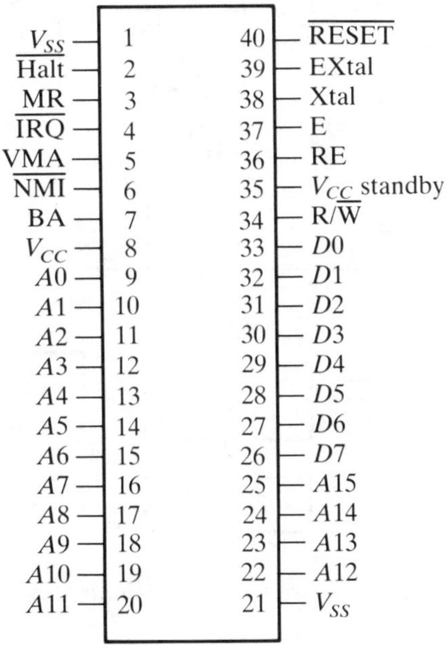

13-8. Structure of the MEK6802D5 Evaluation Kit

```
          ┌─────────┐
   Vss ──┤ 1    40 ├── RESET
  Halt ──┤ 2    39 ├── EXtal
    MR ──┤ 3    38 ├── Xtal
   IRQ ──┤ 4    37 ├── E
   VMA ──┤ 5    36 ├── RE
   NMI ──┤ 6    35 ├── Vcc standby
    BA ──┤ 7    34 ├── R/W
   Vcc ──┤ 8    33 ├── D0
    A0 ──┤ 9    32 ├── D1
    A1 ──┤10    31 ├── D2
    A2 ──┤11    30 ├── D3
    A3 ──┤12    29 ├── D4
    A4 ──┤13    28 ├── D5
    A5 ──┤14    27 ├── D6
    A6 ──┤15    26 ├── D7
    A7 ──┤16    25 ├── A15
    A8 ──┤17    24 ├── A14
    A9 ──┤18    23 ├── A13
   A10 ──┤19    22 ├── A12
   A11 ──┤20    21 ├── Vss
          └─────────┘
```

Fig. 13-6. Pin diagram of the M6802 microprocessor.

instructions.[3] A pin diagram of the M6802 is shown in Fig. 13-6. Some of the expected connections are the 5 V power supply and ground, eight data lines, and 16 address lines. On the M6802 the address and data lines are completely dedicated to their functions, in contrast to the 8080 microprocessor where eight of the address lines are multiplexed with the data lines. The microprocessor is paced by an internal clock that is timed by a crystal connected between pins 38 and 39. The microprocessor can operate with frequencies up to about 2 MHz. Pin 34 is the $\overline{\text{READ/WRITE}}$ command going out to memories and I/O's. The reset is $\overline{\text{RES}}$ at pin 40, which is an input to the microprocessor that is automatically activated on a powerup and also under user control from the keyboard. Pin 5 is a VMA or valid memory address signal that the microprocessor sends out as a high state to indicate to external components that the address currently on the bus is valid.

13-8 Structure of the MEK6802D5 Evaluation Kit

This elementary microcomputer has a structure similar to that shown in Fig. 13-4. A 16-bit address bus and an 8-bit data bus, shown in Fig. 13-7, con-

Chapter 13. The Structure of an Elementary Microcomputer

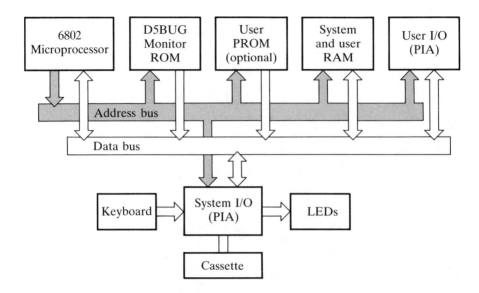

Fig. 13-7. Bus structure of the D5 elementary microcomputer.

nect the microprocessor to the memories and the I/O's. The monitor ROM is called the D5BUG, and both the user and system I/O's are called peripheral interface adapters, or PIAs. Each PIA incorporates two 8-bit parallel ports as well as interrupt lines. One of the PIAs is dedicated to the keyboard and LED display as well as an interface to an audio cassette recorder.

The overall operation suggested by the bus structure is that the microprocessor picks up an instruction as data from a RAM or ROM, executes that instruction by activating a component through the appropriate address, either transmits or receives data, and then asks for the next instruction in the program.

13-9   Memory Map of the D5 Evaluation Kit

The assigned locations of the major blocks of memory in the D5 unit are shown in Fig. 13-8. The D5BUG monitor ROM occupies 2K of memory. The system communication interface adapter is a chip that can be added to the board when serial data (in contrast to parallel data) are to be transmitted in and out of the microcomputer. The user PIA for I/O is treated by the microprocessor as several memory locations. Some sections of the system RAM are used by the microcomputer for certain operations, but the remainder is available for the user as are the 128 RAM memory locations contained on the 6802 chip.

13-10. *Common Features of an Elementary Microcomputer*   271

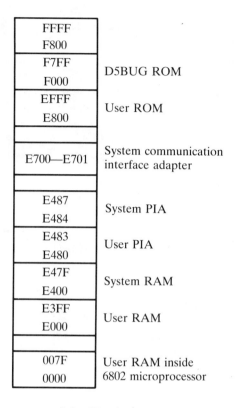

Fig. 13-8. Memory map of the D5 elementary microcomputer.

13-10 Common Features of an Elementary Microcomputer

The two elementary microcomputers investigated in this chapter serve several important purposes. One purpose is that they are learning tools to facilitate an understanding of microcomputers that is a step toward familiarity with more powerful computers. The structure of the microcomputer in which certain major components—the microprocessor, memories, and I/O's—are connected through data and address buses is established. Some insight into the functioning of these components together is provided.

It is also realistic to pause and regain a perspective of the extent to which an elementary microcomputer represents a realistic controlling computer. Indeed, such microcomputers as the SDK-85 and MEK6802D5 have been used and are used as controllers in industrial situations. Their applications are not frequent, however, and commercial controllers are likely to branch in one of two directions from these design kits. Dedicated microprocessor-based controllers serving specific, limited functions are likely to be simplified versions of the microcomputers studied in this chapter. These dedicated microcomputers might not possess the keyboard and LEDs, and incorporate only the I/Os that are needed for the application. Small size is often a goal, so chips that combine several required functions are chosen. The other direction to progress from these elementary microcomputers is toward more powerful computers, perhaps similar to personal computers, which provide the flexibility to the user of changing, adding, or deleting control functions.

References

1. R. L. Tokheim, *Theory and Problems of Microprocessor Fundamentals*, Schaum's Outline Series, McGraw-Hill Book Co., New York, 1983.

2. *SDK-85 System Design Kit User's Manual*, Intel Corporation, Santa Clara, CA, 1978.

3. R. Bishop, *Basic Microprocessors and the 6800*, Hayden Book Company, Inc., Rochelle Park, NJ, 1979.

Problems

13-1. When an elementary microcomputer is in a monitor program in the ready state, what action by the user is necessary to shift the microcomputer to any other operation?

13-2. Some addresses are stored in the subroutines within the monitor ROM. To transmit such an address to the microprocessor, over which bus would it be communicated?

13-3. In what block of memory must a program developed by a user be stored in (a) the SDK-85, and (b) the D5 kit?

13-4. What is the procedure used by the 8085 microprocessor to handle 16 address lines and eight data lines using a total of only 16 pins?

Chapter 14

Parallel Input/Output and Interrupts

14-1 Parallel Input/Output

A microcomputer with no input/output (I/O) would be useless, because there would be no way of communicating with it. The two classes of communication used by microcomputers are parallel and serial. In parallel communication, which will be explored in this chapter, eight lines (for an 8-bit computer) are available to the outside. The voltage levels on these lines are either 0 or 5 V, and the level of all lines is sensed simultaneously. Parallel I/O is thus conveying information to or from the world external to the microcomputer in the same manner as the data bus communicates within the microcomputer. In serial communication, on the other hand, only two lines participate in the process, and one of them is normally constant at ground potential. The voltage on the active line is high at one instant and low at another, so only one signal is transmitted at a time. Both parallel and serial communication are vital, and each has its place. Serial communication will be addressed in the next chapter.

Another important means that the microcomputer has of communicating with the external world is through interrupts. As its name implies, when an interrupt is received from a peripheral, the program on which the computer is operating is suspended and shifts to a special subroutine. Operation may or may not automatically return to the original program.

Facilitation of parallel I/O and interrupts is achieved through the functioning of special chips. This chapter first describes a generic chip that incorporates the minimum needed for parallel I/O, then explores in more detail the Motorola MC6821 as it works on the D5 kit and two Intel chips, the 8155 combination of a RAM with I/O ports that is used on the SDK kit and the M8212 8-bit I/O port.

14-2 A Generic Parallel I/O Chip

Some of the pin connections that would be expected on a parallel I/O chip are shown in Fig. 14-1. It is fundamental that eight lines of the data bus of the microcomputer be connected to the chip so that the microprocessor can send or receive data from the port. Also, one or two ports with their eight lines connect with the peripheral. A direct connection between the data bus and the I/O device cannot be permitted, because the data bus must be free to shift to signals other than that being sent out or received on the port. A capability of I/O chips is that the port is latched, so that when a parallel signal is laid on the port to be sent out, it is held until changed to a new one by the microprocessor.

There are two lines designated interrupt—one receiving an interrupt from the outside and another conveying an appropriate signal to the microprocessor. I/O chips are usually equipped so that the user, through software, can block an interrupt or transmit it immediately to the microprocessor.

Routine connections to the parallel I/O chips are the power and ground lines and some means of selecting and/or enabling the chip when it is to be activated. The chip is inherently selected by means of the address bus, and two different concepts are reflected in the D5/MC6821 and the SDK/8155 procedures. The MC6821 has three chip-select pins that are hardwired to certain address lines. On the SDK the address lines connect to the address decoder, which then sends only one line to the designated chip—in this case the 8155.

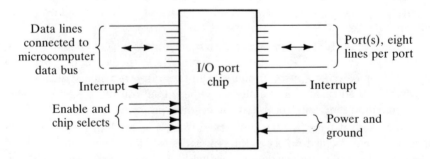

Fig. 14-1. Some pin connections expected on a parallel I/O chip.

14-3 Processing Interrupts

Almost all microcomputers use procedures in responding to an interrupt similar to those shown in Fig. 14-2. When the path through the I/O port is clear, an interrupt from the outside signals the microprocessor, which first completes the operation on which it is working and then processes the interrupt. Separate from the operating program is the interrupt service routine (ISR) located in another section of RAM or ROM. The location of the ISR must be known to the microprocessor. Perhaps the function of the interrupt is to turn on an alarm. In this case the ISR executes that task and then returns to the operating program. In some other cases the alarm may notify a human operator, who must take action before restoring control to the operating program.

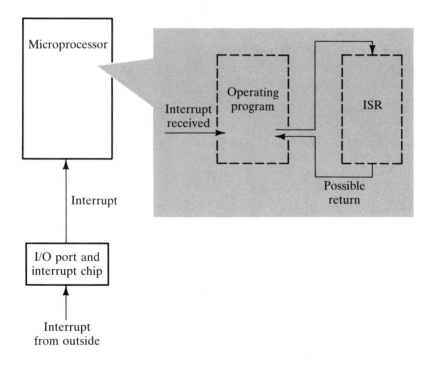

Fig. 14-2. Structure of an interrupt.

14-4 The Motorola Peripheral Interface Adapter (PIA)

The D5 elementary microcomputer is equipped with two PIAs, chips that bear the number MC6821. One of these PIAs is dedicated to the keyboard input and the LED output. The other PIA is available for use by the user for 8-bit I/O. The pinout of the PIA is shown in Fig. 14-3, and the pins may be grouped in the following categories.

1. Power and ground, $V_{SS} = 0$, and $V_{CC} = 5\,\text{V}$.

2. Data bus connections, D0 to D7, that communicate with the microprocessor.

3. Ports A and B. These 8-bit ports, PA0 to PA7 and PB0 to PB7, can be either input or output.

4. Interrupt inputs from the outside CA1 and CB1.

5. Interrupts and peripheral controls, CA2 and CB2, which can convey interrupts from the outside or can transmit interrupts to peripherals.

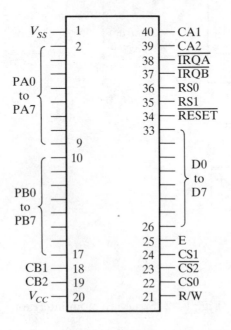

Fig. 14-3. Pinout of the MC6821 PIA.

14-5. Registers in the PIA   277

6. Control and miscellaneous pins. These include interrupt requests IRQA and IRQB; register selects RS0 and RS1; the RESET; the enable E; chip selects CS0, CS1, and CS2; and read/write, R/W.

Another approach to sorting out the functions of the pin connections is to show, as in Table 14-1, which pins are connected to peripherals and which to the microprocessor. The only pins not included in the table are the power supply and ground and the enable pin E. The R/W connection controls the direction of the D0 to D7 pins, thus whether the microprocessor is to read from the PIA or write to the PIA. The direction of ports A and B is controlled by other means that will be explained in Sec. 14-5. The RESET pin, upon receiving a drop in voltage, resets all registers in the PIA.

Table 14-1. Pins on PIA Connected to Peripherals and to the Microprocessor

Connected to Peripherals	Connected to Microprocessor
PA0 to PA7	D0 to D7
PB0 to PB7	IRQA and IRQB
CA1 and CA2	RS0 and RS1
CB1 and CB2	Reset
	CS0, CS1, and CS2
	R/W

14-5 Registers in the PIA

Like the microprocessor the PIA contains several registers—six of them. The three registers associated with each port are (1) the control register, (2) the data direction register, and (3) the peripheral register that is directly connected to the port. The microprocessor has access to these registers by calling specified memory locations. These memory locations on the D5 unit are

$E480 Either DDRA (data direction register for port A) or port A

$E481 CTLA (control register for port A)

$E482 Either DDRB (data direction register for port B) or port B

$E483 CTLB (control register for port B)

Whether $E480 accesses port A or the DDRA is determined by the way the CTLA is set. If CTLA=$x0, addressing $E480 accesses DDRA, or if CTLA is

$x4$, addressing $E480 accesses port A. The contents of the left-most hex digit in CTLA is of no concern. Each bit of the data direction register (DDRx) determines whether the corresponding bit in the associated port is an input or output:

- 0 in a bit of the DDR sets the corresponding bit of the port to be an input, and

- 1 in a bit of the DDR sets the corresponding bit of the port to be an output.

It is possible for some bits of a port to be inputs and the remainder to be outputs.

14-6 Preparing the PIA to Send and Receive Data

Two examples will show the steps necessary to set the ports to receive and transmit data.

Example 14-1. Set port A to receive data.

Procedure

Step 1. Store $$x0$ in $E481. This action causes memory location $E480 to have access to the data direction register DDRA.

Step 2. Store $00 in $E480. All bits are zero in $00, which assigns all bits of Port A to be inputs.

Step 3. Store $$x4$ in $E481. The first two steps accomplished the purpose of setting the desired direction of the data flow at the port, so now this last step gives the microcomputer access to Port A through $E480.

Example 14-2. Set port B to output data.

Procedure

Step 1. Store $$x0$ in $E483, which permits addressing $E482 to access DDRB.

Step 2. Store $FF in $E482. All the bits in $E482 will be ones, which assigns all the bits of the port to be outputs.

Step 3. Store $$x4$ in $E483, which gives the microcomputer access to port B through $E482.

14-7. Interrupt from a Peripheral—An Overview

In Step 2 of both the examples, the entire port A was set to receive data and the entire port B to send data. A port can also serve a divided function. Suppose that the three lowest-order bits of the port are to have input capability and the five highest-order bits are to be equipped to output data. Figure 14-4 shows the setting of the DDR, which is $F8.

$$\underbrace{\begin{matrix} \text{MSB} \\ 1\ 1\ 1\ 1 \end{matrix}}_{F} \quad \underbrace{\begin{matrix} \text{LSB} \\ 1\ 0\ 0\ 0 \end{matrix}}_{8}$$

Fig. 14-4. Setting of DDR for mixed input/output.

It is appropriate at this point to comment on the function of the RESET. When the D5 unit powers up there is an automatic resetting of the registers. Another act that accomplishes the reset is to press the RESET key on the keyboard. Pressing the RESET key would have executed the first two steps in Example 14-1, but it is good practice to carry out the complete procedure anyway. The steps in setting up the PIA will normally be a preliminary portion of the program, the latter part of which performs a control function.

14-7 Interrupt from a Peripheral—An Overview

The previous discussion outlined how the ports would be used to receive and transmit data. The control functions of the microcomputer in certain assignments may not be required to do any more than receive data from sensors, make decisions based on the magnitudes from those sensors, and then output data to regulate the actuators. The program for this control assignment when once set into operation may continue indefinitely.

Another important situation is for the microcomputer to carry out the control function until it receives a signal from outside—an interrupt—whereupon the computer jumps to another program to carry on a different assignment. The microcomputer is capable of continuing in this second assignment until the action required by the interrupt is complete, then returning to the original control assignment. The schematic diagram of the components and operation of an interrupt are shown in Fig. 14-5. The interrupt begins when one of the interrupts CA1, CA2, CB1, or CB2 rises or falls. Whether the PIA responds to a rise or to a fall depends on how the control register is set. The PIA transmits the signal to $\overline{\text{IRQA}}$ if the interrupt comes from CA1 or CA2.

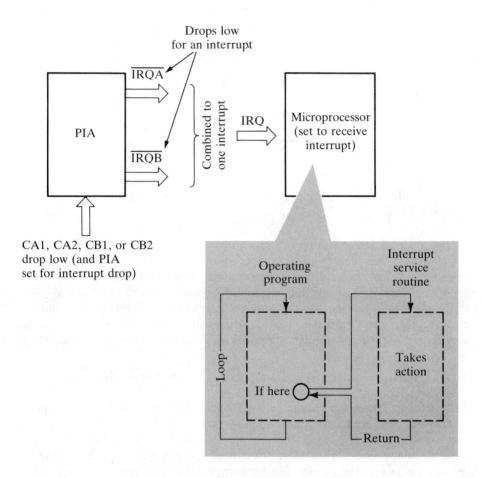

Fig. 14-5. Overview of an interrupt operation.

The signal is transmitted to $\overline{\text{IRQB}}$ if the interrupt comes from CB1 or CB2. $\overline{\text{IRQA}}$ or $\overline{\text{IRQB}}$ drops when it receives an interrupt, and this message is passed on to the microprocessor. The signals from $\overline{\text{IRQA}}$ and $\overline{\text{IRQB}}$ are combined such that only one IRQ passes to the microprocessor. Let us assume that the microcomputer is in the midst of an operating program that has a loop structure. Upon receiving the interrupt, the program counter jumps from wherever it is in the operating program to the interrupt service routine (ISR). If the programs are structured so that the ISR is to take one action and then return to the operating program, the return to the operating program is to the point where control left the operating program, just as occurs for other types of subroutines.

There are four interrupt pins on the PIA, but the microprocessor receives only one interrupt message. A reasonable question is how the microprocessor knows which of the interrupts should be serviced if more than one is active. It is possible to include in the first part of the ISR a test to determine which of the control lines (CA1, CA2, CB1, or CB2) originally caused the interrupt.

14-8 The Control Register and the Control Lines

The control registers, CTLA (location \$E481) and CTLB (location \$E483), have already been encountered as the means of deciding whether \$E480 or \$E482 was accessing a data direction register or a port. The fact that we loaded values of $\$x0$ or $\$x4$ in the control register implied that CTLA and CTLB are 8-bit registers. The function of these bits is shown in Fig. 14-6. The designations in Fig. 14-6 apply to the CTLA. For the CTLB the functions would apply to IRQB1, IRQB2, CB2, etc. So far we have used only bit 2, because when we entered $\$x0$ into \$E481 we assigned bit 2 to 0 and when we entered $\$x4$ into \$E481 we assigned bit 2 a value of 1.

The CA1 control and the interrupt request $\overline{\text{IRQA}}$ will be the only additional bits of the CTLA examined in this chapter, and the CA2 interrupt will not be considered. While IRQA1 and IRQA2 are two bits of the control register, they combine to send out one signal, $\overline{\text{IRQA}}$, at pin 38 on the PIA. As Fig. 14-5 illustrates, the $\overline{\text{IRQA}}$ and $\overline{\text{IRQB}}$ further combine to one interrupt signal that passes to the microprocessor. What is the purpose of having both IRQA1 and IRQA2, then? If both CA1 and CA2 are available for interrupts, the computer must have some way of determining whether it was CA1 or

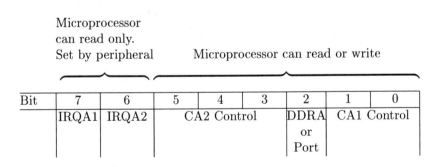

Fig. 14-6. The bits in the control register, CTLA.

CA2 that caused the interrupt, so following an interrupt the microprocessor can read the control register and determine from the setting of bits 6 and 7 whether CA1 or CA2 caused the interrupt.

To transmit an interrupt signal on to the microprocessor the $\overline{\text{IRQA}}$ pin, number 38 (or the $\overline{\text{IRQB}}$ pin 37), must drop from a high to a low. The settings of bits 0 and 1 determine how the $\overline{\text{IRQA}}$ responds to CA1, and the four possible conditions are shown in Table 14-2.

Table 14-2. Response of IRQA to CA1 as a Function of the Settings of Bits 0 and 1 of the Control Register

Bit 1	Bit 0	$\overline{\text{IRQA}}$ (Pin 38)
0	0	Disabled, remains high
0	1	Goes low on ↓ of CA1
1	0	Disabled, remains high
1	1	Goes low on ↑ of CA1

Example 14-3. Extend Example 14-1 to set port A to receive data and also prepare the PIA so that a rise in CA1 will send an interrupt to the microprocessor.

Procedure

Step 1. Store $x0 in $E481 (as in Example 14-1).

Step 2. Store $00 in $E480 (as in Example 14-1).

Step 3. Store $x7 in $E481, which assigns the lowest three bits as follows:

Bit	2	1	0
Value	1	1	1

The 1 in bit 2 accesses port A through $E480. The setting of bits 1 and 0, according to Table 14-2, causes a drop in $\overline{\text{IRQA}}$ on a rise in CA1.

14-9 Setting the Microprocessor to Receive an Interrupt

We have now achieved a drop in $\overline{\text{IRQA}}$ or $\overline{\text{IRQB}}$, which in turn results in a drop in the one IRQ line reaching the microprocessor. The question now is whether the microprocessor will accept the interrupt request. The microprocessor accepts the interrupt request if the interrupt mask in the microprocessor is cleared. The I bit of the condition code discussed in Sec. 11-21 is the interrupt mask.

- If $I = 0$, the microprocessor will respond to a falling edge of IRQ.
- If $I = 1$, the microprocessor will not respond to a change in IRQ.

A powerup or pressing RESET clears the interrupt mask, thus I will normally be 0 and the microprocessor responds to an interrupt. As a precaution the CLI instruction could be used in the program.

14-10 Structure of an Interruptible Program

The four chief sections of a program capable of receiving and acting on an interrupt are:

1. Setting the PIA and the microprocessor
2. Specification of the location of the ISR
3. Operating program
4. ISR

These sections are shown symbolically in Fig. 14-7.

1. **Setting the PIA and the microprocessor.** The settings of the PIA for the desired direction of data and setting the control register for the desired interrupt control line and whether it is a rise or fall that activates the interrupt have been discussed in Secs. 14-6 to 14-8. Section 14-9 explained how the microprocessor is prepared.

2. **Specifying the location of the ISR.** To jump to a usual type of subroutine the operation of the main program had to reach a certain position in the program. Upon receipt of the interrupt signal, on the other hand, the microprocessor finishes the instruction it is currently executing and then jumps to the ISR regardless of where in the main program the microprocessor is executing. The computer must be informed where the ISR begins so that the jump can be made. In the D5

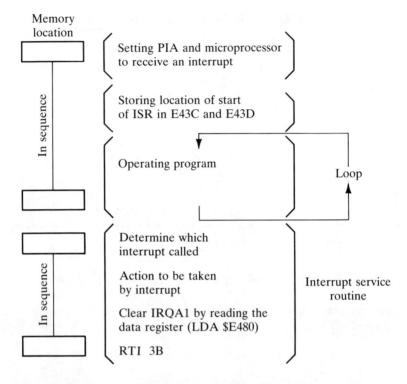

Fig. 14-7. Major sections of an interruptible program.

unit the memory location of the start of the ISR is stored in $E43C and $E43D. For example, if the ISR is to begin at $E100, it is necessary to load:

Memory location	Contents
E43C	E1
E43D	00

3. **Operating program.** The operating program may perform some continuing control functions that can be interrupted, or it may only be a time killer that does nothing but wait for the interrupt. The operating program will normally operate in a loop until interrupted.

4. **ISR.** The structure of the ISR is shown in Fig. 14-7. If there are several interrupts (CA1, CA2, CB1, and CB2) that could be expected, the first step is to poll the PIA control registers to determine which line called the interrupt. We are limiting ourselves to the case where only one interrupt is active, so the polling process would not be necessary.

14-11. User I/O Socket

The next section in the ISR takes the action called for by the interrupt, which may be a task such as changing the value of a temperature setting, sounding an alarm, etc.

If after taking action the program operation returned immediately to the operating program it would find that bit 7 in the control register shown in Fig. 14-6 still indicates an interrupt, $\overline{\text{IRQA}}$ (pin 38) is still low, and the microprocessor would think it had another interrupt. It is therefore necessary to clear IRQA1, which the PIA has the capability of doing by reading the data register. For example, LDA $E480 (B6 E4 80) clears bit 7 in the CTLA.

If the function of the interrupt is to demand some operator intervention, the ISR may be structured as a loop that does not return to the operating program until the user manually makes that transfer. Typically, however, the interrupt causes some action to be taken and the operation then returns to the operating program to carry out the assigned functions there and to wait for the next interrupt.

The return from interrupt, RTI, is called by op-code 3B, and execution returns to the instruction in the operating program that would have been the next one had the interrupt not occurred.

14-11 User I/O Socket

The D5 kit provides a 24-pin socket through which the user can make I/O connections. This socket, numbered U-19 on the board, has a pinout as shown in Fig. 14-8. The connections to the socket are for port *A*, port *B*, the interrupts (CA1, CA2, CB1, CB2), and the power connections of 5 V and

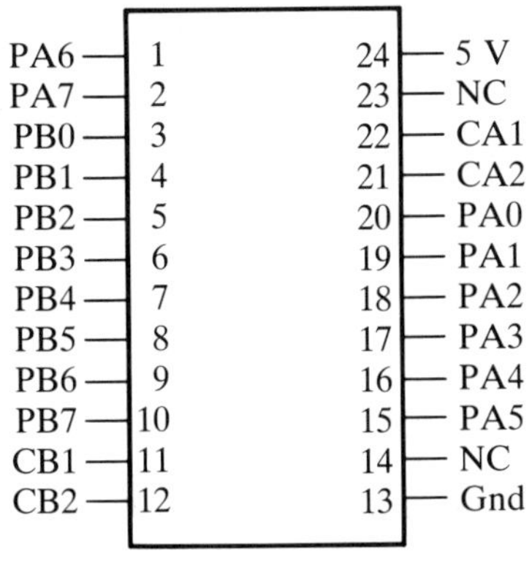

```
PA6 — 1        24 — 5 V
PA7 — 2        23 — NC
PB0 — 3        22 — CA1
PB1 — 4        21 — CA2
PB2 — 5        20 — PA0
PB3 — 6        19 — PA1
PB4 — 7        18 — PA2
PB5 — 8        17 — PA3
PB6 — 9        16 — PA4
PB7 — 10       15 — PA5
CB1 — 11       14 — NC
CB2 — 12       13 — Gnd
```

Fig. 14-8. User I/O socket on D5 board.

ground which are supplied from the board. The port and interrupt connections are tied directly to the PIA, and as Table 14-1 stated, are the pins of the PIA that are accessible to the peripherals.

14-12 Intel 8155/8156 RAM with I/O

This chip combines the function of a 256-byte RAM with three I/O ports having the pin diagram shown in Fig. 14-9. Ports A and B are 8-bit, and port C is 6-bit.

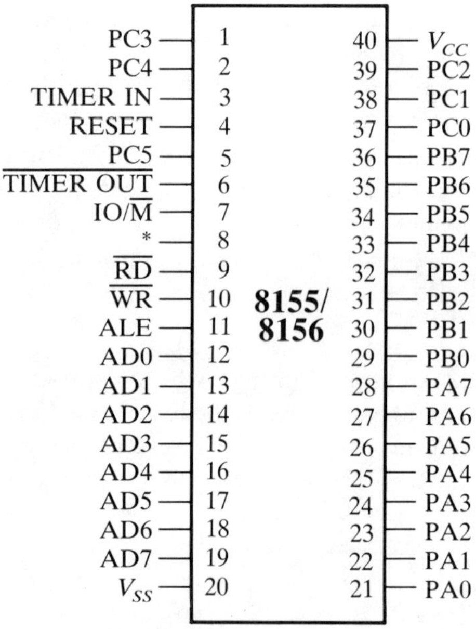

```
        PC3 ──┤ 1        40 ├── V_CC
        PC4 ──┤ 2        39 ├── PC2
    TIMER IN ─┤ 3        38 ├── PC1
       RESET ─┤ 4        37 ├── PC0
         PC5 ─┤ 5        36 ├── PB7
   TIMER OUT ─┤ 6        35 ├── PB6
        IO/M ─┤ 7        34 ├── PB5
           * ─┤ 8        33 ├── PB4
          RD ─┤ 9        32 ├── PB3
          WR ─┤ 10 8155/ 31 ├── PB2
         ALE ─┤ 11  8156 30 ├── PB1
         AD0 ─┤ 12       29 ├── PB0
         AD1 ─┤ 13       28 ├── PA7
         AD2 ─┤ 14       27 ├── PA6
         AD3 ─┤ 15       26 ├── PA5
         AD4 ─┤ 16       25 ├── PA4
         AD5 ─┤ 17       24 ├── PA3
         AD6 ─┤ 18       23 ├── PA2
         AD7 ─┤ 19       22 ├── PA1
        V_SS ─┤ 20       21 ├── PA0
```

Fig. 14-9. Pin diagram of 8155 combination RAM and parallel I/O chip.

14-13 Intel 8212 I/O Chip

The 8212 chip, whose pin diagram is shown in Fig. 14-10, is more nearly comparable to the PIA in that it combines an 8-bit port with interrupt capabilities. The strobe pin is the interrupt from outside, and the INT pin signals the microprocessor that an interrupt has occurred.

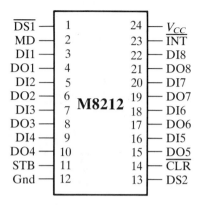

Fig. 14-10. The 8212 combination I/O port and interrupt.

14-14 Rudimentary Control Capability Now Available

The significant achievement that has now been reached through the work of the preceding chapters is that the entire control loop has been traversed in an elementary manner. The route has covered the path from sensors through transducers and signal conditioning to multiplexing and analog-to-digital conversion to the microcomputer. On the basis of the information from the sensors, the computer, acting on its program, decides what control measures to take. Commands are sent out in digital form from the microcomputer, converted to analog values, demultiplexed, and sent to the appropriate actuator(s). The control loop is complete.

From this point on the efforts can be devoted to improving the utility and convenience of the programming process. Knowledge of the microcomputer hardware and experience in machine language programming is instructive, because the steps required are all ones that occur within the microcomputer. It is not economical in user time, however, to continue working on such a primitive level. For future work the engineer will want to use high-powered microcomputers that have large memories and that can be programmed in a more convenient manner. A further capability that enhances the application of the microcomputer is serial data transmission. In order to tie microcomputers together and transmit data over long distances the 8-bit or 16-bit parallel data transmission must be replaced by voltage or current pulses along two conductors.

Even after shifting to commercial microcomputer controllers, there will be operations and requirements that are reminiscent of the basic behavior of the elementary microcomputer.

Problems

14-1. The eight lines of a parallel port are connected directly to the input of a DAC. What is the signal to the DAC when the I/O chip is not selected?

14-2. Write an assembly language program that sets up a PIA as follows: (a) an interrupt occurs when CA1 falls, (b) bits 0 to 3 of port A are input, and bits 4 to 7 are output.

14-3. In what situation would an interrupt line from the outside world (with proper signal characteristics, of course) connected directly to the microprocessor function satisfactorily, and in what situation would it be too restrictive?

14-4. Where is the starting address of the interrupt service routine stored?

14-5. With a PIA as the I/O chip, write in assembly language one part of a program that determines whether it was CA1 or CA2 that caused an interrupt. Already available:

1. Program that prepares port A and the microcomputer for interrupts and designates the ISR to start at $E090

2. Subroutine starting at $E010 if CA1 was the interrupt

3. Subroutine starting at $E020 if CA2 was the interrupt

Chapter 15

Serial Input/Output and Modems

15-1 Serial Data Transmission

Chapter 14 explained how data in parallel form may be transmitted from and received by a microcomputer. All the lines of a port are simultaneously sending or receiving data. In contrast, the serial mode of data transmission sends bits of data one at a time along a single conductor, as in Fig. 15-1a. A high voltage may appear at one time and a low voltage an instant later, as shown in Fig. 15-1b. If the voltage is high for 1/300 s, in one type of serial data transmission, for example, that signal could mean a 1 and if low it could mean a zero. A high voltage for 1/150 s is interpreted as two 1's in succession. Instead of a voltage signal, pulses of current and the lack of them

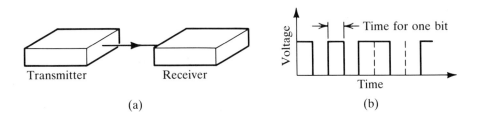

Fig. 15-1. (a) Serial data transmission between two microcomputers, and (b) voltage-time pattern.

could indicate 0's and 1's. Sending parallel data requires eight conductor lines, while a rudimentary system transmitting serial data requires only one line plus a ground line.

Both parallel and serial data transmission have unique roles to play. Within the microcomputer and to closely coupled peripherals parallel data transmission has the advantage of speed. For long distances it would be possible to bundle the eight parallel lines and enclose them with a shield, but such cable would be expensive. Serial data communication is mandatory for transferring data on and off tapes or disks and for sending data over telephone lines. Several disadvantages of serial data transmission are its slow speed relative to parallel communication and the complexity of receiving serial data.

This chapter first proposes a generic chip as the serial I/O port for an elementary computer. The necessary functions will be enumerated as though the chip were being designed. A portion of the chapter is devoted to a Motorola chip and an Intel chip that implement serial data transmission. The emphasis then shifts to the more rugged form of serial transmission, RS-232-C, and finally to modems for transmission over telephone lines.

15-2 Mark, Space, and Baud Rate

Several terms that will appear frequently should be defined. A "mark" is a logical 1, and a "space" is a logical zero. These marks and spaces are grouped into a "character," which might be one byte of data consisting of eight marks and spaces. In addition, one character may be separated from another character by other bits, such as a start bit, a stop bit, and a parity bit, all of which will be explained later. We will call the assembly of the start bit, the character, the stop bit, and parity bit a "cluster." The standard convention is:

- Start bit is a space.

- Stop bit is a mark.

It should not be assumed that a mark is always represented by 5 V and a space by 0 V, as has been true with parallel data transmission. The convention for marks and spaces is a function of the protocol—for example, the passage of a 20 mA pulse of current may indicate a mark, and the lack of current flowing may indicate a space. In another convention the existence of +12 V may indicate a space and −12 V a mark.

The "baud rate" is the number of bits per second being transmitted. Typical rates are 75, 110, 150, 300, 600, 1200, 2400, 4800, 9600, and 19,200 baud.

15-3 Synchronous and Asynchronous Communication

The need for some synchronization between the transmitter and the receiver is obvious. If the transmitter sends bits when the receiver is not expecting them, errors in interpretation are likely. There are two concepts in timing—one is synchronous and the other asynchronous communication.

In both timing methods the transmitter and receiver are regulated by clock pulses—the transmitter laying bits on the communication line according to its clock. In synchronous communication once the transmitter begins sending data it continues at a regular pace until all the data it has available have been sent. The transmitter and receiver must be timed by the same clock or by clocks of identical frequency.

In asynchronous communication, data flow along the line in spurts—one cluster at a time, as illustrated in Fig. 15-2. The transmitter sends a character (eight bits, for example) bounded by the start and stop bits and then may wait a period at its own discretion. The convention is that a mark is used for the stop bit and a space for the start bit. Following transmission of a cluster, the final stop bit leaves the line in a mark status, a condition in which it remains until the next start bit switches the line to a space status. The waiting time may be zero, in which case the clusters are transmitted continuously. An important characteristic of asynchronous communication is that the transmitter and receiver coordinate themselves at the start of each cluster. Both the transmitter and receiver need essentially the same clock frequency so that the character will not be misunderstood, but there can be a slight deviation between their two frequencies without disturbing the accuracy of transmission, since the transmitter and receiver synchronize at the start of each new cluster.

A comparison of the two methods is that synchronous transmission requires extra provisions to provide identical clocking for both the transmitter and receiver, while in asynchronous transmission the two clocking rates need only be very close to each other. Synchronous communication does not require the start and stop bits, since once the coordinated transmission and

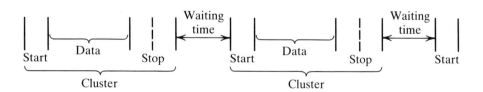

Fig. 15-2. Asynchronous communication.

reception begin the transmitter and receiver are in agreement where the characters start and end. An advantage of synchronous transmission is therefore that more data can be transmitted even at the same bit rate. Asynchronous communication is widely used, is more convenient, and will be the only type considered hereafter.

15-4 Parity

The transmission of data is not always flawless, and it is possible that a bit in a character may be distorted in transmission such that it is received as 0 when it was sent as 1 and vice versa. A procedure to detect such an error is the use of a parity bit. A parity bit is an extra bit added to each cluster, as shown in Fig. 15-3. Odd parity or even parity may be chosen. If odd parity is chosen, the sum of the 1's in the combination of data and parity bit is an odd number. If even parity is chosen, the sum of the 1's is an even number. It may be noted that the use of parity will detect an error in one bit of data, but not errors in two bits.

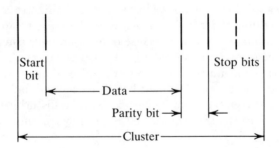

Fig. 15-3. A parity bit in a cluster.

15-5 Shift Register

A fundamental component in the conversion between parallel and serial data is the shift register. In the conversion from parallel to serial, the eight bits of parallel data must be pushed one at a time onto the serial line. During the conversion from serial to parallel, the marks and spaces coming from the serial line must be loaded in sequence in a memory location or register. The conversion by a shift register from parallel to serial is shown schematically in Fig. 15-4.

15-6. A Generic UART

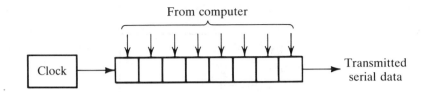

Fig. 15-4. A shift register for conversion from parallel to serial.

The steps in the operation are first to write the eight bits of data into the shift register. Upon a signal, the shift register sends out its rightmost bit to the line and simultaneously shifts all the other bits one position to the right. The position just vacated at the leftmost location is loaded with a 0. The successive shifting of characters from the register to the line proceeds at a pace controlled by the clock.

Individual chips are available for parallel-to-serial conversion, such as the SN74165, and for conversion in the opposite direction from serial to parallel, such as the SN74164. Rather than explore the details of these chips we will move directly to components that include the shift functions with the other required operations.

15-6 A Generic Universal Asynchronous Receiver/Transmitter (UART)

Imagining what capabilities would be needed by a UART is one approach to understanding its structure and operation. First identify the minimum connections needed for the chip, as in Fig. 15-5. Fundamental requirements include the power supply of 5 V, ground, and eight lines to the data bus. It is expected that the chip will be capable of communicating serial data, either in or out, so two serial lines would be provided—the transmit (Tx) and the receive (Rx) pins. The UART incorporates a shift register and, as Fig. 15-4 indicated, there is a clock to time the transmission or reception of each bit. One or more chip-select pins are needed to activate the UART when desired. One chip-select pin could suffice if the microcomputer is equipped with an address decoder (as in Fig. 13-4). If certain address lines are connected directly to the UART, there would normally be several chip-select pins to offer flexibility in selecting various other chips as well. The remaining pin connections provide means of informing the UART whether data are to arrive from the microprocessor and to be transmitted serially, or whether the data are flowing in the opposite direction. One possibility for specifying the direction of data flow as shown in Fig. 15-5 is a combination of two-pins—the R/W and

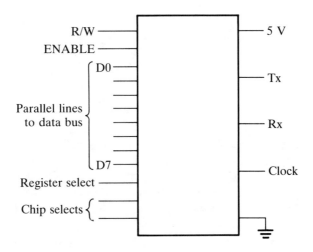

Fig. 15-5. Minimum pin connections on a generic UART.

the enable. If eight bits of parallel data are to be extracted from the UART by the microprocessor, the R/W pin would be set for R, following which the enable signal is activated.

The external connections illustrated in Fig. 15-5 may be adequate for an elementary UART. There are additional demands placed on the UART that lead to some internal requirements of the chip. Three operations might be imagined: (1) structuring the cluster, (2) temporary storage of data during the transmit mode and some indication of when the microprocessor could send another byte, and (3) temporary storage of data during the receive mode and some indication of when a new byte is ready to be read.

Figure 15-3 showed a cluster consisting of a start bit, the data, two stop bits, and a parity bit. Some variation in the structure of the cluster should be provided; for example, perhaps only seven bits of data will be included, and perhaps only one stop bit is to be used. Also, it is necessary to choose odd or even parity on both the communicating microcomputers. With respect to operations 2 and 3, an important requirement is that the shift register not be read during the reception process until it is entirely filled and that a partially empty shift register during transmission not be overwritten by a new cluster. These requirements can be met by providing internal registers within the UART, as illustrated schematically in Fig. 15-6. The six registers proposed are the transmit-shift, transmit-data, receive-shift, receive-data, control, and status registers.

15-6. A Generic UART

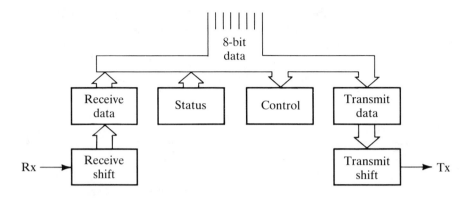

Fig. 15-6. Internal registers in a generic UART.

When the microprocessor reads the UART, the eight bits may come from either the receive-data register or from the status register, and some means must be provided for the microprocessor to specify which. When the microprocessor writes a byte to the UART the destination could conceivably be either the control register or the transmit-data register, and here again the UART must be equipped with means of specifying to which register the byte is intended. It is here that the register-select pin, shown in Fig. 15-5, performs a function. By the combination of the signals to the R/W and the register-select pins, the intent could be specified as in Fig. 15-7.

A preliminary step before transmitting or receiving any data will be to write into the control register of both communicating microcomputers the combination of bits that specify the structure of the cluster, including the number of data, start, and stop bits, and whether the parity is to be odd or even.

The status register will assist in preventing disruption of the receive-shift or the transmit-shift operations until they are complete. One bit in the sta-

R/W	Register Select	
	High	Low
R	Read status register	Read receive-data register
W	Write to control register	Write to transmit-data register

Fig. 15-7. Combinations of R/W and register-select signals to determine internal function of UART.

tus register will indicate a "receive-data-register-full" (RDRF) condition and another bit a "transmit-data-register-empty" (TDRE) condition.

Listed below are the individual steps, first for a transmission and then for the reception of a cluster. A preparatory operation is to access the control registers of both microcomputers and establish identical forms of clusters.

Transmission Assume that both the transmit-data and transmit-shift registers are empty. By the proper choice of R/W and register-select signals (Fig. 15-7), load the byte into the transmit-data register. The UART automatically appends the start, stop, and parity bits to the data and transfers the full cluster to the transmit-shift register. Immediately upon receipt of the cluster, the transmit-shift register begins sending out the cluster serially on the Tx line.

Assume that the microcomputer has a series of bytes to transmit. Before the next byte is written to the transmit-data register, the program should read the TDRE bit in the status register to assure that the transmit-data register is empty and can receive another byte. condition. The transfer of the first byte from the transmit-data register to the transmit-shift register is fast, but the shifting out of bits onto the Tx line is slow (300 or 1200 baud) relative to operations within the microcomputer . The transmitting microcomputer may have to spend most of its time checking the TDRE bit waiting for the last byte to clear the transmit-shift register.

Reception Assume that the receive-shift and receive-data registers are initially empty. The UART stands in a ready state, and as soon as bits begin to arrive on the Rx pin they shift into the receive-shift register. When the complete cluster is lodged in the receive-shift register, the start, stop, and parity bits are stripped from the cluster and the byte of data is transferred to the receive-data register. The above operations are automatic for the UART, and meanwhile the microcomputer has been checking the RDRF bit in the status register to determine when the byte is transferred from the receive-shift register. When the RDRF condition is verified, the microcomputer then reads the receive-data register by choosing the combination of pin signals indicated in Fig. 15-7.

The microcomputer would be able to read the receive-data register repeatedly at a rapid rate but must wait for the slow process in the sequence—the reception of bits into the receive-shift register. Anticipating that the shift operation will be an order of magnitude slower than internal microcomputer operations allows the user to program other tasks between occasional checks of the RDRF.

15-7 The MC6850 Asynchronous Receiver/Transmitter (ACIA)

The elements of the generic UART will be distinguishable in two commercial UARTs that will be explored, the first being the Motorola MC6850. This version of the UART is called an ACIA which is the abbreviation of "asynchronous communications interface adapter." The pinout of the MC6850 ACIA is shown in Fig. 15-8, and many of the pins are already familiar from the generic UART. Several that have not yet been mentioned are Request To Send (pin 5), Data Carrier Detect (pin 23), and Clear To Send (pin 24). These connections fit in the category of "handshake" signals whose use can be bypassed when using the elementary microcomputer but become crucial when using personal computers and computers of even higher levels.

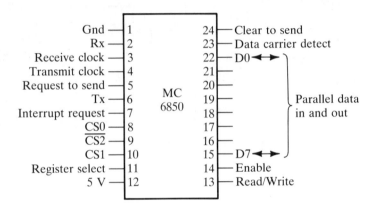

Fig. 15-8. Pinout of MC6850 ACIA.

15-8 Registers in the ACIA

Figure 15-9 shows a diagram of the registers in the ACIA as well as the schematic connections with the pins. The six registers found in the generic UART—transmit-data, transmit-shift, receive-data, receive-shift, control, and status registers—are components of the ACIA as well. The clock signals are shown feeding to the transmit-shift and receive-shift registers. Designation of which register is being accessed in the ACIA is determined by a similar process to that explained in Fig. 15-7 for the generic UART. Specifically for the ACIA, Fig. 15-10 shows the combinations of signals supplied to pins 11

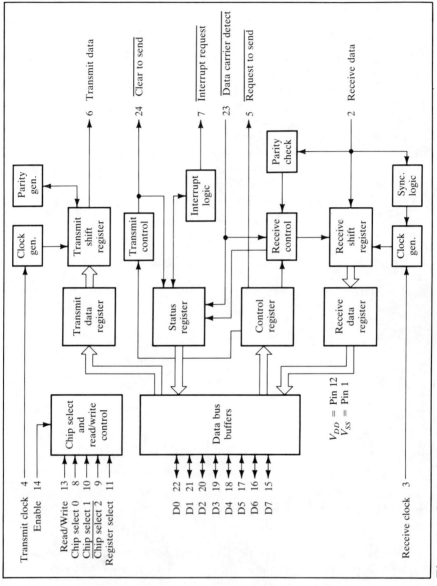

Fig. 15-9. Internal arrangement of the ACIA.

	Pin 11,	Pin 13, R/W	
Address	Register Select	Low (Write)	High (Read)
$E700	Low	Control register	Status register
$E701	High	Transmit data	Receive data

Fig. 15-10. Access to registers on the ACIA.

and 13 that specify both the register and whether the read or write operation is desired. On the elementary microcomputer the register-select pin is tied to one of the address lines, so the control and status registers, for example, appear as the same memory location. The read/write signal, pin 13, is hard-wired to the R/W coming from the microprocessor. Pin 14 is the enable signal that is usually connected to a clock line from the microcomputer, and the edge of this pulse triggers the transfer of data between the data bus and whichever register has been selected of the four in Fig. 15-10.

15-9 The Control Register

When accessed, the microprocessor can write eight bits of data into the control register. The bits in the control register CR0 to CR7 are grouped into several blocks:

CR0 and CR1	Counter divide and reset
CR2, CR3, and CR4	Word selects
CR5 and CR6	Transmit control
CR7	Receive control

The specific conditions controlled by the combination of CR0 and CR1 are shown in Fig. 15-11. Pins 3 and 4 on the ACIA are connected to an external clock, such as a baud-rate generator. The sampling rate used by the transmit and receive-shift registers may be chosen the same as the clock rate or divided

CR1	CR0	Function
0	0	Divide by 1
0	1	Divide by 16
1	0	Divide by 64
1	1	Master reset

Fig. 15-11. Counter division and reset controlled by CR0 and CR1.

by 16 or 64, depending on the values of CR0 and CR1. When CR0 and CR1 are both equal to 1, the ACIA is reset. This resetting must take place after a power-off condition, so an expected sequence upon startup is to first use CR0 and CR1 to reset, and thereafter use CR0 and CR1 to select the desired divide function.

Figure 15-12 shows the role of CR2, CR3, and CR4 in establishing the structure of the cluster. The combinations of these bits determine the number of data bits, number of stop bits, and whether a parity check will be made, and if so, whether parity is odd or even.

Bits 5, 6, and 7 of the control register adjust the Request-To-Send (RTS) and whether the interrupt is enabled. Explanation and use of these features will be bypassed.

CR4	CR3	CR2	Structure		
			Bits of data	Parity	Stop bits
0	0	0	7	even	2
0	0	1	7	odd	2
0	1	0	7	even	1
0	1	1	7	odd	1
1	0	0	8	—	2
1	0	1	8	—	1
1	1	0	8	even	1
1	1	1	8	odd	1

Fig. 15-12. Structure of clusters.

15-10   The Status Register

The status of the ACIA can be read by the microprocessor, as indicated in Fig. 15-10, by providing a low level to the register select, pin 11, and a high level to the R/W, pin 13. The status register contains eight bits that indicate the condition of the transmit-data register, receive-data register, and the inputs from the peripherals. The status register also indicates that certain errors have been detected.

Bit 0 (Receive-Data Register Full, RDRF) When this bit is high it indicates that the receive-data register is full, which means that a complete transfer from the shift register has occurred. After the contents of the receive-data register are read by the microprocessor, bit 0 drops to a low level.

15-11. Transmitting and Receiving with the ACIA 301

Bit 1 (Transmit-Data Register Empty, TDRE) When this bit is high it indicates that the transmit-data register is empty, thus the data have been transferred from the transmit-data register to the transmit-shift register and new data may be entered. The low state of bit 1 indicates that the transmit-data register is full.

Bit 2, (Data Carrier Detect, DCD) This bit goes high when pin 23 goes high. This capability will not be discussed further in this chapter.

Bit 3 (Clear-To-Send, CTS) This bit is the Clear-To-Send bit which indicates the signal from the peripheral. A low level means that the peripheral is ready.

Bit 4 (Framing error) When the received cluster is improperly framed by a start and at least one stop bit, the framing error bit goes high.

Bit 5 (Receiver overrun) This bit goes high if one or more characters were received by the receive-data register before the last byte was read from this register.

Bit 6 (Parity error) This flag is set (bit goes high) if the parity of the received character does not agree with the parity selected by CR2 through CR4.

Bit 7 (Interrupt request) Pin 7 on the ACIA is an interrupt request that may be connected to the microprocessor. When an interrupt is being requested by a low value on pin 7, bit 7 in the status register is high.

15-11 Transmitting and Receiving with the ACIA

The steps in sending out a byte of data are:

1. Reset the registers by writing ones into CR0 and CR1.

2. Select the dividing constant of the input clock pulses and assign CR0 and CR1 appropriately.

3. Choose the structure of the cluster and assign CR2, CR3, and CR4 accordingly.

4. Access the transmit-data register as in Fig. 15-10. When a byte of data is written into the register, the process is triggered that adds the framing bits and sends the cluster out over the transmit data line, pin 6.

To receive a byte with the ACIA the first three steps are identical to those for transmitting. Step 4 is:

4. If a cluster of serial data has been previously received, the ACIA will have stripped off the start, stop, and parity bits and stored the data in the receive-data register. Access the receive-data register as specified by Fig. 15-10, and read the register. This operation readies the receive-data register to accept the next character transferred from the receive-shift register.

15-12 The Intel 8251A Programmable Communication Interface

Another commercial version of a UART is the 8251A, which is actually a USART because it can transmit and receive both synchronous and asynchronous serial data. To retain consistency in the scope of this chapter, only the asynchronous serial data communication will be examined. This chip has 28 pins with the labels indicated in Fig. 15-13.

Certain of the pins and groups of pins can be anticipated from knowledge of the generic UART, Fig. 15-5:

- Power (+5 V) and ground, pins 26 and 4

- Parallel lines to the data bus, pins 27, 28, 1, 2, 5, 6, 7, and 8

- Transmit and receive serial data, pins 19 and 3, respectively

- Clock, pin 20

- Chip select, pin 11

- Read and write, pins 13 and 10, respectively

There is only one chip-select pin, which suggests that an address decoder (see Sec. 13-5) translates the signals from several address lines such that a proper combination of them activates the 8251A chip. The write operation sends a byte either to be shifted out as serial data or to establish the structure in the control register. The read operation calls for either a byte from the status register or a byte that has been restructured from serial data that have been received by the chip. The combination of signals on four pins determines which of these operations occurs, as shown in Fig. 15-14. The $C/\overline{D}$ pin is called the control/$\overline{\text{data}}$ signal in that in combination with $\overline{\text{RD}}$, $\overline{\text{WR}}$, and $\overline{\text{CS}}$ determines whether the byte is control/status in which case $C/\overline{D} = 1$, or whether the byte is data in which case $C/\overline{D} = 0$.

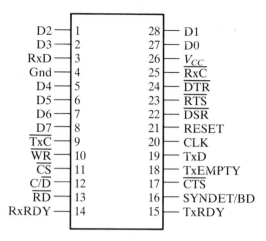

Fig. 15-13. Pin diagram of the 8251A communication interface.

C/$\overline{\text{D}}$	$\overline{\text{RD}}$	$\overline{\text{WR}}$	$\overline{\text{CS}}$	Function
0	0	1	0	Reads a byte of received data
0	1	0	0	Writes a byte of data for transmission
1	0	1	0	Reads status register
1	1	0	0	Establishes the control register

Fig. 15-14. Combinations of C/$\overline{\text{D}}$, $\overline{\text{RD}}$, $\overline{\text{WR}}$, and $\overline{\text{CS}}$ to read data, read the status register, transmit data, or set the control register.

15-13 The Control and Status Register on the 8251A

A preliminary step before transmission or reception is to write a byte into the control register to establish the structure of the cluster. The reset operation must precede a change of the control register, which is achieved by feeding a high pulse to the RESET, pin 21. The cluster can be structured by writing to the control register by setting the appropriate combination of signals specified in Fig. 15-14, and then writing a byte from the microprocessor. The byte written to the 8251A determines the structure according to the map shown in Fig. 15-15. Bits 0 and 1 assign the dividing factor. If the clock rate at TxC is 4800 Hz, for example, and a dividing factor of 16 is chosen, the transmission

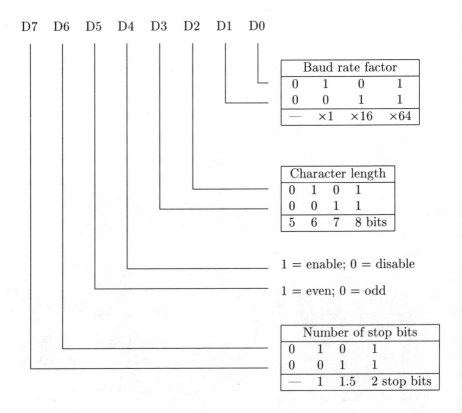

Fig. 15-15. Structure of the cluster as determined by the byte written to the control register.

on TxD will occur at $4800/16 = 300$ baud. Bits 2 and 3 specify the character length, bit 4 the parity enable, and bit 5 whether parity is even or odd. Bits 6 and 7 designate the number of stop bits in the cluster.

The meaning of the bits in the status register is shown in Fig. 15-16. A check of Fig. 15-13 shows that several functions seem to be duplicated by pin connections, namely, TxRDY, RxRDY, and TxEMPTY. Indeed the TxEMPTY and RxRDY bits in the status register have the same status as the corresponding pins on the chip, which permits the microprocessor to read the pin (if suitable hardwiring is provided) directly without the process of following the multiple-step operation of reading the status register and determining whether TxEMPTY $= 1$ so that another byte may be transmitted or RxRDY $= 1$ so that an incoming byte can be read.

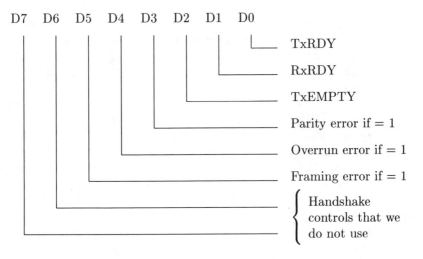

Fig. 15-16. Status register of the 8251A USART.

15-14 Communicating Using RS-232-C and Modems

This chapter has explained the need of converting between the parallel communication, which is the mode within the computer, and serial communication, which is the practical method outside the computer. The concept of shifting a byte of parallel data into serial form, as well as the reverse process, has been explored. Commercial chips are equipped with the capabilities of providing associated operations needed for reliable serial data transmission.

The serial data leaving the UART will be of the TTL (0 and 5 V) level, which can then be processed further by techniques that enhance the ruggedness of the signal and facilitate the transmission of data over long distances. It would be possible to use the 0 and 5 V convention for short distances, but for long distances the signals would be susceptible to distortion caused by electrical noise, differences in ground potential, and damage from short circuits. The RS-232-C convention, which will be described in Sec. 15-15, provides a greater voltage difference between the marks and spaces.

When the need arises to send data over a telephone line, either because of convenience or out of necessity, modulator-demodulators, called "modems," are needed. The modem at the transmitting end of the telephone line converts the incoming RS-232-C signal to sine waves of frequencies compatible with the telephone system. A modem at the opposite end of the phone line reconverts the sine waves back to an RS-232-C signal.

15-15 RS-232-C Interface

The most commonly used serial interface specification is RS-232-C, which has been established by the Electronic Industries Association.[1] Newer standards have been developed, such as the RS-422 and RS-423, which may one day supplant the RS-232-C standard, but RS-232-C is currently the dominant one. The standard is lengthy, and while it specifies necessary electrical, mechanical, and functional characteristics of the data exchange, it includes many specifications that the user usually does not need to know.

Briefly, RS-232-C specifies that the driver circuits must be able to withstand open circuits and shorts without damage. The receiver circuit must be able to withstand an input signal of up to ±25 V without damage.

The typical RS-232-C signal is shown in Fig. 15-17 corresponding to the TTL levels from which the RS-232-C signal was generated. Typically the high voltage of RS-232-C is of the order of 12 to 15 V and the low voltage is −12 to −15 V, with a mark represented by the negative voltage and a space by the positive voltage. The difference in voltage between the mark and space is thus between 24 and 30 V.

The RS-232-C standard also specifies a connector, which is a 25-pin D-shell connector as shown in Fig. 15-18. Some of the more frequently used pins in the connector are given in Table 15-1.[2]

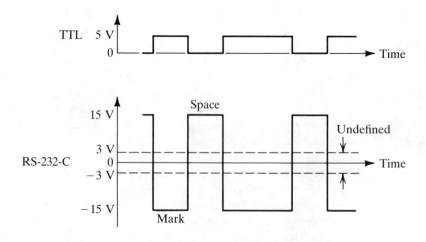

Fig. 15-17. Marks and spaces in the RS-232-C convention.

15-15. RS-232-C Interface

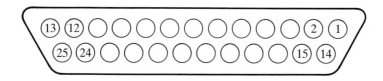

Fig. 15-18. Front view of the standard female connector used for RS-232-C.

Table 15-1. Some Pin Assignments in the RS-232-C Connector

Pin Number	Function	Abbreviation
1	Protective ground	
2	Transmitted Data	TxD
3	Received Data	RxD
4	Request To Send	RTS
5	Clear To Send	CTS
6	Data Set Ready	DSR
7	Signal ground	
8	Data Carrier Detect	DCD
20	Data Terminal Ready	DTR

The RS-232-C standard was developed specifically for the interconnection of equipment such as a display terminal and a modem. In the terminology of the standard, these devices are known as the data terminal equipment (DTE) and data communications equipment (DCE), respectively. The signal names are given from the perspective of the DTE, so, for example, the transmitted data are sent on pin 2 by the DTE and received by the DCE. Pin 7 is the signal ground, and the line that connects the number 7 pins of the receiver and transmitter provides a common ground level reference. The RTS, CTS, DSR, DCD, and DTR pins are "handshaking controls." Reference to the pinout of the ACIA in Chapter 15 shows that the DCD, CTS, and RTS controls were available on this chip. For the circuit used on the D5 kit the RTS signal was active, but the DCD and CTS signals were tied to ground, signifying that this ACIA is always assumed ready to receive incoming data. The CTS, RTS, and DCD signals are necessary when the data transmission passes through a modem. Our first assignment will be to transmit signals in RS-232-C levels over a direct line.

15-16 Level Conversion Between RS-232-C and TTL

Sometimes the user must provide the facilities to convert between the ±12 to ±15 V pulses of an RS-232-C data line and 0/5 V pulses at a PIA or an ACIA. Integrated circuits are available for this task, and two of the possibilities are:

MC1488 Quad MDTL line driver for conversion to RS-232-C

MC1489 Quad MDTL line receiver for conversion from RS-232-C

The pinout of the MC1488 line driver is shown in Fig. 15-19. This chip incorporates four converters as inverters such that if the input is 0 V the output is 15 V, and if the input is 5 V the output is −15 V. The converter between pins 2 and 3 has one input while the others have two inputs. Unless some auxiliary control is to be used on the two-input converters, the two inputs can be tied together, and they function as a single-input converter.

For converting in the opposite direction, the MC1489, whose pinout is shown in Fig. 15-20, serves the purpose. The RS-232-C signal is the input to one of the NAND gates, and the response control is supplied with 5 V through a 13 kΩ resistor. The magnitude of the resistance in the response control line shifts the input voltage threshold at which the output changes between 0 and 5 V. The 13 kΩ choice causes the shift to occur at approximately 0 V input voltage.

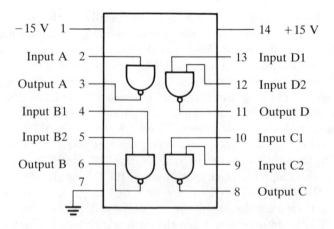

Fig. 15-19. Pinout of MC1488 quad line driver.

15-16. Level Conversion Between RS-232-C and TTL

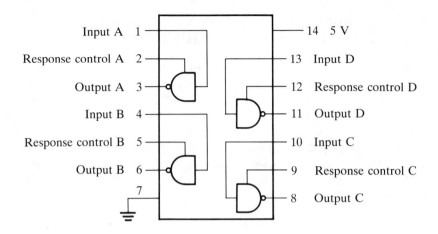

Fig. 15-20. Pinout of MC1489 quad line receiver.

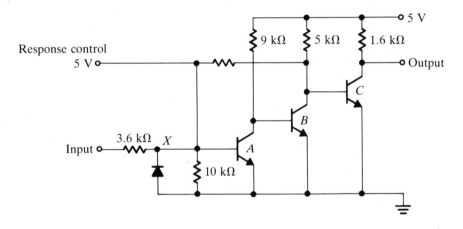

Fig. 15-21. Circuit diagram of MC1489 line receiver.

The schematic diagram of one of the four receiver circuits in the MC1489 is shown in Fig. 15-21. The +15 V and −15 V signals are received by the input, and the diode in the reverse direction to ground prevents the voltage from dropping lower than −0.6 V. The series of three transistors results in a reversal of the signal at X to the output. Thus, a high voltage at X results in transistor C grounding the output, but a low voltage at X causes transistor C to develop a high resistance between the collector and emitter, which provides close to 5 V at the output.

15-17 Communicating Between Two Elementary Microcomputers Using RS-232-C

Chapter 15 outlined how to communicate between two elementary microcomputers using TTL. The step needed to communicate by means of RS-232-C is the simple one illustrated in Fig. 15-22. The TTL signal from the transmitting microcomputer is converted to RS-232-C by the level converter that uses the MC1488. It would be expected that if the signal is to be transmitted any appreciable distance the long length would be traversed with the RS-232-C lines. At the destination the level converter using the MC1489 converts the RS-232-C signal back to TTL for acceptance by the serial data receiver in the microcomputer.

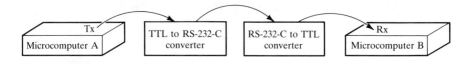

Fig. 15-22. Transmitting between two elementary microcomputers using RS-232-C.

15-18 Transmission over Telephone Lines Using Modems

Transmitting RS-232-C over long distances poses some problems. The maximum distance for transmitting RS-232-C signals is officially about 15 m, but practically RS-232-C is used for much longer distances. At some length, however, degradation of the signal occurs because of noise. Furthermore, connections by hard wires become prohibitive if the lines must run through other people's property, across the city, or from one city to another. Surmounting the transmission problem leads to using conductors that already exist, namely, telephone lines.

The telephone system is set up to accommodate conversation between people, and telephone companies speak of "voice grade" channels that transmit signals satisfactorily in the 300 to 3000 Hz frequency range. Telephone lines are equipped with amplifiers to build up the strength of the signal along the line to compensate for losses. The telephone system has difficulty sending one constant voltage for a long time as may be needed in transmitting RS-232-C, so it is necessary to convert RS-232-C to a more amenable form.

15-18. Transmission over Telephone Lines Using Modems

The device that converts between RS-232-C and the sine waves in the voice frequency range is called a modulator-demodulator, or modem for short. The term "data set" is sometimes used synonymously with modem. The modem appears in several forms, as illustrated in Fig. 15-23. Figure 15-23a shows the modem using an acoustic coupler and the telephone receiver as the interface with the telephone line. When a modular phone plug is available, the modem can be connected directly to the phone line, as in Fig. 15-23b. The acoustic couplers permit the phone to be used for other purposes when not transmitting data, but the hardwired arrangement results in more reliable transmission.

One of the classifications of transmission is "half-duplex" or "full-duplex." The half-duplex modem can transmit or receive but cannot do both at the same time. Full-duplex transmission can send and receive simultaneously.

When two modems communicate with one another over the phone lines, one must be in the "originate" mode and the other in the "answer" mode. The originate modem is usually lower in cost than the answer/originate modem, so the originate-type would usually serve, for example, as the modem at the remote terminal of a multiterminal computer installation, while the answer-type modem would be located at the central computer.

The convention chosen for frequencies to transmit spaces and marks in the widely used Bell System 103F series of modems is shown in Table 15-2. The originate modem transmits either 1070 or 1270 Hz, and the answer modem transmits either 2025 or 2225 Hz.

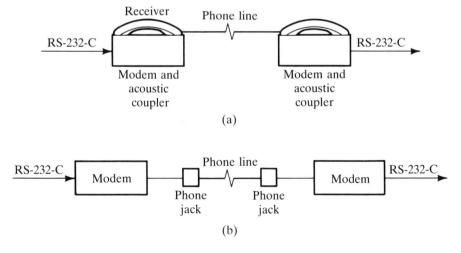

Fig. 15-23. (a) Modem combined with an acoustic coupler, and (b) modem hardwired to the telephone line.

Table 15-2. Frequencies Used by Modems

	Space (Logical Zero)	Mark (Logical One)
From originate modem	1070 Hz	1270 Hz
From answer modem	2025 Hz	2225 Hz

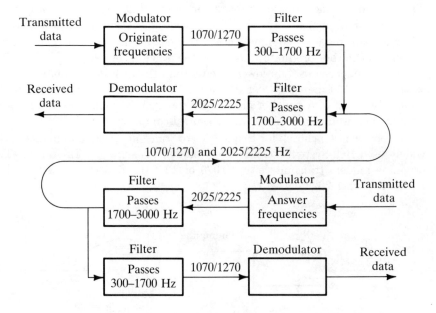

Fig. 15-24. Structure of a full-duplex modem operation.

The functional diagram[3] of two modems in communication is shown in Fig. 15-24. The originate modulator sends frequencies of 1070 and 1270 Hz, and the answer modulator sends 2025 and 2225 Hz. The two filters at the originate location direct the 2025/2225 Hz signals to the demodulator, and the two filters at the answer location pass the received frequencies of 1070/1270 Hz to its demodulator.

The frequencies shown in Table 15-2 are adequate for transmission of data at a 300 baud rate.

15-19 Dial-Up Modems

The modems described in the previous section expect some outside control, such as by a human, to set the transmission into operation. Following the exchange of data between the modems, human intervention is probably the means of terminating the communication. Another type of modem, called the dial-up modem, allows one computer to initiate and terminate the interchange. Upon the proper instruction from the initiating computer, the modem dials the telephone number associated with the other modem. The dial-up modem at that computer performs the equivalent of lifting the receiver, and communication begins. When the interchange has been completed the addressed modem hangs up the receiver, following which the originating modem does the same. Dial-up modems are widely used, for example, when a central controller periodically polls a remote controller and inquires of the status at the remote location and perhaps updates instructions to the remote computer.

15-20 ASCII Characters

A standard that will be encountered when working with such devices as terminals and printers is the American Standard Code for Information Interchange (ASCII). The ASCII code is a group of 7-bit binary numbers, shown in Fig. 15-25, representing all the letters of the alphabet, numbers, punctuation, and certain carriage control instructions. In order to transmit the letter B in ASCII, for example, the binary number 1000010 is communicated. Characters numbered 0011111 and below are not printable but act as delimiters or modify the positioning of subsequent text. The most commonly used control codes are CR (carriage return) and LF (line feed), which position printing at the beginning of the current line or advance to the next line, respectively.

15-21 One-on-One Communication

The serial data transmission described in this chapter consisted of communicating between two parties or computers that are connected by direct lines. The listener is assumed to be courteous and to accept the data transmitted by the talker. If a central computer needs to communicate with two microcomputers, the central computer could speak through separate UARTs, one dedicated to each of the two microcomputers. This practice works satisfactorily if the number of microcomputers or peripherals is small.

Another approach is to construct a party line where the addressed microcomputer could listen when spoken to and respond as directed, and meanwhile the other microcomputers or peripherals on the line would not participate in the conversation. The concept to achieve this type of communication is the bus structure typified by the IEEE-488 protocol.

NUL 0000000	SOH 0000001	STX 0000010	ETX 0000011	EOT 0000100	ENQ 0000101	ACK 0000110	BEL 0000111
BS 0001000	HT 0001001	LF 0001010	VT 0001011	FF 0001100	CR 0001101	SO 0001110	SI 0001111
DLE 0010000	DC1 0010001	DC2 0010010	DC3 0010011	DC4 0010100	NAK 0010101	SYN 0010110	ETB 0010111
CAN 0011000	EM 0011001	SUB 0011010	ESC 0011011	FS 0011100	GS 0011101	RS 0011110	US 0011111
space 0100000	! 0100001	" 0100010	# 0100011	$ 0100100	% 0100101	& 0100110	' 0100111
(0101000	) 0101001	* 0101010	+ 0101011	, 0101100	- 0101101	. 0101110	/ 0101111
0 0110000	1 0110001	2 0110010	3 0110011	4 0110100	5 0110101	6 0110110	7 0110111
8 0111000	9 0111001	: 0111001	; 0111011	< 0111100	= 0111101	> 0111110	? 0111111
@ 1000000	A 1000001	B 1000010	C 1000011	D 1000100	E 1000101	F 1000110	G 1000111
H 1001000	I 1001001	J 1001010	K 1001011	L 1001100	M 1001101	N 1001110	O 1001111
P 1010000	Q 1010001	R 1010010	S 1010011	T 1010100	U 1010101	V 1010110	W 1010111
X 1011000	Y 1011001	Z 1011010	[1011011	\ 1011100	] 1011101	^ 1011110	_ 1011111
` 1100000	a 1100001	b 1100010	c 1100011	d 1100100	e 1100101	f 1100110	g 1100111
h 1101000	i 1101001	j 1101010	k 1101011	l 1101100	m 1101101	n 1101110	o 1101111
p 1110000	q 1110001	r 1110010	s 1110011	t 1110100	u 1110101	v 1110110	w 1110111
x 1111000	y 1111001	z 1111010	{ 1111011		1111100	} 1111101	~ 1111110	DEL 1111111

Fig. 15-25. ASCII characters.

References

1. *Interface Between Data Terminal Equipment and Data Communication Equipment Employing Serial Binary Interchange*, Standard RS-232-C, Catalog 3, Electronic Industrial Association, 2001 Eye Street, NW, Washington, DC 20006.

2. J. E. Oleksy and G. B. Rutkowski, *Microprocessor and Digital Computer Technology*, Prentice-Hall, Englewood Cliffs, NJ, 1981.

3. J. E. McNamara, *Technical Aspects of Data Communication*, Digital Press, Bedford, MA, 1977.

Problems

15-1. The generic UART of Fig. 15-5 is connected to lines of the address and data bus of an elementary microcomputer as shown in Fig. 15-26, where the signal to $\overline{CS0}$ must be low and that to CS1 high, in order to select the chip. The register select operates as shown in Fig. 15-26. Show the assembly language statements, including the appropriate addresses that (a) read the status register of the UART and place contents into accumulator A, and (b) write a byte from accumulator B into the data register for transmission.

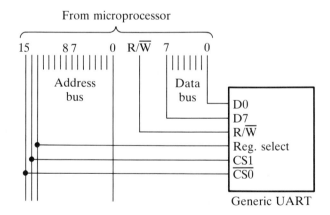

Fig. 15-26. Connections of the generic UART to the address bus and data bus in Prob. 15-1.

15-2. The parity check uses one bit to detect a reversal of one bit (actually an odd number of bits) in serial transmission. Various other techniques are available, such as double transmission and transmitting several bits that are processed by an algorithm to attempt to detect errors. If you were given the opportunity of transmitting two bits as error detectors, propose how they might be used to detect errors in a way that is superior to using the one parity bit.

15-3. Write the assembly language program of a loop that repeatedly checks the status of the RDRF bit in the ACIA. Then show the program statements to read the byte. Assume that the address $E700 accesses the status register and $E701 the data register.

15-4. Write the assembly language statements to structure the control register on the 8251A USART such that the cluster incorporates eight bits of data, uses 1 stop bit, chooses odd parity, and uses a baud rate factor of ×16. Assume that the status register is accessed through address $E700.

15-5. Comparison of the MC6850 ACIA with the 8251A shows that the former devotes one pin for a combination of read/write, while the latter has separate pins for the read and the write instructions. Consult the pin diagrams of the Motorola 6800 and Intel 8080/8085 microprocessors to explain why the pin assignments can be different but both still work.

15-6. What is the voltage of each bit in a cluster of asynchronous data on an RS-232-C line when the letter A in ASCII is being transmitted? The cluster is structured for one start bit, two stop bits, seven data characters, and even parity.

15-7. Explain a difficulty that will arise when attempting to transmit 9600 baud using modems with the Bell System 103F series.

15-8. What is the voltage on an RS-232-C line when it is in its idle state between the transmission of two asynchronous clusters?

Chapter 16

Dynamic Behavior of Systems Under Computer Control

16-1 Returning to the Thermal and Mechanical System

The subject of this book, as reflected in its title, is the control of thermal and mechanical systems. The controller to be used is of the computer type, and the overall objective is to learn the characteristics of computer controllers in order to apply them properly to these systems. The objective of Chapters 16 and 17 is to study the interface of the computer control and the thermal and mechanical system.

The two major topics covered in this chapter are (1) on/off control and (2) modulating control of the proportional and proportional-integral types. An engineer should be equipped to perform mathematical analyses of control loops when the situation warrants. The need for analysis of on/off controllers is probably rare, and in practice the setpoints, dead bands, and sampling intervals are usually chosen empirically based on the way the system behaves. Sections 16-2 to 16-4 are included for completeness with the understanding that a computer may be used for the controller and tuned in the field. A different situation applies to modulating controls, however, where analysis helps in the understanding of the loop characteristics and may be useful in setting the control constants prior to the start of operation. A portion of this chapter is nothing more than classical control theory applicable to continuous variables (in contrast to the sampling typical of computer control). The two modes of control to be examined are proportional and proportional-integral control. The integral mode is important in computer control because it can often be incorporated in software at a low cost compared to buying hardware to provide it. When both proportional and integral modes are employed, there is a need to tune the constants in relationship to one another, so a thorough grounding in the behavior of these control loops is valuable.

Analog electric and pneumatic controllers process continuous signals, in contrast to computer controllers, which intermittently sense variables and send signals to the actuators. Chapter 17 concentrates on the sampling nature of computer controllers and explores how the computer achieves proportional and proportional/integral control. The sampling interval is of concern in computer control because when the interval is too long the loop may go unstable. The final sections of Chapter 17 explore the mathematical basis of stability analysis using the z-transform.

16-2 On/Off Controls

Most of the later sections in this chapter concentrate on modulating controls where a valve, damper, voltage regulator, speed regulator, or other actuator is positioned through a range of positions in an effort to hold the controller variable at the setpoint. Another type of control that is widely used and is simple in operation is the on/off control wherein a signal from the controller changes the status of the actuator from one extreme position to the other. The burner in a residential gas furnace, for example, is an on/off control where the solenoid valve controlling the flow of gas is either in its completely open or completely closed position.

Some very effective energy management is achieved through on/off controls regulated by timers. Hardly any other action is as effective in conserving energy as turning off the device if it is not needed. Computer controllers[1] are used extensively as time clocks wherein daily, weekly, and annual schedules can be programmed into the memory of the computer to turn on and off lights, motors, heaters, air conditioners, and other energy consumers. Conceptually, in terms of a control block diagram the timers are open-loop controllers, as shown in Fig. 16-1, in that they do not regulate the magnitude of a variable. The only feedback in this loop is an occasional revision of the timing program based on operating experience.

An example of an on/off controller of a variable is the water-level controller shown in Fig. 16-2. The water flow into the tank is controlled by a solenoid valve that is either completely open or completely closed.

Two of the available level sensing possibilities are (1) make/break sensor, and (2) analog sensor. The signal from each of these two types of sensors must be handled differently by the microcomputer controller, and the way these signals are accommodated represents our first serious effort at interfacing the microcomputer controller with the system.

16-3. Make/Break Sensor with On/Off Actuator

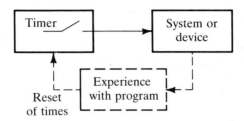

Fig. 16-1. Block diagram of an on/off controller regulated by a time clock.

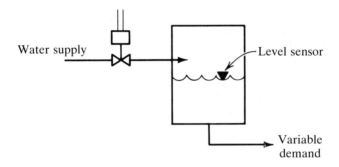

Fig. 16-2. On/off controller of the water level in a tank with a make/break level sensor

16-3 Make/Break Sensor with On/Off Actuator

Using the level controller in Fig. 16-2 as an example, let us suppose that the sensor closes a contact (makes) when the water level drops below the setpoint, and breaks contact at a level above the setpoint. It is also assumed that the flow rate of water supplied to the tank when the solenoid valve is open is greater than the maximum demand, and thus the water level could be elevated even during maximum demand.

The principal interfacing consideration that arises in this application is that the microcomputer can react several orders of magnitude faster than is appropriate for the sensor and actuator hardware. If the microcomputer instructed the valve to open each time the sensor made contact and instructed the valve to close each time the sensor broke contact, the control system might not work at all. If the microcomputer were dedicated to this control function and had no other duties, it could respond within microseconds. The attempt to open and close the valve in cycles of microsecond or even millisecond duration is incompatible with the 60 Hz power supplied to the valve.

With the make/break sensor, time delays must be built into the program. Were the microcomputer to be dedicated to only this control function and change the valve position each time the sensor reversed, the valve would open and close so rapidly it would probably chatter. Furthermore, waves on the liquid surface could open or close the valve.

A typical flow diagram of the computer control program incorporating time delays is shown in Fig. 16-3. Even time delays of several seconds might result in the valve cycling so frequently that its life would be short. The maximum time delay should be based on the permissible variations in level. The greatest drop in level occurs at full demand when the valve is off. The greatest rise in level occurs at no demand when the valve is open.

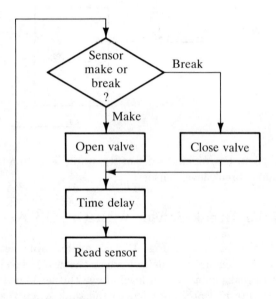

Fig. 16-3. Flow diagram of program that provides time delays when a make/break sensor is used.

16-4 Analog Sensor with On/Off Actuator

If an analog sensor is available that sends out a variable voltage (or current) as a function of the level (Fig. 16-4), comparable procedures must be provided to prevent the actuator from rapidly switching on and off. The strategy is

16-5. Modulating Control Strategies

to incorporate a "deadband." Suppose that the analog sensor in Fig. 16-4 delivers 9 V when the level is at the top extremum of its sensing element and 4 V when the level is at the bottom. A deadband could be incorporated into the control loop, as shown in the flow diagram of Fig. 16-5, which opens the valve whenever the voltage from the sensor is below 6 V and closes the valve whenever the voltage is above 7 V.

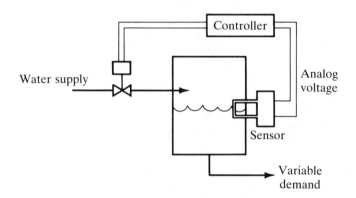

Fig. 16-4. On/off controller of water level in a tank with an analog sensor.

16-5 Modulating Control Strategies

The remainder of this chapter concentrates on modulating control in which both the sensor (as in Sec. 16-4) and the actuator are of the modulating type. Modulating types of actuators are ones in which there is an infinite gradation of adjustments between the extremum positions. Pneumatic actuators, many electrically motorized actuators, and magnetic actuators (Sec. 6-8) are examples of modulating actuators.

In the next pages the important industrial control modes of proportional, proportional-integral (PI), and proportional-integral-derivative (PID) will first be described. Initially these modes will be considered to perform in a quasi-steady-state manner that ignores stability considerations. Thereafter comes a review of elementary control theory to analyze the stability of proportional and PI loops.

The ultimate objective is to study computer control, which differs from classical control in that instead of the variables throughout the control loop

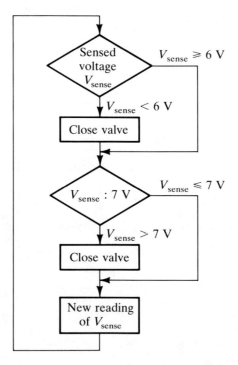

Fig. 16-5. Flow diagram of program that provides a deadband when an analog sensor is used.

being continuous with respect to time, the computer only intermittently receives a message from the sensor and only intermittently sends a signals to the actuator. This type of analysis, called sampled data analysis, will be examined in Chapter 17.

16-6 Proportional Control

Proportional controllers are ones in which the action is proportional to the error between the sensed and set values of the controlled variable. Several examples of proportional controllers are shown in Figs. 16-6, 16-7, and 16-8, representing mechanical, pneumatic, and electric control loops, respectively. The mechanical controller in Fig. 16-6 regulates the level of water in a tank. As the water level drops, the valve opens. The upper limit of the water level occurs when there is no demand, and the level rises enough to close the valve completely. As the demand increases, the level drops and the valve opens.

16-6. Proportional Control

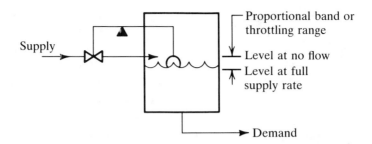

Fig. 16-6. A mechanical proportional controller regulating the water level in a tank.

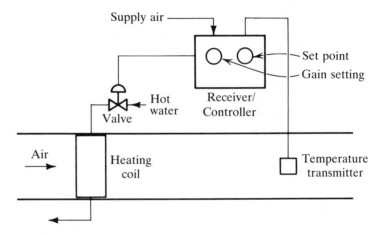

Fig. 16-7. A pneumatic proportional controller.

Demand rates higher than the maximum inflow would result in a drop in water level below the controlled range. The difference in water levels between the closed and fully open valve positions is called the proportional band or the throttling range.

The standard hardware in a pneumatic proportional controller regulating the discharge temperature of air from a heating coil in an air-conditioning system is shown in Fig. 16-7. The major elements in the control loop are a temperature transmitter, a receiver/controller, and the valve. The temperature transmitter regulates a pressure that is conveyed to the receiver/controller, and this pressure varies linearly with the temperature. Temperature transmitters are available in several ranges of temperature, such as -10 to $30°C$ or 10 to $65°C$, during which the output pressure varies from 20 to 100 kPa. The receiver/controller picks up a narrow range of the output of the temperature transmitter and amplifies this pressure, sending it to the valve. The extent of amplification is regulated by the gain setting on the receiver/controller. The particular narrow range selected from the temperature transmitter output is decided by the setpoint adjustment on the receiver/controller.

Example 16-1. Determine the throttling range of a discharge air temperature controller, as in Fig. 16-7, if the spring range of the valve is 15 to 50 kPa (completely open when the pressure supplied to the valve is 15 kPa and completely closed when the pressure is 50 kPa), the gain setting of the receiver/controller is 10:1, and the pressure range of the temperature transmitter varies from 20 to 100 kPa as the temperature varies from 10 to 65°C.

Solution. The throttling range is the change in controlled temperature needed to drive the valve from one limit position to the other. In order to develop the change of pressure between 15 and 50 kPa a pressure difference of 35 kPa is required. With the gain setting of 10:1, the controller works from a pressure increment of 3.5 kPa from the temperature transmitter, which translates into a throttling range of temperature of

$$\text{Throttling range} = (3.5\,\text{kPa})(65 - 10°\text{C})/(100 - 20\,\text{kPa})$$
$$= 2.4°\text{C}$$

A circuit providing proportional control through an electric analog circuit is shown in Fig. 16-8 regulating the speed of a motor through a variable-frequency inverter. The tachometer provides a voltage to the control circuit that is linearly proportional to the motor speed. As the load on the motor increases, the voltage difference across the motor must be increased by dropping the voltage at point 1, which is accomplished by delivering higher current into the base of the power transistor. This higher current is provided by increasing the voltage at point 2, as shown in Fig. 16-9a. The voltage at the inverting input of the op amp is essentially the same voltage as at the non-inverting input, V_3, which is adjusted by the speed setting. The increase in voltage at point 2 is accompanied by a decrease in voltage at 4, which results in, as Fig. 16-9b shows, a drop in the speed.

16-7 Proportional-Integral Control

A characteristic of any proportional controller, as was illustrated in the previous section, is the existence of a proportional band or throttling range.

16-7. Proportional-Integral Control

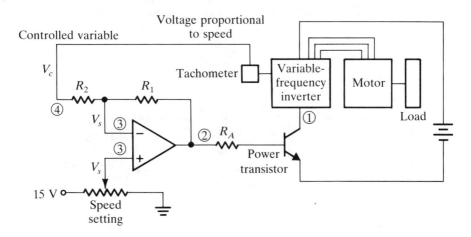

Fig. 16-8. An electric analog speed controller of a motor through a variable-frequency inverter.

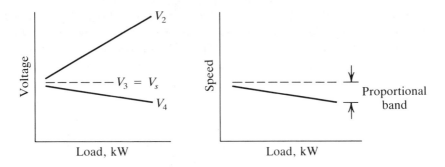

Fig. 16-9. (a) Variations in controller voltages, and (b) variation in speed as the load changes in the system of Fig. 16-8.

Except at one operating point the controlled variable is always offset from the setpoint. When this deviation is unacceptable, an additional feature can be incorporated, namely the integral mode. The integral mode integrates with respect to time the deviation between the controlled variable and the setpoint, and based on the magnitude of this integral it determines the adjustment to the control signal.

An electric analog proportional-integral (PI) control loop can be constructed by combining an integrating op amp through a summing circuit with the basic proportional control circuit. In the integrator (Fig. 3-14), a capacitor

326   Chapter 16. Dynamic Behavior of Systems

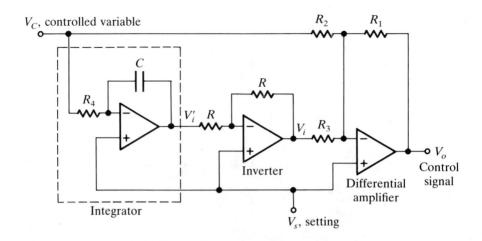

Fig. 16-10. An electric analog proportional-integral control circuit.

replaces the forward resistor in an inverting op amp circuit. A PI controller circuit is shown in Fig. 16-10. The integrator, shown enclosed in the dashed lines, integrates the deviation between V_c and V_s and feeds the result to an inverter. Integrating a revised form of Eq. (3-7) yields

$$\int \frac{V_c - V_s}{R_4 C}\, dt = -(V_i' - V_s) = V_i - V_s \qquad (16\text{-}1)$$

The summing portion of Fig. 16-10 sums and proportions the inputs V_c and V_i according to an equation comparable to Eq. (3-5):

$$V_o - V_s = \frac{R_1}{R_2}(V_s - V_c) + \frac{R_1}{R_3}(V_s - V_i) \qquad (16\text{-}2)$$

Substituting Eq. (16-1) into Eq. (16-2) gives

$$V_o - V_s = \underbrace{\frac{R_1}{R_2}(V_s - V_c)}_{\text{proportional}} + \underbrace{\frac{R_1}{R_3 R_4 C}\int (V_s - V_c)\, dt}_{\text{integral}} \qquad (16\text{-}3)$$

If V_c for some reason has been greater than V_s for a period of time, the integral of the final term in Eq. (16-3) will deduct from V_o, which progressively reduces the offset of the controlled variable from the setpoint.

16-8 Proportional-Integral-Derivative (PID) Control

Still another mode of control that is available is the derivative control, which senses the rate of change of the controlled variable and contributes a component to the control signal that is proportional to the derivative. The derivative mode provides the equivalent of how an alert human would correct the control signal in response to a change in the controlled variable with respect to a setting. If a change in the controlled variable were observed, the human operator would make a greater adjustment of the control signal if the rate of change were high than if the rate of change were low.

The incorporation of the derivative mode to form the PID control is especially necessary where rapid changes may occur, such as in some applications in the chemical and process industries. We will not consider the derivative mode further, however, but will concentrate instead on the widely used P and PI controls.

16-9 Dynamic Analysis

The technical challenge in controls lies especially in dynamic behavior (changes with respect to time), particularly when interfacing microcomputer controls with thermal and mechanical systems. The next several sections review a few major points from classical control theory that are applicable to continuous controllers. Thereafter the investigation moves into the considerations that arise with the discrete-change type of control provided by microcomputers. Readers who have the principles of automatic control theory at their fingertips may be able to move rapidly through the remainder of this chapter.

Representation of the behavior of components and systems that are changing with respect to time calls for the use of differential equations. An important tool to help solve differential equations is the Laplace transform, and this tool has power in addition to solving differential equations when dealing with control systems. Much of the investigation of control systems, such as stability analysis of a loop, can be performed in the transformed domain, which is much simpler to work with than operating in the physical domain.

16-10 Laplace Transforms

The Laplace transform is an operation that converts a function of t, called $F(t)$, to a function of the transformed variable s, called $f(s)$. The definition of the Laplace transform is

$$\mathcal{L}\{F(t)\} = \int_0^\infty F(t)e^{-st}\,dt = f(s) \qquad (16\text{-}4)$$

To find $f(s)$, substitute $F(t)$ into Eq. (16-4) and integrate. For example, if $F(t) = a$, where a is a constant,

$$f(s) = a\int_0^\infty e^{-st}\,dt = a\left[\left(-\frac{1}{s}\right)e^{-st}\right]_0^\infty = \frac{a}{s}$$

If $F(t) = t$,

$$f(s) = \int_0^\infty te^{-st}\,dt = -\frac{d}{ds}\int_0^\infty e^{-st}\,dt = -\frac{d\left(\frac{1}{s}\right)}{ds} = \frac{1}{s^2}$$

To facilitate working with Laplace transforms, tables of pairs of $F(t)$ and $f(s)$ are available; a sample of such a table is presented in Table 16-1.

16-11 Inverting a Transform

The inverse of the transform $f(s)$ is the function $F(t)$ that provided $f(s)$ in the first place. Thus from Table 16-1 te^{at} is the inverse of $1/(s-a)^2$. When inverting complicated expressions, the usual strategy is to break the expression into components that can be inverted using tables of pairs of transforms.

Example 16-2. Invert

$$\frac{1}{s^2 - s - 6}$$

by decomposing into fractions, then inverting the individual terms.

Solution.

$$\frac{1}{s^2 - s - 6} = \frac{A}{s-3} + \frac{B}{s+2}$$

One technique for determining A and B is

$$A = |[f(s)](s-3)|_{s=3}$$

and

$$B = |[f(s)](s+2)|_{s=-2}$$

16-11. Inverting a Transform

Table 16-1. Table of Laplace Transforms

$f(s)$	$F(t)$	$f(s)$	$F(t)$
$\dfrac{1}{s}$	1	$\dfrac{1}{(s-a)(s-b)}$	$\dfrac{1}{a-b}\left(e^{at}-e^{bt}\right)$
$\dfrac{1}{s^2}$	t	$\dfrac{1}{(s-a)(s-b)(s-c)}$	$\dfrac{(b-c)e^{at}+(c-a)e^{bt}}{(a-b)(b-c)(c-a)}$ $-\dfrac{(a-b)e^{ct}}{(a-b)(b-c)(c-a)}$
$\dfrac{1}{s^n}$, n=integer	$\dfrac{t^{n-1}}{(n-1)!}$	$\dfrac{s}{(s-a)(s-b)}$	$\dfrac{1}{(a-b)}\left(ae^{at}-be^{bt}\right)$
$\dfrac{1}{\sqrt{s}}$	$\dfrac{1}{\sqrt{\pi t}}$	$\dfrac{b}{(s+a)^2+b^2}$	$e^{-at}\sin bt$
$\dfrac{1}{s-a}$	e^{at}	$\dfrac{s+a}{(s+a)^2+b^2}$	$e^{-at}\cos bt$
$\dfrac{1}{(s-a)^2}$	te^{at}	$\dfrac{1}{s(s^2+a^2)}$	$\dfrac{1}{a^2}(1-\cos at)$
$\dfrac{1}{(s-a)^n}$, n=integer	$\dfrac{t^{n-1}e^{at}}{(n-1)!}$	$\dfrac{1}{s^2(s^2+a^2)}$	$\dfrac{1}{a^3}(at-\sin at)$
$\dfrac{1}{s^2-a^2}$	$\dfrac{1}{a}\sinh at$	$\dfrac{1}{(s^2+a^2)^2}$	$\dfrac{1}{2a^3}(\sin at - at\cos at)$
$\dfrac{s}{s^2-a^2}$	$\cosh at$	$\dfrac{s}{(s^2+a^2)^2}$	$\dfrac{t}{2a}\sin at$
$\dfrac{1}{s^2+a^2}$	$\dfrac{1}{a}\sin at$	$\dfrac{1}{(s+a)(s^2+b^2)}$	$\dfrac{1}{a^2+b^2}\left[e^{-at}\right.$ $\left.+\dfrac{\sqrt{a^2+b^2}}{b}\sin(bt-\theta)\right]$ where $\theta=\tan^{-1}\left(\dfrac{b}{a}\right)$
$\dfrac{s}{s^2+a^2}$	$\cos at$		

Then

$$A = \left\|\left[\frac{1}{(s-3)(s+2)}\right](s-3)\right\|_{s=3} = \frac{1}{5}$$

and

$$B = -\frac{1}{5}$$

$$f(s) = \frac{1}{5(s-3)} - \frac{1}{5(s+2)}$$

so inverting the individual terms gives

$$F(t) = \frac{e^{3t}}{5} - \frac{e^{-2t}}{5}$$

which is confirmed by the inversion shown by Table 16-1 for $1/(s-a)(s-b)$.

16-12 Transforms of Derivatives

The Laplace transform of a derivative, $dF(t)/dt$, or $F'(t)$, can be derived by substituting $F'(t)$ into Eq. (16-4) and integrating by parts,

$$\mathcal{L}\left\{F'(t)\right\} = \int_0^\infty F'(t)\,dt = s\,f(s) - F(0) \qquad (16\text{-}5)$$

The expression for the transform of the second derivative can be developed by cascading Eq. (16-5),

$$\mathcal{L}\left\{F''(t)\right\} = s^2\,f(s) - s\,F(0) - F'(0) \qquad (16\text{-}6)$$

16-13 Solving Differential Equations by Means of Laplace Transforms

The steps in solving a differential equation using Laplace transforms are:

1. Transform the differential equation to the s domain.

2. Substitute the boundary conditions that apply to $t = 0$.

3. Invert the $f(s)$ expression.

Example 16-3. Using Laplace transforms, solve the differential equation

$$\frac{dY}{dt} + 3Y = 2$$

16-13. Solving Differential Equations

subject to the boundary condition of $Y = 4$ when $t = 0$.

Solution. The transform of the differential equation is

$$s\,y(s) - Y(0) + 3y(s) = \frac{2}{s}$$

Substituting $Y(0) = 4$ and rearranging,

$$y(s) = \left(\frac{2}{s} + 4\right)\left(\frac{1}{s+3}\right) = \frac{2+4s}{s(s+3)}$$

Inverting,

$$Y(t) = \frac{2}{3} + \frac{10}{3}e^{-3t}$$

Example 16-4. Using Laplace transforms, solve the differential equation

$$\frac{d^2Y}{dt^2} - 4Y = 3$$

subject to the boundary conditions $Y(0.5) = 0$ and $Y'(0) = 1$.

Solution. First invert the differential equation,

$$s^2\,y(s) - s\,Y(0) - Y'(0) - 4y(s) = \frac{3}{s}$$

The value of $Y'(0) = 1$ can be substituted into the transform, but $Y(0)$ is as yet unknown, so assign it a symbol A.

$$s^2\,y(s) - As - 1 - 4y(s) = \frac{3}{s}$$

so

$$y(s) = \frac{As^2 + s + 3}{s\,(s^2 - 4)}$$

Inverting,

$$Y(t) = \frac{1}{8}\left[-6 + (4A+5)e^{2t} + (4A+1)e^{-2t}\right]$$

Now the boundary condition, $Y(0.5) = 0$, may be substituted to find A to be -0.6446. The solution is

$$Y(t) = -0.75 + 0.3027\,e^{2t} - 0.1975\,e^{-2t}$$

The validity of the solution may be checked by substituting $Y(t)$ back into the differential equation and also verifying the boundary conditions.

16-14 Transfer Functions

The transfer function of a process or system is defined as the Laplace transform of the output divided by the transform of the input, as shown in Fig. 16-11. Transfer functions are elements of control block diagrams, which will be introduced in the next section. The transfer function derives from the differential equation by first transforming the equation and then solving for the ratio $\mathcal{L}\{\text{output}\}/\mathcal{L}\{\text{input}\}$.

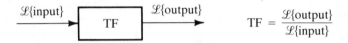

Fig. 16-11. A transfer function.

Example 16-5. The torque θ developed by the dc motor in Fig. 16-12 is proportional to the applied voltage Y, $\theta = aY$. This torque overcomes the load represented by $f\omega$, where f is constant and ω is the rotative speed. Another component of the torque is absorbed by the rate of change of rotative members, $I(d\omega/dt)$, where I is the moment of inertia. Determine the transfer function with Y the input and ω the output.

Solution. The differential equation is

$$\theta = aY = f\omega + I\frac{d\omega}{dt}$$

Transforming the differential equation,

$$a\,\mathcal{L}\{Y\} = f\,\mathcal{L}\{\omega\} + I\left[s\,\mathcal{L}\{\omega\} - \omega(0)\right]$$

It is at this point that a policy decision is made: the value of the output variable at $t = 0$ is set to zero. Then the transfer function TF can easily be solved,

$$\text{TF} = \frac{\mathcal{L}\{\omega\}}{\mathcal{L}\{Y\}} = \frac{a}{Is + f}$$

Fig. 16-12. Developing the transfer function of a motor driving a load.

16-15 Feedback Loops

The requirement that $\omega(0) = 0$ in Example 16-5, and in general that the output variable is zero at time $= 0$, looms important whenever numerical calculations are being made in the time domain. The stipulation requires that the variables must be normalized with respect to a steady state.

16-15 Feedback Loops

A fundamental control arrangement is the feedback loop, two examples of which are presented in Fig. 16-13a and 16-13b. The summing point, represented by the circle, computes the difference between the setpoint $R(s)$ and the controlled variable $C(s)$ in Fig. 16-13a. This error regulates the actuator position, which in turn influences the status of the system whose transfer function is represented by $G(s)$. In the non-unity feedback loop of Fig. 16-13b, a time-dependent process, whose transfer function is $H(s)$, translates the controlled variable to a variable $B(s)$ that is compared to the setpoint. An example of $H(s)$ would be a temperature transducer with some thermal capacity.

To compute the response and analyze for stability of the loop, the transfer function of the loop is needed,

$$\text{TF}_{\text{loop}} = \frac{C(s)}{R(s)}$$

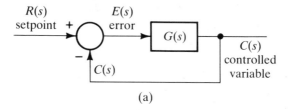

(a)

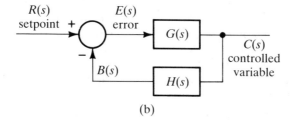

(b)

Fig. 16-13. (a) A unity feedback, and (b) a non-unity feedback control loop.

For the unity feedback loop of Fig. 16-13a,

$$\text{TF}_{\text{loop}} = \frac{C(s)}{R(s)} = \frac{E(s)G(s)}{R(s)} = \frac{[R(s) - C(s)]\,G(s)}{R(s)}$$

and $\text{TF}_{\text{loop}} = G(s) - G(s)\,(\text{TF}_{\text{loop}})$ so

$$\text{TF}_{\text{loop}} = \frac{G(s)}{1 + G(s)} \tag{16-7}$$

In a similar manner, the TF_{loop} of the non-unity feedback loop of Fig. 16-13b can be derived:

$$\text{TF}_{\text{loop}} \frac{G(s)}{1 + G(s)H(s)} \tag{16-8}$$

16-16 Stability Criteria for a Feedback Control Loop

One of the strengths of the Laplace transform representation of a control loop is that stability analyses can be performed without ever inverting back to the time domain. The reason for this ease of analysis is that the TF_{loop}, for example,

$$\frac{as + b}{cs^3 + ds^2 + es + f}$$

can be decomposed into components,

$$\frac{A}{s - r_1} + \frac{B}{s - r_2} + \frac{C}{s - r_3} \tag{16-9}$$

The nature of the roots, the r values, indicates stability or instability, because the inverses of the terms in Eq. (16-9) are exponents as shown in Fig. 16-14. If a root is less than zero, the contribution of that term decays, indicating a stable response. If, on the other hand, the root is positive as shown in Fig. 16-14c, the magnitude grows with time so the response is unstable. When the root is zero, the response is a constant value and hence is neutral.

Figure 16-14 implies that the roots are real numbers, but roots that are imaginary numbers appear frequently as well. If the root is $a + ib$, for example, the inverse is $e^{at}e^{ibt}$. Since $e^{ibt} = \cos(bt) + i\sin(bt)$, this contribution is oscillatory, neither growing nor decaying with time. The foregoing observations lead to the Nyquist criterion,[2] which states that a control loop is unstable if the real part of any of the roots of the denominator of the TF_{loop} is positive (Fig. 16-15).

16-16. Stability Criteria for a Feedback Control Loop

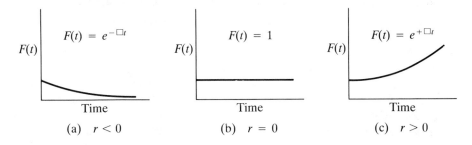

Fig. 16-14. Inverses $F(t)$ are dependent upon the sign of the roots in $(\text{const})/(s - r)$ terms.

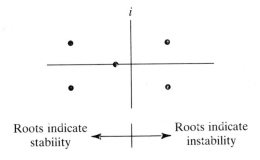

Fig. 16-15. Location of roots of the loop transfer function indicates stability or instability.

Example 16-6. Is the control loop having the transfer function

$$\frac{s+2}{s^2 + 2s + 3}$$

stable or unstable?

Solution. The roots of the denominator are

$$s = \frac{-2 \pm \sqrt{2^2 - (4)(3)}}{2} = -1 + i\sqrt{2} \quad \text{and} \quad -1 - i\sqrt{2}$$

The real parts are negative, so the response of the loop is stable.

16-17 A Proportional Controller Regulating the Pressure in an Air-Supply System

Another step will now be taken in the progression to hardware and the representation of a physical situation in symbolic and mathematical form. The first experiences that this example will provide are (1) developing a transfer function for another process, (2) constructing a control block diagram, and (3) analyzing a loop for stability. The air-supply system, shown in Fig. 16-16, consists of a motor-driven compressor that delivers a mass-flow rate of air, w_i kg/s, that is proportional to the speed ω rev/s. The speed is regulated with the objective of maintaining a constant pressure in the tank. The rate of air flow drawn by the users is w_o kg/s, a quantity that may vary with time. The controller is of the proportional type and supplies a voltage to the motor, $Y = k(p_{set} - p)$. The torque, θ N-m, developed by the motor is proportional to the applied voltage, $\theta = c_1 Y$. The moment of inertia of the motor and compressor is I kg-m^2, and the torque required for compression of the gas is a function of the speed, $\theta_{comp} = c_3 \omega$.

Example 16-5 has already expressed the transfer function from the voltage Y to the rotative speed ω as

$$\frac{c_1}{Is + c_3}$$

Another dynamic process that exists in this system has an input of the difference in flow rates and an output of pressure p. The differential equation

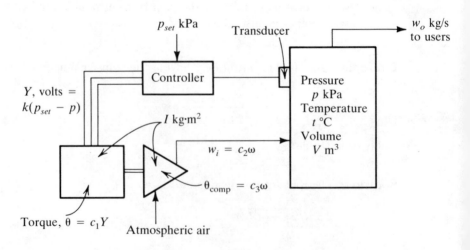

Fig. 16-16. Pressure control in an air-supply system.

16-17. A Proportional Controller Regulating an Air-supply System

representing the process is

$$w_i - w_o = \frac{dm}{dt} = \frac{d(pV/RT)}{dt} \quad (16\text{-}10)$$

where

- m = mass in tank, kg
- R = gas constant for air = 0.287 kJ/(kg-K)
- V = volume of tank, m³
- T = absolute temperature of air, K

Assuming that the temperature of air remains constant,

$$w_i - w_o = \frac{V}{RT}\frac{dp}{dt}$$

The transfer function derived from this differential equation is

$$\text{TF} = \frac{RT/V}{s} \quad \text{if } p(0) = 0$$

One version of the control block diagram representing the system is shown in Fig. 16-17. The transfer function of this loop having p_{set} as the input, p as the output, and assuming w_o constant is

$$\text{TF} = \frac{\mathcal{L}\{p\}}{\mathcal{L}\{p_{\text{set}}\}} = \frac{\dfrac{kc_1c_2RT/V}{Is^2+c_3s}}{1+\dfrac{kc_1c_2RT/V}{Is^2+c_3s}}$$

or

$$\text{TF} = \frac{kc_1c_2RT/V}{Is^2+c_3s+kc_1c_2RT/V} \quad (16\text{-}11)$$

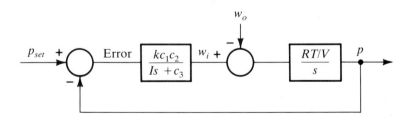

Fig. 16-17. Version 1 of the control block diagram of the system in Fig. 16-16.

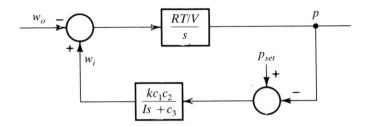

Fig. 16-18. Version 2 of the control block diagram of the system in Fig. 16-16.

Equation (16-11) is the pertinent transfer function to analyze the response of the controlled pressure p to a change in the pressure setting. While indeed the setpoint may be changed on occasion, a far more frequent occurrence is a change in the demand w_o. To analyze the response of p to that change, the form of the block diagram in Fig. 16-18 is chosen. The signs of variables no longer correspond to the standard of Fig. 16-13, and the transfer function of the loop shown in Fig. 16-18 is the negative of that in Eq. (16-8). With a constant pressure setting the transfer function of this loop is

$$\text{TF} = \frac{\mathcal{L}\{p\}}{\mathcal{L}\{w_o\}} = \frac{-(RTI/V)s - RTc_3/V}{Is^2 + c_3 s + kc_1 c_2 RT/V} \qquad (16\text{-}12)$$

The roots of the denominator of the transfer function in Eq. (16-12) indicate, by means of the Nyquist criterion, whether the loop is stable. All the coefficients in the denominator are positive, so for convenience we set a comparable expression

$$As^2 + Bs + C$$

equal to zero to solve for the roots,

$$s = \frac{-B \pm \sqrt{B^2 - 4AC}}{2A}$$

Since A, B, and C are all positive, there is no way in which s can be positive. If the quantity within the radical is negative, the square root is imaginary, which does not influence stability.

A fact to file for future reference is that this loop is invariably stable, in contrast to what will be discovered later when using a computer controller. The loop may go unstable if the sampling interval is too long.

16-18 Response of a Proportional Air-Pressure Controller to a Disturbance in Air-Flow Rate

The response of the pressure controller (Fig. 16-16) to a disturbance will now be quantified. In particular, the variation of the tank pressure following a step change in the outlet flow rate will be computed. Because numerical values will be expressed in the time domain (in contrast to the transformed domain) it will first be necessary to normalize the variables shown in Fig. 16-18. This normalization is one of relating the variables to their values at zero time, the need for which was first pointed out in Sec. 16-14. The revised variables on the control loop are shown in Fig. 16-19 with w^* representing the steady-state flow rate (both in and out of the tank) and p^* the steady-state tank pressure at zero time.

Suppose that the following parameters apply:

$$
\begin{aligned}
c_1 &= 3.2 \, \text{N-m/V} \\
c_2 &= 0.02 \, \text{kg/s per rev/s} \\
c_3 &= 12 \, \text{N-m per rev/s} \\
I &= 10 \, \text{kg-m}^2 \\
\text{Temperature} &= 26.85°\text{C} \; (T = 300 \, \text{K}) \\
V &= 3 \, \text{m}^3 \\
R &= 0.287 \, \text{kJ/(kg-K)} \\
p_{\text{set}} &= 800 \, \text{kPa}
\end{aligned}
$$

Several different proportionality constants, the k's, will be explored. Suppose that w_o has been 0.8 kg/s for a long time such that steady-state conditions

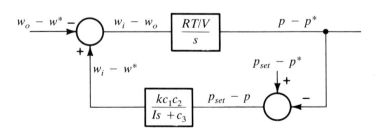

Fig. 16-19. Control loop with variables normalized so that their values are zero at zero time.

prevail, and at time $t = 0$ the outlet flow rate makes a step decrease to 0.7 kg/s. The transform of this step change, $\mathcal{L}\{w_o - w^*\} = -0.1/s$, so from Eq. (16-12),

$$p - p^* = \mathcal{L}^{-1}\left\{\frac{-0.1}{s}\left[\frac{-\overbrace{(RTI/V)}^{D}s - \overbrace{RTc_3/V}^{E}}{\underbrace{I}_{A}s^2 + \underbrace{c_3}_{B}s + \underbrace{kc_1c_2RT/V}_{C}}\right]\right\} \quad (16\text{-}13)$$

Three different solutions are possible from Eq. (16-13), depending on the relationships of $A, B,$ and C. The character of these solutions is summarized in Table 16-2. The three solutions are given in Eqs. (16-14) to (16-16), and corresponding exercises are presented in Probs. 16-7, 16-8, and 16-9, respectively. The terms r_1 and r_2 are the roots of the quadratic expression in the denominator,

$$A(s - r_1)(s - r_2) = As^2 + Bs + C$$

$$p - p^* = \frac{0.1}{A}\left(\frac{E}{r_1 r_2} + \frac{(D + E/r_1)e^{r_1 t} - (D + E/r_2)e^{r_2 t}}{r_1 - r_2}\right) \quad (16\text{-}14)$$

$$p - p^* = \frac{0.1}{A}\left[\frac{E}{r^2} + \left(D + \frac{E}{r}\right)te^{rt} - \frac{E}{r^2}e^{rt}\right] \quad (16\text{-}15)$$

$$p - p^* = \frac{0.1}{C}\left[E - Ee^{-at}\cos(bt) + \left(\frac{DC}{Ab} - \frac{Ea}{b}\right)e^{-at}\sin(bt)\right] \quad (16\text{-}16)$$

where

$$a = B/(2A) \quad \text{and} \quad b = \sqrt{4AC - B^2}/(2A)$$

Table 16-2. Three Solutions to Eq. (16-13)

Characterization	Name	Equation Number Giving Solution	Constant, k
$B^2 > 4AC$	Overdamped	(16-14)	< 1.961
$B^2 = 4AC$	Critically damped	(16-15)	1.961
$B^2 < 4AC$	Underdamped	(16-16)	> 1.961

16-18. Response of a Proportional Air-pressure Controller

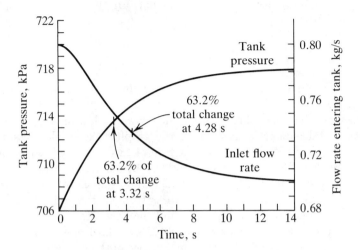

Fig. 16-20. Response of the tank pressure and the inlet air-flow rate to a step change in outlet air-flow rate from 0.8 to 0.7 kg/s. This is the overdamped case with $k = 1.6$.

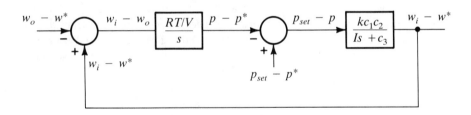

Fig. 16-21. Control loop with $w_i - w^*$ as the output.

The history of the tank pressure and the inlet flow rate to the tank in response to the step change in outlet flow rate are shown in Fig. 16-20 for the overdamped case ($k = 1.6$). The tank pressure, which is computed from Eq. (16-14), starts at 706.3 kPa and increases to the final pressure of 717.9 kPa. To compute the inlet flow rate to the tank, the control loop of Fig. 16-19 can be restructured to the form of Fig. 16-21, which shows w_i as the output of the loop. The transfer function of this loop is

$$\frac{\mathcal{L}\{w_i - w^*\}}{\mathcal{L}\{w_o - w^*\}} = \frac{kc_1c_2RT/V}{Is^2 + c_3s + kc_1c_2RT/V} \qquad (16\text{-}17)$$

If w_o experiences the step change from 0.8 to 0.7 kg/s, inversion of Eq. (16-17) yields the time-domain equation for w_i:

$$w_i = w^* - \frac{0.1C}{A}\left(\frac{1}{r_1 r_2} + \frac{r_2 e^{r_1 t} - r_1 e^{r_2 t}}{r_1 r_2 (r_1 - r_2)}\right) \qquad (16\text{-}18)$$

where the roots r_1 and r_2 are the same as in Eq. (16-14).

A few observations may be made on the trends in Fig. 16-20. To express a measure of the speed of response of the two variables, the time for 63.2 percent of the total change is marked on the graph. The tank pressure responds more rapidly (3.32 s) than the flow rate (4.28 s). The quicker response of the pressure may be explained qualitatively by referring to the diagram of the physical system in Fig. 16-16. A change in w_o impacts the tank pressure first, and then this change in pressure works through the controller to ultimately revise the speed of the compressor and the flow rate.

The pressures between 706 and 718 kPa are far from the pressure setting of 800 kPa. This large offset is the consequence of the low proportional constant, k. Not until the flow is reduced to zero does the pressure in the tank reach the set pressure of 800 kPa. Would the operator be satisfied with this magnitude of pressure if he or she really wanted 800 kPa? Probably not, and if the typical demand on the system was in the range of 0.7 to 0.8 kg/s the operator would readjust the setpoint to 880 or 890 so that with the 82 to 94 kPa offset the tank pressure would remain near 800 kPa. In fact, many controllers do not have a scale on the setpoint adjustment, with the expectation that the adjustment would be made to bring the controlled variable into the desired range.

Another approach to reducing the offset would be to increase the gain, which is an appropriate strategy for this system since it is always stable. For the first adjustment of the gain, increase k to 1.961, a unique value that results in a critically damped control. Equation (16-15) expresses the response of the tank pressure. The r value in Eq. (16-15) is the root of the equation $(s - r)^2 = As^2 + Bs + C$. The response of the pressure is plotted in Fig. 16-22 and has a similar shape to the overdamped case shown in Fig. 16-23, but the pressure changes from 723.5 to 733.1 kPa, and 2.64 s is required for 63.2 percent of the total travel. Thus, the offset is reduced and the speed of response increased. From a mathematical standpoint the critically damped case is represented by a unique equation, Eq. (16-15), but the major physical significance of the response is that it is the borderline before overshoot occurs in response to a step change.

Increasing k appreciably—to a value of 12, for example—shifts to an underdamped response as shown in Fig. 16-23. From the initial steady-state pressure of 787.5 kPa the pressure rises to a maximum value of 790.6 and then undergoes a slight oscillation until settling to its new steady-state value of 789.1 kPa. The appearance of the trigonometric functions in Eq. (16-16)

16-18. *Response of a Proportional Air-pressure Controller* 343

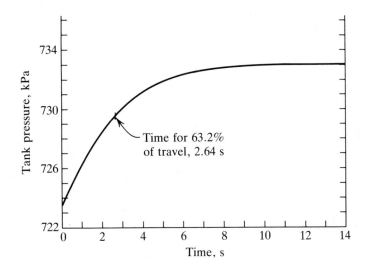

Fig. 16-22. Response of tank pressure to a step change in outlet air-flow rate from 0.8 to 0.7 kg/s—critically damped with $k = 1.961$.

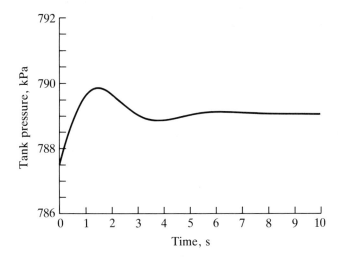

Fig. 16-23. Response of tank pressure to a step change in outlet air-flow rate from 0.8 to 0.7 kg/s—underdamped case with $k = 12$.

hints that there will be an oscillation of the tank pressure. The high gain setting is clearly preferable for this control, because the response is rapid and the offset from the setpoint of 800 kPa is the smallest of any of the cases considered.

16-19 The Integral Mode of Control

Section 16-7 presented an example of a controller that incorporated the integrating mode along with the proportional mode of control. The function of the integral mode was to eliminate the offset associated with the proportional mode by integrating the error and applying a correction proportional to that integral.

The reason for analyzing the classical control topic of the integral mode in this book on computer control should be clarified. Analog electric and pneumatic controllers that provide the integral mode are available on the market, but adding the I mode to the P mode results in an increased cost in the controller. The strategy to be pursued with computer control is to introduce the I mode in the software of the computer—hopefully with little extra cost. In this section we shall examine the integral-only mode, even though this choice is rarely found in practice. Our reason for treating the I mode separately is twofold: to show why it is not used alone, and to identify its characteristics more distinctly.

The mathematical expression for the action of the I mode is

$$\text{Control signal} = K_I \int (\text{error})\, dt \qquad (16\text{-}19)$$

The transfer function of the I mode is K_I/s, which may be developed as in Fig. 16-24. In the time domain, if the input experiences a step input of Δ at time zero, the integration gives an output of $K_I(\Delta)(t)$. The transfer function is the transform of the output divided by the transform of the input,

$$\text{TF} = \frac{K_I(\Delta)/s^2}{\Delta/s} = \frac{K_I}{s} \qquad (16\text{-}20)$$

The block diagram for the air-pressure controller, Fig. 16-19, could be revised to show an I-mode controller as in Fig. 16-25. The form of the diagram presents the outlet air-flow rate as the input variable and the tank pressure p as the output variable. The transfer function of the control loop in Fig. 16-25 is

$$\frac{\mathcal{L}\{p - p^*\}}{\mathcal{L}\{w_o - w^*\}} = \frac{-\dfrac{RT/V}{s}}{1 + \left(\dfrac{RT/V}{s}\right)\left(\dfrac{c_1 c_2}{Is + c_3}\right)\left(\dfrac{K_I}{s}\right)}$$

16-19. The Integral Mode of Control

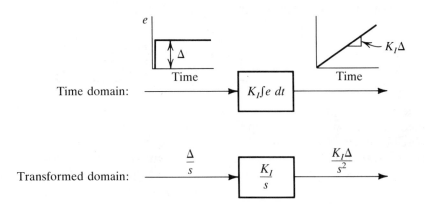

Fig. 16-24. Transfer function of the I mode.

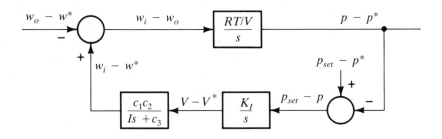

Fig. 16-25. Block diagram of the air-pressure controller operating with an I-mode controller.

which can be restructured to the form:

$$\frac{\mathcal{L}\{p - p^*\}}{\mathcal{L}\{w_o - w^*\}} = \frac{-(RT/V)s^2 - (RTc_3/IV)s}{s^3 + (c_3/I)s^2 + (RT/IV)(c_1c_2)K_I} \quad (16\text{-}21)$$

We next examine the response of the integral-only control to the same disturbance chosen in the previous section, namely, from a steady-state operating condition with a flow rate in and out of the tank of 0.8 kg/s, the outlet flow rate, w_o, undergoes a step change to 0.7 kg/s at zero time. The flow rate w^* thus equals 0.8 kg/s, and we shall assume that the controller had indeed eliminated the offset so that $p^* = 800$ kPa. Since $\mathcal{L}\{w_o - w^*\} = -0.1/s$, the transform of the tank pressure is

$$\mathcal{L}\{p - p^*\} = -0.1 \left[\frac{\overbrace{-(RT/V)s}^{D} \overbrace{-RTc_3/IV}^{E}}{s^3 + \underbrace{(c_3/I)}_{A} s^2 + \underbrace{0}_{B} s + \underbrace{(RT/IV)(c_1c_2)K_I}_{C}} \right] \quad (16\text{-}22)$$

One of the first problems to be confronted is to determine the roots of the denominator in Eq. (16-22)—a cubic equation. There are numerous approaches possible, many of which consist of iterative sequences. One iterative method takes advantage of the knowledge that there will always be at least one real root of a cubic equation. One real root may be found using a technique such as Newton-Raphson where a value of s is sought that results in $y = 0$, where

$$y = s^3 + As^2 + Bs + C$$

The method consists of choosing a trial value of s, then applying repeated corrections according to the algorithm

$$s_{new} = s_{old} - \frac{y}{y'} = s_{old} - \frac{s^3 + As^2 + Bs + C}{3s^2 + 2As + B} \quad (16\text{-}23)$$

Once one root, r_1, has been determined, y can be divided by $(s - r_1)$ to obtain a second-degree equation that can be solved by the quadratic formula.

An alternative closed-form approach to finding the roots is to use the following formulas for the roots r_1, r_2, and r_3:

$$\begin{align} r_1 &= p + q - \frac{A}{3} \\ r_2 &= -(p+q)/2 - \frac{A}{3} + i\sqrt{3}(p-q)/2 \quad (16\text{-}24) \\ r_3 &= -(p+q)/2 - \frac{A}{3} - i\sqrt{3}(p-q)/2 \end{align}$$

where

$$p = \sqrt[3]{-\frac{v}{2} + \sqrt{\frac{v^2}{4} + \frac{u^3}{27}}}$$

$$q = \sqrt[3]{-\frac{v}{2} - \sqrt{\frac{v^2}{4} + \frac{u^3}{27}}}$$

where u and v are

$$u = \frac{1}{3}(3B - A^2)$$

$$v = \frac{1}{27}(2A^2 - 9AB + 27C)$$

16-19. The Integral Mode of Control

The nature of the roots is controlled by the magnitude of the group, $(v^2/4 + u^3/27)$, namely,

Case 1: $(v^2/4 + u^3/27) > 0$; one real root and two roots that are a conjugate pair

Case 2: $(v^2/4 + u^3/27) = 0$; three real roots, two of which are equal

Case 3: $(v^2/4 + u^3/27) < 0$; three real roots, all different

Cases 1 and 2 only require straightforward substitution of the coefficients of the cubic equation into Eqs. (16-24). In solving for p and q in case 3, the cube root must be extracted of an imaginary number. If the term within the radical sign is designated p^3, which is shown in Fig. 16-26, the cube root of that imaginary number may be extracted by using the polar representation. Thus

$$p = (\text{absolute magnitude of } p^3)^{1/3} e^{i\theta}$$

where θ is one-third the polar angle of p^3. The expressions for the sum and difference of p and q resolve into simple expressions:

$$p + q = 2|p| \cos\theta \qquad \text{and} \qquad p - q = 2|p| \sin\theta$$

where $|p|$ = absolute magnitude of p.

Before attempting to invert Eq. (16-22), the nature of the roots of the denominator will be examined. The same constants used for the proportional

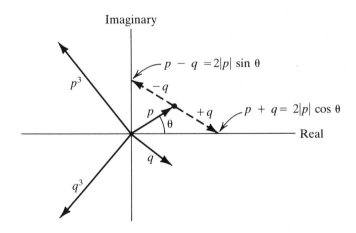

Fig. 16-26. Extraction of cube root of imaginary numbers to obtain p and q.

Table 16-3. Some Roots of the Denominator of Eq. (16-22)

K_I	A	B	C	Roots	
0.01	1.2	0	0.0018368	-1.201,	$0.00063647 \pm i\ 0.039098$
0.1	1.2	0	0.018368	-1.212,	$0.006247 \pm i\ 0.1229$
1.0	1.2	0	0.18368	-1.307,	$0.053726 \pm i\ 0.37095$
10	1.2	0	1.8368	-1.7798,	$0.2899 \pm i\ 0.97363$

controller in Sec. 16-18 will apply, namely: $R = 0.287$, $T = 300$, $I = 10$, $V = 3$, $c_1 = 3.2$, $c_2 = 0.02$, and $c_3 = 12$. Table 16-3 shows the roots of the denominator for four values of K_I. Several predictions can be made about the response of the tank pressure in the time domain. The complex roots indicate that the tank pressure will assume an oscillatory form, regardless of the value of K_I. For the values of K_I shown in Table 16-3, the response is unstable because the real part of the complex roots is always positive.

Inverting Eq. (16-22) to obtain the response in the time domain provides some insight into the nature of the I mode controlling this particular system. Performing the inversion in terms of the symbols for the variables would be cumbersome, so the inversion will be made for a specific case, namely, $K_I = 0.1$. The expression to invert is

$$-0.1 \left(\frac{-28.7s - 34.44}{(s + 1.212)(s - 0.006247 - i\,0.1229)(s - 0.006247 + i\,0.1229)} \right)$$

which may be divided into two fractions

$$0.1 \left(\frac{C_1}{s + 1.212} + \frac{C_2 s + C_3}{s^2 - 0.01249s + 0.1514} \right) \quad (16\text{-}25)$$

for which $C_1 = -0.2616$, $C_2 = 0.2616$, and $C_3 = 28.333$.

Substituting those constants into Eq. (16-25) and subdividing the second group further yields

$$0.1 \left(\frac{-0.2616}{s + 1.212} + 0.2616 \frac{s - 0.006245}{(s - 0.006245)^2 + 0.12289^2} \right.$$

$$\left. - \frac{28.335}{(s - 0.006245)^2 + 0.12289^2} \right) \quad (16\text{-}26)$$

16-19. The Integral Mode of Control

The parts of Eq. (16-26) now fit several of the standard forms of Table 16-1, so can be inverted to

$$p - p^* = 0.1\Big[-0.2616\,e^{-1.212t} + 0.2616\,e^{0.002645t}\cos(0.3506t)$$
$$+ 28.335\,e^{0.002645t}\sin(0.3506t)\Big] \tag{16-27}$$

A qualitative examination of Eq. (16-27) shows the first term to be a transient that dies out quickly. Of the remaining two trigonometric terms, the sine term, because of its large coefficient, dominates. The response, as shown in Fig. 16-27, is essentially a sine wave having a period of 17.9 s with a steadily increasing amplitude.

From time zero, where the output flow rate drops to 0.7 kg/s, the pressure in the tank initially begins to rise at a rate of $(0.8 - 0.7)(RT/v) = 2.87$ kPa/s shown by the dashed line in Fig. 16-27. The controller begins integrating the error, progressively decreasing $V - V^*$ in Fig. 16-25, which steadily decreases w_i. At point u, $w_i = w_o = 0.7$ kg/s, but the pressure is still above p_{set} so the integration of the errors continues and w_i sinks lower than 0.7 kg/s. At point v the tank pressure has been brought back to the setpoint, but w_i is below 0.7 kg/s and furthermore the controller has accumulated a positive integral shown by area A that it seeks to cancel out by dropping the pressure below

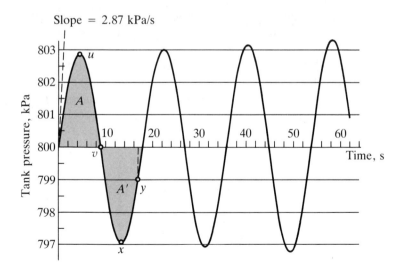

Fig. 16-27. Response of an air-pressure controller when using an I mode with $K_I = 0.1$.

800 kPa. Beyond v the flow rate starts to increase and at x has returned to 0.7 kg/s. The positive integral A will be canceled out by negative integral A' at y, but unfortunately the tank pressure is slightly below 800 kPa and the flow rate is greater than 0.7 kg/s. The oscillation continues, and because of the positive exponential, $e^{0.002645t}$, the amplitude progressively increases.

Three observations from the foregoing I-mode example on this particular system are: (1) the controlled variable oscillates, (2) the mean value of the controlled variable equals the setpoint, and (3) the control is unstable. While it cannot be concluded from this specific example that all I-mode controllers are unstable, in comparison to the I-mode controller of this air-supply system, the P mode was always stable.

16-20 The Proportional-Integral (PI) Mode of Control

The last two sections have explored the characteristics of the proportional and the integral modes of control when functioning separately. The proportional controller has the desirable feature of changing the control signal rapidly when the error changes but always regulates the controlled variable to a value that is offset from the setpoint. The integral mode, on the other hand, holds the mean value of the controlled variable to the setpoint but gives an oscillatory control signal. A widely used control concept is a combination of these two modes in a way that selects the advantage of each mode and deemphasizes its disadvantage. That combination is the proportional-integral, or PI, mode.

The two modes are additive and thus form a combination in a block diagram as shown in Fig. 16-28. The addition of the P and I effects were demonstrated in the PI controller of Fig. 16-10 through the summing circuit of the differential amplifier.

When the PI controller is applied to the compressed air system of Fig. 16-16, the control diagram may be represented as in Fig. 16-29. The

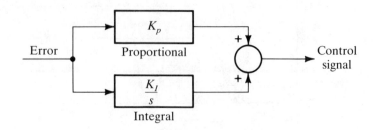

Fig. 16-28. Combination of the proportional and integral modes into a PI control.

16-20. The Proportional-Integral (PI) Mode of Control

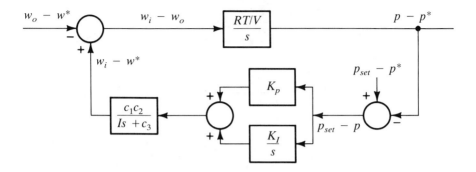

Fig. 16-29. Block diagram of air-supply system operating under PI control.

parallel paths of Fig. 16-28 now represent the controller. The transfer function of the loop is

$$\frac{\mathcal{L}\{p-p^*\}}{\mathcal{L}\{w_o-w^*\}} = \frac{-\frac{RT/V}{s}}{1+\left(\frac{RT/V}{s}\right)\left(\frac{c_1c_2}{Is+c_3}\right)\left(K_P+\frac{K_I}{s}\right)}$$

which can be restructured to

$$\frac{\mathcal{L}\{p-p^*\}}{\mathcal{L}\{w_o-w^*\}} = \frac{-(RT/V)s^2 - (RTc_3/IV)s}{s^3 + (c_3/I)s^2 + (RT/IV)c_1c_2K_Ps + (RT/IV)c_1c_2K_I} \quad (16\text{-}28)$$

Now apply the step decrease in outlet flow rate from 0.8 to 0.7 kg/s. The transform of the response is

$$\mathcal{L}\{p-p^*\} = -\frac{0.1}{s}\left[\frac{\overbrace{-(RT/V)s^2}^{D} - \overbrace{(RTc_3/IV)s}^{E}}{\underbrace{s^3 + (c_3/I)s^2}_{A} + \underbrace{(RT/IV)c_1c_2K_Ps}_{B} + \underbrace{(RT/IV)c_1c_2K_I}_{C}}\right] \quad (16\text{-}29)$$

It is the combination of K_P and K_I that will determine the response to the disturbance. Furthermore, the roots of the denominator, shown in Table 16-4, indicate the nature of the response.

One combination—$K_P = 2$ and $K_I = 10$—might be ruled out immediately, because the real parts of the imaginary roots are positive, which indicates instability. The large value of the integral constant, $K_I = 10$, introduces the instability that Fig. 16-27 showed to be the tendency of the I mode alone.

Table 16-4. Roots of the Denominator of Eq. (16-29) for Several Combinations of K_P and K_I

K_P	K_I	A	B	C	Roots
2	0.1	1.2	0.3674	0.01837	$-0.06185, -0.7329, -0.40524$
2	1.0	1.2	0.3674	0.1837	$-1.016, -0.091826 \pm i\,0.4151$
2	10	1.2	0.3674	1.837	$-1.6512, 0.22562 \pm i\,0.030338$
10	1.0	1.2	1.837	0.1837	$-0.10679, -0.5466 \pm i\,1.19226$

The remaining three K_P-K_I combinations are reasonable, so Eq. (16-29) will be inverted for those cases:

$$K_P = 2,\ K_I = 0.1$$

$$p - p^* = 14.18e^{-0.06185t} + 6.097e^{-0.7329t} - 20.27e^{-0.40524t} \tag{16-30}$$

$$K_P = 2,\ K_I = 1$$

$$p = p^* = 0.51472e^{-1.016t} - 0.51472e^{-0.091825t}\cos(1.00378t)$$
$$+ 3.333e^{-0.091825t}\sin(1.00378t)$$
$$\tag{16-31}$$

$$K_P = 10,\ K_I = 1$$

$$p - p^* = 1.9428e^{-0.10679t} - 1.9428e^{-0.54661t}\cos(1.1923t)$$
$$+ 1.6904e^{-0.54661t}\sin(1.1923t)$$
$$\tag{16-32}$$

The responses of these three combinations are shown in Fig. 16-30. Equation (16-30) with $K_P = 2$ and $K_I = 0.1$ shows no oscillation, experiences an appreciable deviation of the pressure from the setpoint in the early stages, and is relatively slow to drive back to the setpoint. When $K_P = 2$ and $K_I = 1$ the deviation from the setpoint is never large, and this deviation shrinks rather quickly. The oscillation does persist for a prolonged period. The third response shown in Fig. 16-30 displays Eq. (16-32) for $K_P = 10$ and $K_I = 1$. The initial deviation is approximately the same as for Eq. (16-31), and thereafter the pressure slowly drives to the setpoint.

16-20. The Proportional-Integral (PI) Mode of Control

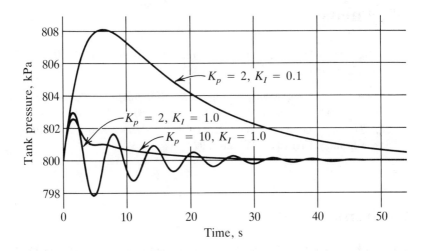

Fig. 16-30. Response of the air-pressure controller to a step decrease in flow rate when using PI modes with various combinations of K_P and K_I.

Choosing the values of K_P and K_I is called tuning. A favorable choice of K_P and K_I is influenced by the response of the system being controlled and the nature of the disturbances. If the disturbances are infrequent and slight, a low K_I provides steady control and has adequate time to erase the offset. If the disturbances are more radical, a higher K_I would be called for in order to wipe out the offset quickly, even though suffering the disadvantage of oscillations about the setpoint.

The mathematical derivations that resulted in being able to show the responses of P, I, and PI controls have served the valuable purpose of showing trends in control behavior. It should not be inferred, however, that such calculations are typically used for tuning a controller in practice. The process is usually an empirical one that is a sequence of introducing a disturbance (usually by changing the setpoint) and then adjusting the K_P and K_I settings. Often K_P is adjusted first to give a slightly underdamped response. Then K_I is adjusted so that the offset is eliminated in a reasonable time period.

The final comment is a reminder that the study of P, I, and PI controls is in preparation for incorporating the control algorithms into the software of a computer controller. The next chapter explains how to implement such control and apply it to systems that exhibit non-ideal behavior.

References

1. Major Suppliers of Energy Management Systems for Small Users, *Energy User News*, p. 13, Nov. 30, 1981.

2. H. Nyquist, "Regeneration Theory," *Bell Systems Technical Journal*, January 1932, pp. 126–147; also in *Automatic Control: Classical Linear Theory*, G. J. Thaler, ed., Dowden, Hutchinson and Ross, Inc., Stroudsburg, PA, pp. 105–126, 1974.

Problems

16-1. A pneumatic control loop like that shown in Fig. 16-7 regulates the air temperature leaving a cooling coil by controlling a valve in the chilled-water supply line. The temperature transmitter has a range of -10 to $30°C$, in which range its output pressure varies linearly from 20 to 100 kPa, respectively. The gain setting of the receiver controller is 8:1, and the setpoint (in the center of the throttling range) is $14°C$. The spring range of the valve is 60 to 88 kPa. At what discharge air temperature does the valve close off completely?

16-2. In the mechanical level controller shown in Fig. 16-6, designate the water level in the tank (which is 0.5 m in diameter) as x, with $x = 0$ at the no-flow level and $x = 100$ mm at the wide-open valve status. The input flow varies from 0 to $0.04 \, \text{m}^3/\text{s}$, respectively, during this valve travel. If the demand had been $0.01 \, \text{m}^3/\text{s}$ for a long period and is suddenly increased to $0.03 \, \text{m}^3/\text{s}$ where it holds steady, develop the equation for x as a function of time. **Ans.:** Exponential from 25 to 75 mm with a time constant of 0.49 s.

16-3. Invert the Laplace transform

$$\frac{2s + 3\sqrt{s} - 2}{\sqrt{s}(s-1)}$$

by first decomposing into fractions.

16-4. Using Laplace transforms, solve the differential equation

$$\frac{dY}{dt} + 2Y + 4t = 0$$

subject to the boundary condition that $Y = 0$ when $t = 1$. **Ans.:** Equation should show $Y(2) = -2.865$.

Problems

16-5. Using Laplace transforms, solve the differential equation

$$\frac{d^2Y}{dt^2} + 2\frac{dY}{dt} + 5 = 0$$

subject to the boundary conditions $Y'(0) = 4$ and $Y(0) = 1$. **Ans.:** Equation should show $Y(0.5) = 1.80$.

16-6. A fluid of temperature T_f flows over a block of metal having a temperature T, as shown in Fig. 16-31. Applicable properties are:

h = heat-transfer coefficient from air to metal, W/(m²-K)
A = heat-transfer area, m²
m = mass of metal, kg
c = specific heat of metal, J/(kg-K)

The metal has a high thermal conductivity, so the temperature in the metal is uniform at any given time.

Develop the transfer function, $\mathcal{L}\{T\}/\mathcal{L}\{T_f\}$, for the process.

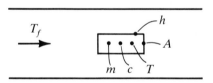

Fig. 16-31. Transfer function of a heat-transfer process.

16-7. Invert Eq. (16-13) for the overdamped case of the air-pressure controller to develop Eq. (16-14).

16-8. Invert Eq. (16-13) for the critically damped case of the air-pressure controller to develop Eq. (16-15).

16-9. Invert Eq. (16-13) for the underdamped case of the air-pressure controller to develop Eq. (16-16).

16-10. An I-mode controller regulates the water level in a tank that has a cross-sectional area of 0.8 m². The controller $K_I = 0.2\,\mathrm{m^2/s^2}$ in the equation

$$w_i - 0.12 = K_I \int (0.6 - x)\,dt$$

A steady state has been achieved with both w_i and $w_o = 0.12\,\text{m}^3/\text{s}$, and as shown in Fig. 16-32 the water level is at the setpoint of 0.6 m. If the draw rate w_o undergoes a step increase to $0.18\,\text{m}^3/\text{s}$, what is the equation for x as a function of time? **Ans.:** Sine wave with period of 12.56 s.

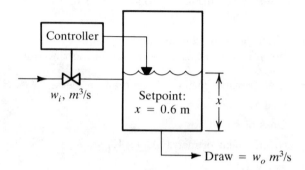

Fig. 16-32. I-mode controller in Prob. 16-10.

Chapter 17

Interfacing the Computer and Its Sampling Processes with the System

17-1 Unique Features of Computer Control

Controlling a system with a computer offers some advantages and a few drawbacks too. Intelligent application of computer control seeks to exploit the strengths and minimize the deficiencies of the computer as a controller. One advantage of computer control is the ability to build into software at a low cost some control features that might be expensive if done in hardware. A characteristic of computer control that at certain times could cause difficulties is the sampling process used in sensing variables and positioning actuators. One of the important pitfalls to avoid in applying computer control to thermal and mechanical systems is instability due to the sampling operation.

This chapter builds on the groundwork laid in Chapter 16 and pursues topics that could be grouped into three categories: (1) computer control modes, (2) types of signals to the actuators, and (3) two frequently encountered non-linearities. Before attacking any of these topics, two analytical tools will be developed: numerical simulation and the z-transform—subjects that will be used frequently as the chapter proceeds. The types of control modes, actuator signals, and non-linearities to be treated are:

1. Computer-control modes

   (a) proportional

   (b) proportional-plus-integral

   (c) sense / fixed change / delay

2. Signals to the actuators
   (a) on/off
   (b) modulated, latched
   (c) pulse-width modulation
3. Nonlinearities
   (a) hysteresis
   (b) transportation lag or dead time

17-2 Numerical Simulation

Inverting the transfer functions of components and loops in order to obtain the response of variables in the time domain, as was done in Chapter 16, served our purposes well. Not only did the resulting closed-form equation afford the means of computing the behavior of the variables with respect to time, but the form of the equations and the magnitudes of constants and coefficients also provided insight into the nature of the response. In this chapter we are forced to abandon closed-form solutions because of complexities appearing principally due to the sampling process and non-linearities. Prediction of response in the time domain remains valuable, however, so another technique will be introduced: numerical simulation.

Numerical simulation will refer here to computing the response of operating variables by subdividing the processes into extremely small time steps for which simple expressions that are often linear apply. As might be expected when subdividing the entire time span of interest into small steps, the number of calculations is large, so performing the simulation on a computer is anticipated. In the dynamic behavior of systems we encounter differential equations, and the fundamental procedure is that of replacing the differential quantities (dy, dt, etc.) by small, but not infinitesimal, magnitudes (Δy, Δt, etc.).

The example of a transient heat-transfer process will illustrate the procedure. The process to be simulated is that of a mass of material (perhaps a temperature-sensing bulb) in a fluid stream, as in Fig. 17-1. The conductivity of the bulb material is assumed to be high so that the temperature throughout the bulb is uniform at any time. The temperature of the bulb θ is to be computed following a step change in the fluid temperature, θ_f. The applicable differential equation developed from an energy balance is

$$(\theta_f - \theta)hA = Mc\frac{d\theta}{dt} \qquad (17\text{-}1)$$

17-2. Numerical Simulation

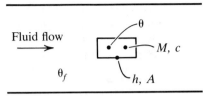

Fig. 17-1. A temperature-sensing bulb in a fluid stream.

where

θ = temperature of bulb, °C
θ_f = fluid temperature, °C
M = mass of bulb, kg
c = specific heat of bulb, kJ/(kg-K)
h = convection heat-transfer coefficient, kW/(m²-K)
A = area for convection heat transfer, m²

Assume that θ_f has been constant at 30°C for a long time such that the temperature of the bulb has also stabilized at 30°C. At time $t = 0$, θ_f undergoes a step change to 40°C. The magnitudes of the other physical terms in Eq. (17-1) are: $M = 0.03$ kg; $c = 0.6$ kJ/(kg-K); $h = 0.09$ kW/(m²-K); and $A = 0.01$ m². With these values, Eq. (17-1) becomes

$$\frac{d\theta}{dt} = \frac{(0.09)(0.01)}{(0.03)(0.6)}(\theta_f - \theta) = \frac{1}{20}(\theta_f - \theta) \tag{17-2}$$

The time constant of the process is 20 s, and to provide the basis of comparison for the numerical simulation this simple differential equation can be solved in closed form:

$$\theta = 30 + (40 - 30)(1 - e^{-t/20}) \tag{17-3}$$

The numerical simulation can be represented by a calculation loop that is similar to a control loop, Fig. 17-2.

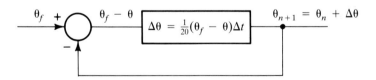

Fig. 17-2. Numerical simulation process for transient heat-transfer process.

A straightforward approach to a numerical simulation is to convert Eq. (17-1) to a finite difference by replacing $d\theta$ with $\Delta\theta$ and dt with Δt. In this technique, called the Euler method, the rate of temperature change at the start of the time step is assumed to prevail throughout the entire step,

$$\frac{\Delta\theta}{\Delta t} = \frac{hA}{Mc}(\theta_f - \theta_{\text{start}}) \tag{17-4}$$

If a time step of 1 s is chosen, the change of temperature in the first time increment will be

$$\Delta\theta = \frac{(0.09)(0.01)}{(0.03)(0.6)}(40-30)(1\,\text{s}) = 0.5°\text{C}$$

The temperature after 1 s is therefore $30 + 0.5 = 30.5°\text{C}$. After 2 s the temperature is

$$\theta = 30.5 + (1/20)(40-30.5) = 30.975°\text{C}$$

Figure 17-3 shows the exact values of θ computed from Eq. (17-3) and the error resulting from the numerical simulation with a 1 s time step. The numerical simulation is not exact, because, as Fig. 17-4 shows, the numerical simulation assumes that $\Delta\theta/\Delta t$ remains constant over the entire step. This assumption is in error because the actual temperature difference, $\theta_f - \theta$, decreases progressively during the step, and the assumption of a constant rate of change overestimates the change in θ. The exact equation for the temperature of the object indicates θ after 1 s to be $30.4877°\text{C}$, rather than $30.5000°\text{C}$ as calculated from the numerical simulation.

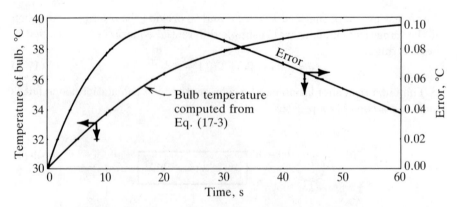

Fig. 17-3. Temperature history of the bulb in Fig. 17-1 following a step change in fluid temperature from 30 to 40°C and error if 1 s time steps are used in the numerical simulation.

17-2. Numerical Simulation

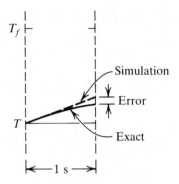

Fig. 17-4. Error resulting from assumption of the rate of change being constant over the entire step.

A direct approach to reducing the error in the numerical simulation is to shrink the size of the time step. If a step size of 0.1 s is chosen, the simulation computes a temperature after 1 s of 30.4889, which is a noticeable improvement over the results of the 1 s step size. The accuracy of the numerical simulation can usually be improved by decreasing the step size, because the integration moves toward the exact situation of the infinitesimal step size.

The objective of numerous possible refinements to the Euler method, such as Runge-Kutta, Adams predictor-corrector, etc., is to maintain good accuracy while using a large step size. An example of how a modest sophistication improves the accuracy of the numerical simulation being explored here is to use the second-degree term in the Taylor's series,

$$y_{n+1} = y_n + \frac{y'}{1!}\Delta t + \frac{y''}{2!}(\Delta t)^2 + \underbrace{\cdots}_{\text{neglect}} \tag{17-5}$$

where

y_n = value of independent variable at current time
y_{n+1} = value of independent variable at one time step in the future
y' = first derivative at n
y'' = second derivative at n

Equation (17-5) is precisely the Euler simulation of Eq. (17-4) with the addition of the second-degree term. To obtain the second derivative of θ in the transient heat-transfer process, differentiate the first derivative

$$\frac{d}{dt}\left(\frac{d\theta}{dt}\right) = \frac{d}{dt}\left(\frac{\theta_f - \theta}{20}\right) = -\frac{1}{20}\frac{d\theta}{dt} = -\frac{1}{20}\left(\frac{\theta_f - \theta}{20}\right) = -\frac{\theta_f - \theta}{400}$$

When the second-degree term is included, the temperature after one 1 s time step is

$$\theta = 30 + \frac{40 - 30}{20}(1\,\text{s}) - \frac{(40 - 30)}{(400)2!}(1\,\text{s})^2 = 30.4875°\text{C}$$

This result is much closer to the exact value of 30.4877°C than the 1 s Euler calculation, which yielded 30.5000°C.

This section on numerical simulation now closes with an evaluation that attempts to put into perspective numerical simulation techniques, at least as we might want to use them for this chapter. An important question is whether or not to program one of the refined integration techniques or simply to shorten the step size to achieve satisfactory accuracy. The trade-off is generally between programming time and computer time. A refined integration technique, while it does introduce a few additional calculations, also permits much longer time steps. The net result is a reduction in computer time. Programming one of the refined techniques requires additional programming time, particularly in the treatment of some non-linearities such as will be encountered later in the chapter.

We propose to use the simple Eulerian integration with short time steps and take the penalty on computer time. What is a short time step? We should be aware that in control loops there may be several time-dependent elements in series, which means that the changes in the input variable to one element may have to pass through other elements before returning to the original one. Also, since the appropriate choice of time step is influenced by the speed of response (the time constant, for example), the choice of time step should adequately simulate the fastest element in the control loop. These influences combine to suggest erring on the side of short time steps.

An engineer in professional practice may need many reruns of a program to simulate a product or a system under study. In that case the investment in programming time to develop a more sophisticated integration routine may pay for itself in reduced computer time and quicker turnaround time following submission of the program.

17-3 Sampled Data

The microcomputer is typically responsible for multiple tasks, including directing several control loops, calculations, and managing displays on a screen

17-3. Sampled Data

or printing. Figure 17-5 shows a microcomputer controller with the responsibility for several sensors and actuators that are distributed throughout several systems.

The sequence of events that pertains to system n might be as shown in Fig. 17-6a, where at one instant the microcomputer devotes itself to sensing a controlled variable of the system, a short time later processes the value of this variable in the control algorithm, and then later transmits an instruction to an actuator in system n. It would be quite orderly to perform all three operations—sensing, deciding, and transmitting—in sequence, as in Fig. 17-6b, before going on to other duties. In either pattern of Fig. 17-6,

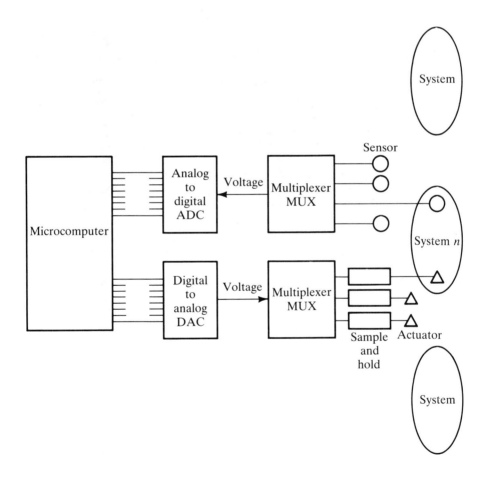

Fig. 17-5. A microcomputer with multiple duties.

364   Chapter 17. The Computer and Its Sampling Processes

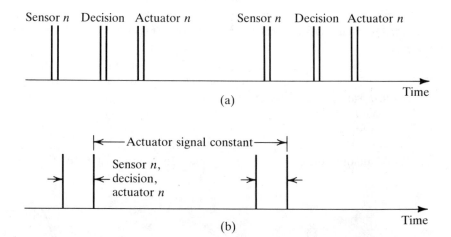

Fig. 17-6. Sequence of sensing, deciding, and transmitting the control signal (a) if the operations are interrupted by other duties of the microcomputer, and (b) if performed in sequence without interruption.

the sample-and-hold operation delivers a constant signal to the actuator until updated by the next cycle, as shown in Fig. 17-7. Holding the signal constant throughout the sampling interval is called zero-order hold and is almost a universal practice in computer control. Other possible holding concepts, as shown in Fig. 17-8, include first-order hold, in which the variable changes linearly during the sampling interval based on an extrapolation from the previous two samples.

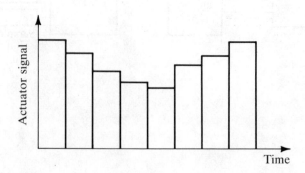

Fig. 17-7. A series of short-term constant signals sent to the actuator.

17-4. Responses to Sampled Values

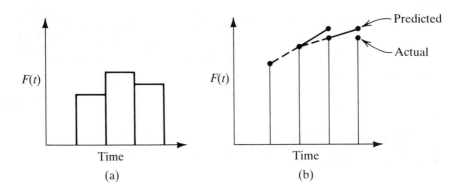

Fig. 17-8. Two different holding operations: (a) zero order and (b) first order.

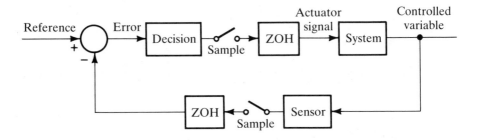

Fig. 17-9. Diagram of a sampled-data control loop corresponding to the sequences of Fig. 17-6.

The conventional symbolism of sampled-data systems corresponding to Fig. 17-6 is shown in Fig. 17-9. The sampling operation is indicated by an open switch which picks off an instantaneous value of a continuous function and feeds it to the holding operation, which here is zero-order (ZOH). The controlled variable can vary continuously, so the value obtained by sampling the sensor may be different in the sequences of Figs. 17-6a compared to Fig. 17-6b.

17-4 Responses to Sampled Values

A feature of sampled data is that its starting point may be at a time other than zero, and the function may not extend to infinity. This section develops the Laplace transforms of a few simple sampled forms and shows an inversion back to the time domain. A fundamental function of time t shown in Fig. 17-10 has a zero value until $t = k$ and then takes on a constant value of r. The Laplace transform of the function can be obtained by substituting into the definition of the Laplace transform,

$$\mathcal{L}\{F(t)\} = \int_k^\infty F(t)\, e^{-st}\, dt = \int_k^\infty r\, e^{-st}\, dt = \frac{r}{s} e^{-ks} \qquad (17\text{-}6)$$

The transforms of this and several other important functions that can be found by integration are shown in Table 17-1.

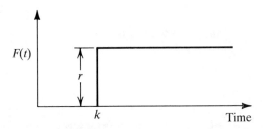

Fig. 17-10. A function of constant value that begins at $t = k$.

Example 17-1. Compute the temperature history of the sensing bulb in Fig. 17-1 if its time constant is 2.5 s and if the temperature of the fluid is at 30°C except for the time between 2 and 3 s when its value is 38°C.

Solution. Figure 17-11 shows the input signal and the transfer function of the process. One approach to solving this simple process is to take advantage of the knowledge of exponential response and predict the behavior of the bulb temperature as shown in Fig. 17-12. The three sections of the response are:

For $0 < t < 2$, $\theta = 30$
For $2 < t < 3$, $\theta = 30 + 8\left(1 - e^{-(t-2)/2.5}\right)$
For $3 < t$, $\theta = 32.6374 - 2.6374\left(1 - e^{-(t-3)/2.5}\right)$

In the time period between 2 and 3 s, the bulb temperature changes exponentially from 30°C toward 38°C, reaching 32.6374°C at 3 s. At this time the fluid temperature drops back to 30°C and the bulb temperature changes exponentially from 32.6374°C toward 30.0°C.

17-4. Responses to Sampled Values

Table 17-1. Laplace Transforms of Several Functions that Start at a Non-zero Time

$F(t)$ (step of height r starting at k)	$\dfrac{r}{s}e^{-ks}$
$F(t)$ (pulse of height r from k to m)	$\dfrac{r}{s}\left(e^{-ks} - e^{-ms}\right)$
$F(t)$ ($re^{-a(t-k)}$ starting at k)	$\dfrac{r}{s+a}e^{-ks}$
$F(t)$ ($r[1 - e^{-a(t-k)}]$ starting at k)	$e^{-ks}\left(\dfrac{r}{s} - \dfrac{r}{s+a}\right) = \dfrac{rae^{-ks}}{s(s+a)}$
$F(t)$ (ramp of slope b starting at k)	$\dfrac{be^{-ks}}{s^2}$
$F(t)$ (ramp of slope b from k to m, then constant)	$\dfrac{b}{s^2}\left(e^{-ks} - e^{-ms}\right)$

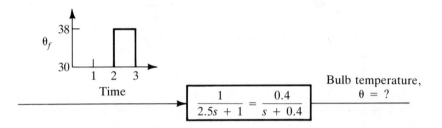

Fig. 17-11. Pulse input of fluid temperature supplied to the temperature-sensing bulb in Example 17-1.

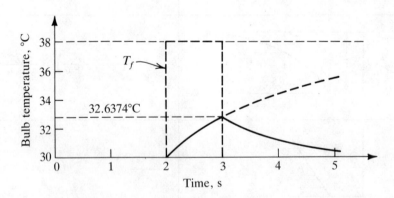

Fig. 17-12. Response of the bulb temperature in Example 17-1.

17-4. Responses to Sampled Values

Table 17-2. Values of θ in Example 17-1 at Several Values of t

Time, s	Physical Model	From Eq. (17-7)
2	30.0000	30.0000
3	32.6374	32.6374
4	31.7679	31.7679
6	30.7944	30.7944
10	30.1604	30.1604

The solution can also be obtained by inverting the transform of the output, which is the product of the transform of the input,

$$\frac{8}{s}\left(e^{-2s} - e^{-3s}\right)$$

and the transfer function $0.4/(s + 0.4)$,

$$\theta - 30 = \mathcal{L}^{-1}\left\{\frac{8}{s}e^{-2s}\left(\frac{0.4}{s+0.4}\right) - \frac{8}{s}e^{-3s}\left(\frac{0.4}{s+0.4}\right)\right\}$$

From Table 17-1,

$$\theta - 30 = \underbrace{8\left[1 - e^{-0.4(t-2)}\right]}_{\text{for } t > 2} - \underbrace{8\left[1 - e^{-0.4(t-3)}\right]}_{\text{for } t > 3} \qquad (17\text{-}7)$$

Table 17-2 shows a comparison of θ at several values of t computed using the physical model of Fig. 17-12 and using Eq. (17-7).

If instead of the input of the single rectangular pulse of Fig. 17-11 the input were two pulses, as shown in Fig. 17-13, the output would be the same as shown in Example 17-1 until $t = 6$ s. Thereafter, because the controlling differential equation is linear, the contribution of the second pulse would be added to the contribution of the first pulse.

The concept of combining rectangular pulses can be extended to a continuous series of pulses, such as those shown in Fig. 17-7. The treatment of such a series reduces to the summation of the influence of each pulse following its occurrence. It will certainly be convenient to develop a closed-form representation of the transform of a series, and this is one of the strengths of the z-transform approach explained in the next section.

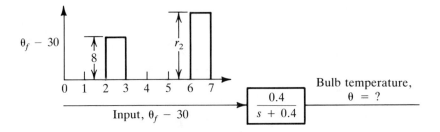

Fig. 17-13. Input signal of two rectangular pulses.

17-5 The z-Transform

The Laplace transform has already proved its value in processing differential equations, including their application to dynamic physical processes and the response of control systems. The classical Laplace transform applies to functions of time that have continuous derivatives. In computer control systems the data are discrete because of the sampling process shown symbolically in Fig. 17-14a. When the sampling process of time interval T is applied to a continuous function in Fig. 17-14b, the result is a series of values shown in Fig. 17-14c. The assembly of the samples is designated $F^*(t)$. We shall only consider the case that is almost universal where the sampling interval is uniform.[1]

To find the Laplace transform of the sample $F^*(t)$ shown in Fig. 17-14c, the first step is to interpret a data sample as an "impulse." If the sample were a finite value lasting for zero time, the integral resulting from the definition of the Laplace transform would be zero. Instead, as Fig. 17-15 shows, the impulse is provided an area equal to the value r of the variable at time t_0 by letting it be a narrow sliver of thickness ϵ and height r/ϵ. Ultimately ϵ will be shrunk to zero to provide the limit value of the transform. Next apply the definition of the Laplace transform to this impulse,

$$\mathcal{L}\{F(t)\} = \int_0^\infty F(t)e^{-st}\,dt = \int_{t_0}^{t_0+\epsilon} \frac{r}{\epsilon} e^{-st}\,dt \qquad (17\text{-}8)$$

$$\mathcal{L}\{F(t)\} = -\frac{r}{\epsilon}\left(\frac{e^{-st_0}e^{-s\epsilon} - e^{-st_0}}{s}\right)$$

Express $e^{-s\epsilon}$ as an infinite series of $s\epsilon$, and cancel where possible:

$$\mathcal{L}\{F(t)\} = -re^{-st_0}\left(-1 + \frac{s\epsilon}{2!} + \cdots\right)$$

17-5. The z-Transform

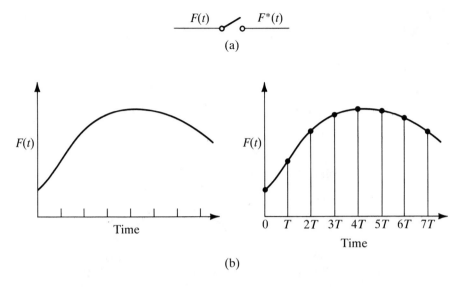

Fig. 17-14. (a) The sampling process, (b) a continuous function, and (c) a series of samples taken from the continuous function.

Finally, letting $\epsilon \to 0$ gives the Laplace transform of the impulse r,

$$\mathcal{L}\{F(t)\} = re^{-st_0} \tag{17-9}$$

Having developed the expression for the Laplace transform of one impulse, the next step is to sum the transforms of all the impulses. For a series of

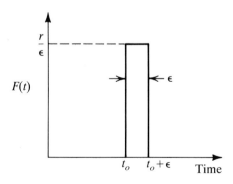

Fig. 17-15. An impulse that provides an area r at time t_0.

impulses whose r values are set by a function, such as in Fig. 17-14c, the summation of the impulses is

$$\mathcal{L}\{F^*(t)\} = \sum_{k=0}^{\infty} F(kT)e^{-kTs} \tag{17-10}$$

Make one cosmetic change, the substitution

$$z = e^{Ts} \tag{17-11}$$

following which Eq. (17-10) leads to the definition of the z-transform

$$Z\{F(t)\} = \sum_{k=0}^{\infty} F(kT)z^{-k} \tag{17-12}$$

In summary, the z-transform of $F(t)$ is the summation of the Laplace transforms of the impulses of magnitude $F(kT)$ occurring at time intervals of T.

For purposes of illustration the z-transform will now be developed for two functions, $F(t) = c$ and $F(t) = bt$. The function that is a constant value c, as shown in Fig. 17-16, has a z-transform of

$$Z\{c\} = \sum_{k=0}^{\infty} cz^{-k} = c\left(1 + z^{-1} + z^{-2} + \cdots\right)$$

The term in the parentheses matches the closed form of the geometric series

$$\frac{1}{1-x} = 1 + x + x^2 + \cdots \qquad \text{if } x < 1$$

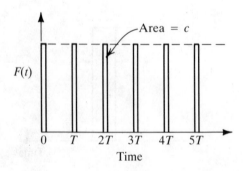

Fig. 17-16. The z-transform of a constant c.

17-5. The z-Transform

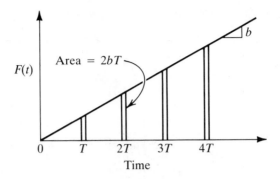

Fig. 17-17. The z-transform of a linear function of time.

so

$$Z\{c\} = c\frac{1}{1-z^{-1}} = \frac{cz}{z-1} \quad (17\text{-}13)$$

The second function for which the z-transform will be determined is bt, as shown in Fig. 17-17.

$$Z\{bt\} = \sum_{k=0}^{\infty} b(kT)z^{-k} = bT\left(0 + \frac{1}{z} + \frac{2}{z^2} + \cdots\right)$$

The closed form of the series in the parentheses is

$$\frac{z^{-1}}{(1-z^{-1})^2} \quad (17\text{-}14)$$

The validity of the closed-form representation can be verified by expanding Eq. (17-14) as a Maclaurin's series,

$$g(z^{-1}) = \frac{Tz^{-1}}{(1-z^{-1})^2} = a_0 + a_1 z^{-1} + a_2 z^{-2} + \cdots$$

Then $a_0 = g\big|_{z^{-1}=0}$, $\quad a_1 = \frac{1}{1!}g'\big|_{z^{-1}=0}$, $\quad a_2 = \frac{1}{2!}g''\big|_{z^{-1}=0}$, $\quad$ etc.

Thus, the z-transform of bt is

$$Z\{bt\} = \frac{bTz^{-1}}{(1-z^{-1})^2} = \frac{bTz}{(z-1)^2} \quad (17\text{-}15)$$

Table 17-3 is a compilation of z-transforms of a number of frequently encountered functions. For convenience the corresponding Laplace transforms are also shown.

Table 17-3. Table of z-Transforms with a Time Interval of T

$f(s)$	$F(t)$	$F(z)$
—	$1, t=0; 0, t \neq 0$	1
—	$1, t=k; 0, t \neq k$	z^{-k}
$\dfrac{1}{s}$	1	$\dfrac{z}{z-1}$
$\dfrac{1}{s^2}$	t	$\dfrac{Tz}{(z-1)^2}$
$\dfrac{1}{s^3}$	$\dfrac{t^2}{2!}$	$\dfrac{T^2 z(z+1)}{2(z-1)^3}$
$\dfrac{1}{s^4}$	$\dfrac{t^3}{3!}$	$\dfrac{T^3 z(z^2+4z+1)}{6(z-1)^4}$
$\dfrac{1}{s^m}$	$\lim_{a \to 0} \dfrac{(-1)^{m-1}}{(m-1)!} \dfrac{\partial^{m-1}}{\partial a^{m-1}} e^{-at}$	$\lim_{a \to 0} \dfrac{(-1)^{m-1}}{(m-1)!} \dfrac{\partial^{m-1}}{\partial a^{m-1}} \dfrac{z}{z - e^{-aT}}$
$\dfrac{1}{s+a}$	e^{-at}	$\dfrac{z}{z - e^{-aT}}$
$\dfrac{1}{(s+a)^2}$	$t e^{-at}$	$\dfrac{Tze^{-aT}}{(z - e^{-aT})^2}$
$\dfrac{1}{(s+a)^3}$	$\dfrac{1}{2} t^2 e^{-at}$	$\dfrac{T^2}{2} \dfrac{e^{-aT} z(z + e^{-aT})}{(z - e^{-aT})^3}$
$\dfrac{1}{(s+a)^m}$	$\dfrac{(-1)^{m-1}}{(m-1)!} \dfrac{\partial^{m-1}}{\partial a^{m-1}} e^{-at}$	$\dfrac{(-1)^{m-1}}{(m-1)!} \dfrac{\partial^{m-1}}{\partial a^{m-1}} \dfrac{z}{z - e^{-aT}}$
$\dfrac{a}{s(s+a)}$	$1 - e^{-at}$	$\dfrac{z(1 - e^{-aT})}{(z-1)(z - e^{-aT})}$

17-5. The z-Transform

Table 17-3. Table of z-Transforms with a Time Interval of T (continued)

$f(s)$	$F(t)$	$F(z)$
$\dfrac{a}{s^2(s+a)}$	$\dfrac{1}{a}(at-1+e^{-at})$	$\dfrac{z\left[(aT-1+e^{-aT})z+(1-e^{-aT}-aTe^{-aT})\right]}{a(z-1)^2(z-e^{-aT})}$
$\dfrac{b-a}{(s+a)(s+b)}$	$e^{-at}-e^{-bt}$	$\dfrac{(e^{-aT}-e^{-bT})z}{(z-e^{-aT})(z-e^{-bT})}$
$\dfrac{s}{(s+a)^2}$	$(1-at)e^{-at}$	$\dfrac{z\left[z-e^{-aT}(1+aT)\right]}{(z-e^{-aT})^2}$
$\dfrac{a^2}{s(s+a)^2}$	$1-e^{-at}(1+at)$	$\dfrac{z\left[z\left(1-e^{-aT}-aTe^{-aT}\right)+e^{-2aT}-e^{-aT}+aTe^{-aT}\right]}{(z-1)(z-e^{-aT})^2}$
$\dfrac{(b-a)s}{(s+a)(s+b)}$	$be^{-bt}-ae^{-at}$	$\dfrac{z\left[z(b-a)-(be^{-aT}-ae^{-bT})\right]}{(z-e^{-aT})(z-e^{-bT})}$
$\dfrac{a}{s^2+a^2}$	$\sin at$	$\dfrac{z\sin aT}{z^2-(2\cos aT)z+1}$
$\dfrac{s}{s^2+a^2}$	$\cos at$	$\dfrac{z(z-\cos aT)}{z^2-(2\cos aT)z+1}$
$\dfrac{s+a}{(s+a)^2+b^2}$	$e^{-at}\cos bt$	$\dfrac{z\left(z-e^{-aT}\cos bT\right)}{z^2-2e^{-aT}(\cos bT)z+e^{-2aT}}$
$\dfrac{b}{(s+a)^2+b^2}$	$e^{-at}\sin bt$	$\dfrac{ze^{-aT}\sin bT}{z^2-2e^{-aT}(\cos bT)z+e^{-2aT}}$

17-6 Response to a Series of Impulses

Before proceeding to the application of z-transforms for predicting the performance in the time domain and determining stability of a control loop, we pause momentarily to seek a physical visualization of the series of impulses. To choose a physical example, suppose that the temperature-sensing bulb of Fig. 17-1 (reproduced in Fig. 17-18a) having a time constant of 2.5 s is subjected to a series of temperature pulses as shown in Fig. 17-18b. The initial temperature of the bulb is 30°C, and the fluid temperature is at 30°C except for the impulses to 38°C every second. The assignment is to develop the time history of the bulb temperature θ.

If a constant value of θ prevailed over the entire time interval, such as in Fig. 17-7, even though θ changes in magnitude every time interval, the physical process could readily be visualized. The challenge in dealing with the impulses is to interpret the meaning of a pulse of 8°C occurring every second. Would a flash of 38°C fluid every second change θ? No, it would not. The definition of the impulse as shown in Fig. 17-15, however, indicates that the impulse is a rectangle with an area for which the base of the rectangle shrinks to zero. As an integral the impulse has units of °C-s and could be envisioned as a pulse of heat Q kJ injected into the fluid stream every second,

$$\text{Impulse, (°C)(s)} = \frac{Q \text{ kJ}}{(m \text{ kg/s})[c \text{ kJ}/(\text{kg-°C})]} = 8°\text{C-s}$$

If the pulse of heat Q were spread out over 1 s as in Fig. 17-19a, the response would be exponential, as in Fig. 17-12, and the temperature after 1 s would be $30 + 2.637 = 32.637°$C. If the pulse of heat were confined to 1/2 s, as in

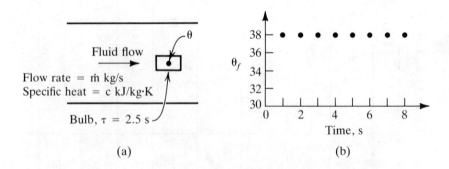

Fig. 17-18. (a) Temperature-sensing bulb subjected to (b) a series of 8°C impulses occurring every second.

17-6. Response to a Series of Impulses

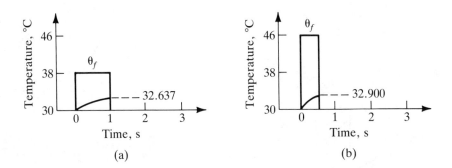

Fig. 17-19. Pulses of heat applied to the fluid stream giving (a) $\theta_f - \theta = 8°C$ for 1 s, and (b) $\theta_f - \theta = 16°C$ for 0.5 s.

Fig. 17-19b, the bulb temperature would be 32.90°C after 1/2 s. In general,

$$\theta - \theta_f = \frac{8}{\epsilon}\left(1 - e^{-\epsilon/\tau}\right) = \frac{8}{\epsilon}\left[1 - \left(1 - \frac{\epsilon}{\tau} + \frac{\epsilon^2}{2\tau^2} - \cdots\right)\right] \qquad (17\text{-}16)$$

where ϵ is the duration of the pulse. If Q is added as an impulse during a negligible time duration, ϵ shrinks to zero, and Eq. (17-16) reduces to

$$\theta - \theta_o = \frac{8}{\tau} = \frac{8}{2.5} = 3.2°C \qquad (17\text{-}17)$$

If these pulses of heat occur every second, the bulb temperature follows the jagged pattern shown in Fig. 17-20, increasing suddenly at the time of the impulse and then decaying exponentially toward 30°C between pulses. After a long period of time the bulb temperature oscillates between 36.505 and 39.706°C.

The intent of the above example is to provide some insight into the impulses that are reflected in the z-transform. The sampler symbol in Fig. 17-14a and the typical representation of the sampled function in Fig. 17-14c imply quick flashes of values. The z-transform, however, is built on a series of areas, as indicated in Figs. 17-16 and 17-17. Not all the questions of interpretation are solved, however, because the sampling process that takes place in the microcomputer control of Fig. 17-5 seems to be properly represented by the symbolism in Fig. 17-14. The major resolution of this conflict is through the realization that in the systems with which we deal some kind of holding operation always follows the sampler. This holding operation will now be combined with the sampling process.

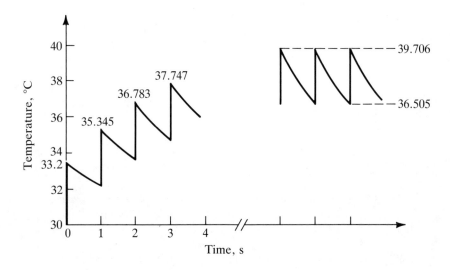

Fig. 17-20. Response of the sensing bulb to 8°C impulses spaced at 1 s intervals.

17-7 The Zero-Order Hold (ZOH)

Two different holding operations were explained in Fig. 17-8, namely, the zero-order and first-order. Almost without exception in computer control work the sampler is followed by the holding operation, and that type of hold is the ZOH. In the functional diagram of the computer controller of Fig. 17-5, the sample-and-hold circuit provides for the actuators, and ZOH of the sensor values occurs because the computer reads or uses the sensor value only once a calculation cycle. The Laplace transform of the function shown in Fig. 17-21, whose samples are subjected to ZOH, is

$$\mathcal{L}\left\{\begin{array}{c} F(t) \\ \text{w/ZOH} \end{array}\right\} = \mathcal{L}\left\{F(t)\right\}\bigg|_{0-T} + \mathcal{L}\left\{F(t)\right\}\bigg|_{T-2T} + \mathcal{L}\left\{F(t)\right\}\bigg|_{2T-3T} + \cdots$$

and

$$\mathcal{L}\left\{\begin{array}{c} F(t) \\ \text{w/ZOH} \end{array}\right\} = \int_0^T F(0)e^{-st}dt + \int_T^{2T} F(T)e^{-st}dt + \int_{2T}^{3T} F(2T)e^{-st}dt + \cdots \quad (17\text{-}18)$$

$$\mathcal{L}\left\{\begin{array}{c} F(t) \\ \text{w/ZOH} \end{array}\right\} = F(0)\left(\frac{1-e^{-sT}}{s}\right) + F(T)\left(\frac{e^{-sT}-e^{-2sT}}{s}\right) + \cdots$$

17-7. The Zero-Order Hold (ZOH)

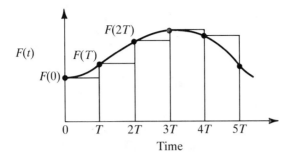

Fig. 17-21. Function with ZOH applied to samples.

$$\mathcal{L}\left\{\begin{array}{c} F(t) \\ \text{w/ZOH} \end{array}\right\} = [F(0) + F(T)e^{-sT} + F(2T)e^{-2sT} + \cdots]\left(\frac{1 - e^{-sT}}{s}\right) \tag{17-19}$$

Comparison of the series in the brackets in Eq. (17-19) with Eq. (17-12) shows it to be the z-transform of $F(t)$, $Z\{F(t)\}$, so

$$\mathcal{L}\{F(t)\} = Z\{F(t)\}\frac{1 - e^{-sT}}{s} \tag{17-20}$$

In Eq. (17-18) the values of $F(0)$, $F(T)$, $F(2T)$, etc. were straightforward numerical values of the function at time intervals that would occur physically with the momentary closing of the sampler. In Eq. (17-19) a term was transferred from each integral to form the series in the brackets that is the sum of areas that is the z-transform.

Example 17-2. What is the Laplace transform of a sampled linear function of time, bt, combined with a ZOH (bt w/ZOH)?

Solution. Figure 17-22 shows the linear function as bt, sampled at intervals of T with these values held until the next sample. One approach is to sum the transforms of the series of pulses of function 2 in Table 17-1:

$$\mathcal{L}\{bt \text{ w/ZOH}\} = \frac{bT}{s}\left(e^{-sT} - e^{-2sT}\right)$$
$$+ \frac{2bT}{s}\left(e^{-2sT} - e^{-3sT}\right) + \cdots$$

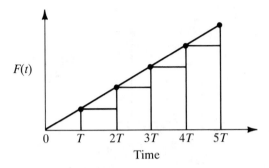

Fig. 17-22. Linear function with ZOH in Example 17-2.

$$\mathcal{L}\{bt \text{ w/ZOH}\} = \frac{bT}{s}\left(e^{-sT} + e^{-2sT} + e^{-3sT} + \cdots\right)$$

which is a series of the form $1/(1-x) = 1 + x + x^2 + \cdots$, so

$$\mathcal{L}\left\{\begin{array}{c} bT \\ \text{w/ZOH} \end{array}\right\} = \frac{bT}{s}\left(\frac{1}{1-e^{-sT}} - 1\right) = \frac{bT}{s}\left(\frac{e^{-sT}}{1-e^{-sT}}\right)$$

$$= \frac{bT}{s(z-1)}$$

The alternative approach is to combine the z-transform of bt from Eq. (17-15) with $(1-e^{-sT})/s$,

$$\mathcal{L}\left\{\begin{array}{c} bT \\ \text{w/ZOH} \end{array}\right\} = \frac{bTz}{(z-1)^2}\left(\frac{1-e^{-sT}}{s}\right)$$

$$= \frac{bTz}{(z-1)^2}\left(\frac{(1-z^{-1})}{s}\right)$$

$$= \frac{bT}{s(z-1)}$$

17-8 Inverting a z-Transform

One of the emphases of both Chapter 16 and this chapter has been to examine the behavior of computer control systems in the time domain to provide physical insight into the behavior of dynamic processes. For a continuous controller this objective necessitates inversion of the Laplace transform, while

17-8. Inverting a z-Transform

for the computer controller with its sampling process, inversion from the z-transform is required. Three procedures to be explained here are (1) table lookup, (2) long division when the transform is a ratio of two polynomials, and (3) decomposition using partial fractions followed by table lookup or long division.

Table lookup is a natural approach to inversion, because if the expression to be inverted is the same as one of the transforms shown in Table 17-3, the corresponding function of time that resulted in the z-transform in the first place is readily available.

The division process of two polynomials should result in a series of terms with ascending powers of z^{-1} that can be translated into numerical values at each of the T intervals.

Example 17-3. Invert by division and subsequent inversion of the series the z-transform $bTz/(z-1)^2$.

Solution. A table lookup would have shown this z-transform to have originated from the linear function of time, bt. The original values at the sampling times were, consequently, 0, bT, $2bT$, etc. For purposes of illustration we shall perform the inversion by division and subsequent inversion. The division is

$$
\begin{array}{r}
bTz^{-1} + 2bTz^{-2} + 3bTz^{-3} + \cdots \\
z^2 - 2z + 1 \overline{\smash{\big)}\, bTz} \\
bTz \quad - \quad 2bT \quad + \quad bTz^{-1} \\
2bT \quad - \quad bTz^{-1} \\
2bT \quad - \quad 4bTz^{-1} \quad + \quad 2bTz^{-2} \\
3bTz^{-1} \quad - \quad 2bTz^{-2}
\end{array}
$$

Reference back to Eqs. 17-9 and 17-11 for the unit impulse on which the z-transform is based indicates that the coefficient of z^{-n} is the value of the original function in the time domain at time nT.

The magnitude of the original function at $t = 0$ is the coefficient of z^0, which is zero. The succeeding magnitudes for $t = T, 2T, 3T$, etc. are $bT, 2bT, 3bT$, etc., respectively. These values correspond to bt at the time intervals $T, 2T, 3T$, etc.

The third method of inversion is to decompose the original transform by partial fractions, developing expressions that appear on Table 17-3.

Example 17-4. Invert

$$\frac{Tz^{-1}}{(1-z^{-1})(1-0.5z^{-1})}$$

by first decomposing using partial fractions.

Solution.

$$\frac{Tz^{-1}}{(1-z^{-1})(1-0.5z^{-1})} = \frac{A}{(1-z^{-1})} + \frac{B}{(1-0.5z^{-1})}$$

$$A - 0.5Az^{-1} + B - Bz^{-1} = Tz^{-1}$$
$$A + B = 0$$
$$-0.5A - B = T$$

so $A = 2T$, and $B = -2T$, and the alternative form of the expression to be inverted is $2T/(1-z^{-1}) - 2T/(1-0.5z^{-1})$, which can be written as two geometric series, $2T(1 + z^{-1} + z^{-2} + \cdots)$ and $-2T(1 + 0.5z^{-1} + 0.25z^{-2} + 0.125z^{-3} + \cdots)$ and combined to $2T(0.5z^{-1} + 0.75z^{-2} + 0.875z^{-3} + \cdots)$. Inverting the series gives the values at $t = 0, T, 2T$, etc. as $0, T, 1.5T, 1.75T$, etc., respectively.

17-9 Cascading z-Transforms and Transforms of a Feedback Loop

In the Laplace transform domain, processes are characterized by their transfer function, as in Fig. 17-23, where the transfer function was defined as the ratio $O(s)/I(s)$. The expression for a TF(s) can be determined by taking

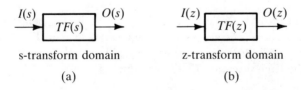

Fig. 17-23. Transfer functions in the s and z planes.

17-9. Cascading z-Transforms

the Laplace transform of the equation (usually a differential equation) that represents the process and solving for $O(s)/I(s)$ as illustrated in Sec. 16-14.

The transfer function in the z-plane is defined in a parallel fashion, namely, the ratio of the output z-transform to the input z-transform, $O(z)/I(z)$. Both the input and output to the process are the series of impulses that make up the z-transform. The next question is how to determine TF(z) for a known process. The answer is that TF(z) is the z-transform that corresponds to the s-transform for the same process. Thus TF(z) for frequently encountered processes appears opposite the TF(s) in Table 17-3.

Example 17-5. For the heat pulses applied to θ_f in Fig. 17-20, invert the z-transform to determine the bulb temperature as a function of time.

Solution. The input $I(z)$ is the transform of the series of 8°C impulses, so from Table 17-3 $I(z) = 8z/(z-1)$. The Laplace transform of the process is $1/(2.5s + 1) = 0.4/(s + 0.4)$, so the corresponding TF(z) from Table 17-3 is $0.4z/(z - e^{-0.4T})$. The z-transform of the output is the product

$$O(z) = I(z)\text{TF}(z) = \frac{8z}{z-1}\left(\frac{0.4z}{z - e^{-0.4T}}\right)$$

The time interval for the process T is 1 s, so

$$O(z) = \frac{3.2z^2}{(z-1)(z-0.6703)}$$

When $O(z)$ is converted to a series of powers of z^{-1} by long division, the inversions of the individual terms are as shown in Table 17-4.

The results shown in Table 17-4 check those presented in Fig. 17-20 for the process that had to be represented by a series of heat pulses in order to obtain a physical interpretation. The z-transform inversion provided the values immediately following the pulsing or sampling, and, as is typical of the z-transform, provided values only at the sampling intervals and supplied no information about what was occurring between samples.

Before leaving this example, it is appropriate to compute the response if the sampling interval were other than the 1 s used in the example. If the sampling interval were 2 s, for example, T would be assigned 2 in Table 17-4, yielding temperatures of 33.2, 34.638,

Table 17-4. Inverting the Transforms in Example 17-5

n	Time, s	Coefficient of z^{-n}	Value if $T = 1$	Temp. at nT
0	0	3.2	3.2	33.2
1	1	$3.2(1 + e^{-0.4T})$	5.345	35.345
2	2	$3.2[(1 + e^{-0.4T})^2 - e^{-0.4T}]$	6.783	36.783
3	3	$3.2[(1 + e^{-0.4T})^3 - 2(1 + e^{-0.4T})e^{-0.4T}]$	7.747	37.747

35.284, and 35.574°C at 0, 2, 4, and 6 s, respectively. These values may be verified by the technique illustrated in Fig. 17-20 with successive 3.2°C jumps followed by 2 s decays.

The next illustration of cascading z-transforms is one that combines a sampler, ZOH, and a time constant block, as shown in Fig. 17-24. The assembly receives a continuous function $I(t)$ and delivers a continuous function $O(t)$, which are exponential responses to a series of rectangles.

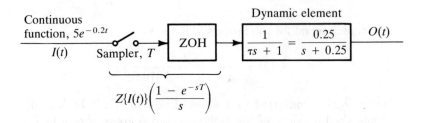

Fig. 17-24. Cascaded transfer function including a sampler and a ZOH.

Example 17-6. What is the output $O(t)$ in the time domain at 0, 2, 4, and 6 s of the cascaded assembly in Fig. 17-24 if $I(t)$ is $5e^{-0.2t}$, the sampling interval $T = 2$ s, and the dynamic element is a time constant block with $\tau = 4$ s.

Solution. The Laplace transform of the combination of the sampler and ZOH is, from Eq. 17-20,

$$Z\{I(t)\}\left(\frac{1 - e^{-sT}}{s}\right)$$

17-9. Cascading z-Transforms

which when combined with the transfer function of the final element gives

$$O(s) = Z\{I(t)\}\left(1 - e^{-sT}\right)\left[\frac{1}{s}\left(\frac{0.25}{s + 0.25}\right)\right]$$

The expression in the brackets may be restructured:

$$\frac{1}{s}\left(\frac{0.25}{s + 0.25}\right) = \frac{A}{s} + \frac{B}{s + 0.25} = \frac{1}{s} - \frac{1}{s + 0.25}$$

The z-transform of the output is

$$O(z) = Z\{I(t)\}\, Z\left\{\left(1 - e^{-sT}\right)\left(\frac{1}{s} - \frac{1}{s + 0.25}\right)\right\}$$

Since the transform of the input is $5z/(z - e^{-0.2T})$,

$$O(z) = \left(\frac{5z}{z - e^{-0.2T}}\right)\left(1 - z^{-1}\right)\left(\frac{z}{z-1} - \frac{z}{z - e^{-0.25T}}\right)$$

or

$$O(z) = \frac{5z(1 - e^{-0.25T})}{(z - e^{-0.2T})(z - e^{-0.25T})}$$

and for $T = 2\,\mathrm{s}$,

$$O(z) = \frac{1.96735 z}{z^2 - 1.27685z + 0.40657}$$

The inversion of $O(z)$ yields values of the output of 0, 1.967, 2.512, and 2.408 for 0, 2, 4 and 6 s, respectively.

The transformation in Example 17-6 emphasizes a precaution. The z-transform of the Laplace transform $0.25/[s(s + 0.25)]$ is

$$Z\left\{\frac{0.25}{s(s + 0.25)}\right\} = \frac{z}{z-1} - \frac{z}{z - e^{-0.25T}} = \frac{z(1 - e^{-0.25T})}{(z-1)(z - e^{0.25T})}$$

a result that is verified by the z-transform corresponding to $a/s(s + a)$ in Table 17-3. Note, however, that

$$Z\left\{\frac{1}{s}\right\} Z\left\{\frac{0.25}{s + 0.25}\right\} = \left(\frac{z}{z-1}\right)\left(\frac{0.25z}{z - e^{-0.25T}}\right)$$

which is not the correct z-transform. Although

$$Z\{G_1(s) + G_2(s)\} = Z\{G_1(s)\} + Z\{G_2(s)\}$$

$$Z\{G_1(s)G_2(s)\} \neq Z\{G_1(s)\} Z\{G_2(s)\} \qquad (17\text{-}21)$$

An apparent violation of Eq. (17-21) is when we say that

$$Z\{(\text{ZOH})G(s)\} = Z\{\text{ZOH}\} Z\{G(s)\}$$

but since $\mathcal{L}\{\text{ZOH}\} = (1 - e^{-sT})/s$ and z-transforms are additive,

$$Z\left\{\frac{1-e^{-sT}}{s}G(s)\right\} = Z\left\{\frac{G(s)}{s}\right\} - Z\left\{\frac{e^{-sT}G(s)}{s}\right\}$$

The time-shift term e^{-sT} can be pulled out as z^{-1}, leaving

$$Z\{(\text{ZOH})G(s)\} = \left(1 - z^{-1}\right) Z\left\{\frac{G(s)}{s}\right\}$$

Transfer functions in the z domain possess the same cascading property as Laplace transforms only when the inputs and outputs are truly z-transforms. A series of Laplace transfer functions can be broken down using partial fractions to expressions that can be converted to the z domain using the table of transforms.

For a feedback loop, such as in Fig. 17-25, the transfer function of the loop in the z domain is

$$\frac{O(z)}{I(z)} = \frac{Z\{(ZOH)G(s)\}}{1 + Z\{(ZOH)G(s)H(s)\}} \qquad (17\text{-}22)$$

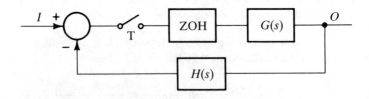

Fig. 17-25. A sampled feedback control loop.

17-10 How a z-Transform Can Indicate Stability of a Control Loop

One of the strengths of the transfer function in the s plane of a continuous controller is that through appropriate interpretation, the transform indicates whether or not the loop is stable. The stability analysis can be made without

17-10. How a z-Transform Can Indicate Stability of a Control Loop

inversion to the time domain. Specifically, if any of the roots of the denominator of the transfer function have a positive real part, the loop is unstable (Sec. 16-16). The z-transform offers similar insight. The stability analysis of a sampled control loop takes on special importance because a loop with sensor-system-actuator characteristics that is stable as a continuous controller may be unstable under certain conditions if converted to a sampled controller. A frequent reason for instability is that the sampling interval is too long.

One of the techniques for inverting a z-transform demonstrated in Sec. 17-8 was decomposition into partial fractions followed by inversion of the individual terms. The z-transform of a control loop would thus be restructured:

$$\frac{Z\{(ZOH)G(s)\}}{1 + Z\{(ZOH)G(s)H(s)\}} = \frac{A}{z-a} + \frac{B}{z-b} + \cdots \qquad (17\text{-}23)$$

where a, b, etc., are roots of the denominator. Next focus on one of the partial fractions and invert it by first finding the coefficients of the negative powers of z. For example,

$$\frac{A}{z-a} = Az^{-1} + Aaz^{-2} + Aa^2 z^{-3} + \cdots \qquad (17\text{-}24)$$

The contributions of $A/(z-a)$ to the entire time-domain representation at 0, T, $2T$, $3T$, etc., are 0, A, Aa, Aa^2, ... respectively. These values increase in an unbounded fashion if $|a| > 1$. So the criterion is: For stability, the absolute magnitude of all roots of the denominator must be less than 1.

Just as roots of s could be complex numbers in Laplace transforms, the roots of z may also be complex in z-transforms.

A simple control assignment will illustrate the difference between the behavior of a continuous controller and a sampled controller. In Fig. 17-26a the air pressure in a tank is regulated by a proportional controller that adjusts the flow rate w_i. The block diagram that represents the system is shown in Fig. 17-26b, with the proportional constant k of 0.2 kg/(s-kPa). The pressure p is a function of the mass in the tank, $2.5M$, and the differential equation for the rate of change of pressure is

$$\frac{dp}{dt} = 2.5\frac{dM}{dt} = 2.5\,(w_i - w_o)$$

which yields the transfer function of the tank pressure, $2.5/s$. The transfer function of the control loop is

$$\text{TF}_{\text{loop}} = \frac{(0.2)(2.5)/s}{1 + (0.2)(2.5)/s} = \frac{0.5}{s + 0.5} = \frac{1}{\tau s + 1} \qquad (17\text{-}25)$$

where $\tau = 2$ s, which is clearly a stable loop.

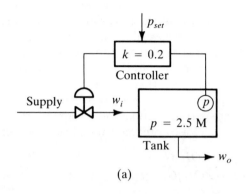

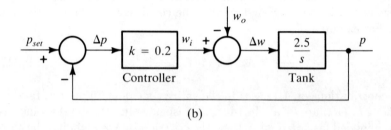

Fig. 17-26. (a) Air-pressure regulator, and (b) block diagram of a continuous controller.

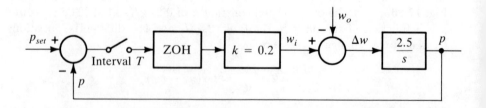

Fig. 17-27. Computer control of the pressure in the air-supply system of Fig. 17-26a.

17-10. How a z-Transform Can Indicate Stability of a Control Loop

Were the continuous controller of Fig. 17-26 to be converted to computer control, the block diagram would be represented by Fig. 17-27. The Laplace transform of the forward path, with a sampling interval of T, is

$$\left(\frac{1-e^{-sT}}{s}\right)(0.2)\left(\frac{2.5}{s}\right)$$

which has a corresponding z-transform of

$$0.5\left(1-z^{-1}\right)\frac{Tz}{(z-1)^2} = \frac{0.5T}{z-1}$$

The z-transform of the loop is

$$\mathrm{TF}(z)_{\mathrm{loop}} = \frac{0.5T/(z-1)}{1+0.5T/(z-1)} = \frac{0.5T}{z-1+0.5T} \quad (17\text{-}26)$$

Next, explore the influence of the sampling interval T. In the system of Fig. 17-26a, which is controlled by the loop represented in Fig. 17-27, suppose that w_o remains constant at 8 kg/s and that steady-state conditions are allowed to develop with $p_{\mathrm{set}} = 260$ kPa, which would result in the tank pressure p stabilizing at 220 kPa. At $t = 0$ a step change of p_{set} to 250 is made. What is the response of p? The response of a continuous controller can be found by inverting Eq. (17-25), giving

$$p = 220 - 10\left(1 - e^{-t/2}\right) \quad (17\text{-}27)$$

which is shown in Fig. 17-28.

The response of the sampled controller can be determined from a numerical simulation. At $t = 0$, p_{set} changes to 250, so $\Delta p = 250 - 220 = 30$ kPa, which translates to a flow rate $w_i = 6$ kg/s, while w_o continues at 8 kg/s. A constant rate of drop in tank pressure occurs during the first time interval. At the end of the first interval the new reading of $(p_{\mathrm{set}} - p)$ translates to a new flow rate. A sampled controller with infinitesimally small values of T approaches the performance of the continuous controller. Even very short values of T, say less than 0.1 s, result in a response of p virtually the same as that of the continuous controller in Fig. 17-28. A longer value of the sampling interval, 3 s for example, which is also shown in Fig. 17-28, gives a response deviating from the continuous controller but eventually settling to the correct steady-state value. If a still longer T were chosen, 5 s for example, the tank pressure at 0, 5, 10, and 15 s would be 220.0, 195.0, 232.5, 176.25 kPa, respectively, indicating unstable control.

Now return to the z-transform of the sampled data loop, Eq. (17-26), and apply the $|z| > 1$ test on the roots of the denominator. The denominator

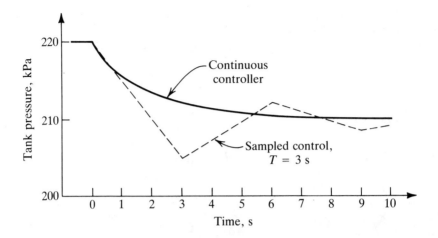

Fig. 17-28. Response of p to a step change in p_{set} for a continuous and sampled data controller for the system in Fig. 17-26a.

shows that values of $T > 4\,\text{s}$ result in $z > 1$, so any sampling intervals longer than 4 s cause instability.

The choice of the sampling interval is under the jurisdiction of the designer with certain limitations. If the computer has few loops to control and also has few other tasks to perform, a short sampling time can be chosen that will provide a stable loop. Very short sampling intervals, however, might require such rapid changes that the actuator may experience excessive wear. For this reason the shortest sampling intervals of which the computer is capable are not necessarily the ones chosen by the control designer.

17-11 Proportional Control

The air-pressure regulator of Fig. 17-26 is a proportional controller, and another example with a modest increase in complexity is the air-temperature controller in Fig. 17-29. A temperature sensor transmits the temperature of discharge air to the controller, which at time intervals of T s compares the sensor temperature θ_s to the reference temperature θ_r, and sends a new message U to the valve. The block diagram of this control loop is shown in Fig. 17-30, where the variables passing between blocks have been normalized with respect to the initial conditions designated by the subscript 0 in anticipation of an inversion to the time domain.

17-11. Proportional Control

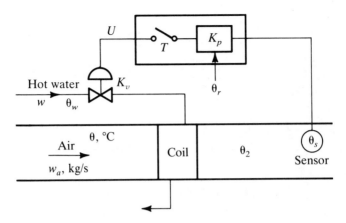

Fig. 17-29. Air-temperature controller.

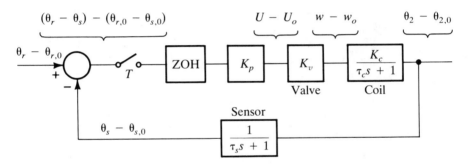

Fig. 17-30. Block diagram of proportional controller of air temperature in Fig. 17-29.

Several linearizations are implicit in the diagram, namely, for the valve and coil. The coil, as is true of most heat exchangers, is complex to model. Even in steady state the outlet air temperature θ_2 is not linear with the flow rate w. Instead, for given values of θ_1, w_a, and θ_w, the relation is as shown in Fig. 17-31. The constant K_c is the slope of the θ_2-w curve in the operating region of interest. The following numerical values apply:

$$
\begin{aligned}
\text{Time constant of sensor, } \tau_s &= 5\,\text{s} \\
\text{Time constant of coil, } \tau_c &= 20\,\text{s} \\
K_c &= 45°\text{C}/(\text{kg-s}) \\
K_v &= 0.012\,\text{kg}/(\text{s-kPa}) \\
K_p &= 8\,\text{kPa per °C error}
\end{aligned}
$$

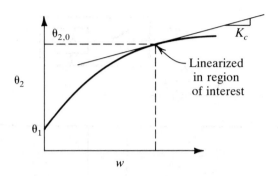

Fig. 17-31. Nonlinear relation of θ_2 to w.

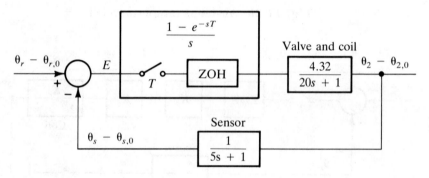

Fig. 17-32. Block diagram with transfer functions shown.

With these numerical values, the block diagram becomes more specific, as in Fig. 17-32.

The Laplace transform of the forward path of the loop is

$$\text{FP}(s) = \left(\frac{1 - e^{-sT}}{s}\right)\left(\frac{4.32}{20s + 1}\right) = (1 - e^{-sT})\left[\frac{1}{s}\left(\frac{0.216}{s + 0.05}\right)\right]$$

and

$$\text{FP}(s) = (1 - e^{-sT})\left(\frac{4.32}{s} - \frac{4.32}{s + 0.05}\right)$$

which when transformed to the z domain becomes

$$\text{FP}(z) = (1 - z^{-1})\left(\frac{4.32z}{z - 1} - \frac{4.32z}{z - e^{-0.05T}}\right)$$

17-11. Proportional Control

$$\text{FP}(z) = \frac{4.32\left(1 - e^{-0.05T}\right)}{z - e^{-0.05T}} \tag{17-28}$$

The denominator D of the loop transfer function in the s domain is

$$D(s) = 1 + \left(1 - e^{-sT}\right)\left[\frac{1}{s}\left(\frac{0.216}{s + 0.05}\right)\right]\left(\frac{0.2}{s + 0.2}\right) \tag{17-29}$$

$$D(s) = 1 + \left(1 - e^{-sT}\right)\left(\frac{4.32}{s} - \frac{5.76}{s + 0.05} + \frac{1.44}{s + 0.2}\right)$$

which when transformed to the z domain is

$$D(z) = 1 + \left(1 - z^{-1}\right)\left(\frac{4.32z}{z - 1} - \frac{5.76z}{z - e^{-0.05T}} + \frac{1.44z}{z - e^{-0.2T}}\right)$$

and

$$D(z) = \frac{z^2 + (0.44M - 6.76N + 4.32)z + 5.32MN - 5.76M + 1.44N}{(z - N)(z - M)} \tag{17-30}$$

where $M = e^{-0.2T}$ and $N = e^{-0.05T}$.

It may be observed that $D(z)$ is not the product of the z-transforms of the groups in Eq. (17-29). Thus, in accordance with Eq. (17-21),

$$Z\left\{\left(\frac{0.216}{s(s + 0.05)}\right)\left(\frac{0.2}{s + 0.2}\right)\right\} \neq Z\left\{\frac{0.216}{s(s + 0.05)}\right\}Z\left\{\frac{0.2}{s + 0.2}\right\}$$

The target of the previous equations has been the loop transfer function, which is Eq. (17-28) divided by Eq. (17-30),

$$\text{TF}(z) = \frac{4.32(1 - N)z + 4.32MN - 4.32M}{z^2 + (0.44M - 6.76N + 4.32)z + 5.32MN - 5.76M + 1.44N} \tag{17-31}$$

Two of the possible uses to which the loop transfer function in Eq. (17-31) can be applied are (1) to analyze the loop for stability and (2) to invert and predict the response of θ_2 to a change in θ_r.

Whether the loop is stable or not will depend upon the magnitude of the sampling interval T. Table 17-5 shows roots of the denominator and the absolute magnitude of these roots at various values of T.

Somewhere between a sampling interval of 17.2 and 17.4 s, the absolute magnitude of z exceeds unity, and the loop goes unstable. This fact can also be verified by imposing a disturbance on θ_r and observing the behavior of θ_2 with various values of T. Two methods available for determining the

Table 17-5. Roots of z in the Loop Transfer Function of Eq. (17-31)

T, s	r_1	r_2	ABS(r_1)	ABS(r_2)
1	$+0.8750 \pm 0.1772i$		0.8928	
10	$-0.1397 \pm 0.7149i$		0.7284	
16	$-0.6502 \pm 0.2947i$		0.7139	
17.2	-0.9512	-0.5223	0.9512	0.5223
17.4	-1.0135	-0.4879	1.0135	0.4879

response of θ_2 in the time domain are (1) through a numerical simulation and (2) through the inversion of the z-transform of θ_2.

Illustrations of the numerical simulation and the z-transform inversion will now be performed. The situation to be considered is where θ_r undergoes a step change such that θ_2 changes from an initial value of 40°C to a final value of 50°C. Examination of the block diagram in Fig. 17-32 shows that to achieve $\theta_2 = 40°$C, the difference $\theta_r - \theta_2$ at the summing point must be $40/4.32 = 9.259$, so the initial θ_r, which is $\theta_{r,0}$, is 49.259°C. By similar logic, θ_r is 61.574 to ultimately give the new steady-state value of $\theta_2 = 50°$C.

Use first the numerical simulation to find θ_2 in the time domain. Before examining the structure of the simulation program, we pause to clarify how to simulate a time-constant block with non-unity gain. For example, convert the transfer function of the coil in Fig. 17-32, $4.32/(20s + 1)$, to a calculation of finite time step. The input is the error E, and the output is $\theta_2 - \theta_{2,0}$. The differential equation from which the transfer function came is

$$4.32E - (\theta_2 - \theta_{2,0}) = \tau \left(\frac{d(\theta_2 - \theta_{2,0})}{dt} \right) \qquad (17\text{-}32)$$

The finite difference form of Eq. (17-32) is

$$\theta_{2,n+1} = \theta_{2,n} + \Delta t \left(\frac{4.32E - (\theta_{2,n} - \theta_{2,0})}{20} \right) \qquad (17\text{-}33)$$

The basic structure of the numerical simulation program is two nested loops, as shown in Fig. 17-33. The outer loop is activated each time the sampling interval T elapses. The inner loop attempts to reproduce the continuous transients of the coil and temperature sensor by using a very small time step, Δt. When the summations of the Δt's in the inner loop have accumulated to T, the value of the error E existing then is brought around to begin a new outer loop.

17-11. Proportional Control

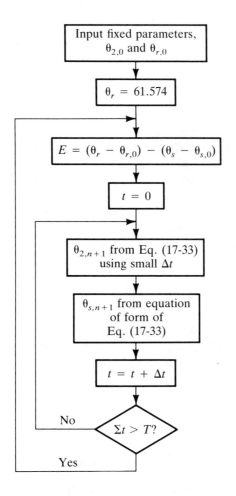

Fig. 17-33. Flow diagram of numerical simulation of control loop in Fig. 17-32.

When choosing a sampling interval of 15 s and a Δt for the inner loop of 0.01 s, the numerical simulation provides the response of the sensed and controlled temperatures shown in two of the columns in Table 17-6. As indicated earlier, the sampling interval of 15 s should still give a stable response, and Table 17-6 indicates this to be the case. There is appreciable oscillation, however, before the temperatures settle to their final values.

The z-transform of the controlled temperature will now be inverted and the results compared with those in Table 17-6 obtained from the numeri-

Table 17-6. Response of the Controlled and Sensed Temperatures of Continuous and Computer Control (with a Sampling Interval of 15 s) of the Air Temperature

Time, s	Continuous control		Numerical simulation		z-transform
	Controlled	Sensed	Controlled	Sensed	Controlled
0	51.71	51.27	40.00	40.00	40.00
15	49.72	49.85	68.08	60.60	68.07
30	50.04	50.01	34.38	42.99	34.42
45	49.99	49.99	58.61	52.58	58.52
60	50.00	50.00	48.18	50.66	48.29
75	50.00	50.00	47.64	47.91	47.55
90	50.00	50.00	53.66	52.07	53.70
105	50.00	50.00	47.01	48.70	47.01
120	50.00	50.00	51.54	50.42	51.50

cal simulation. The z-transform of the output (the controlled temperature) is the product of the z-transform of the input and the TF(z) of the loop, [Eq. (17-31)]. The z-transform of the step change of θ_r from 49.259 to 61.574°C is $12.315[z/(1-z)]$, so $\theta_2(z)$ is

$$12.315\left(\frac{z}{z-1}\right)\left[\frac{4.32(1-N)z + 4.32MN - 4.32M}{z^2 + (0.44M - 6.76N + 4.32)z + 5.32MN - 5.76M + 1.44N}\right] \quad (17\text{-}34)$$

If $T = 15$ s, then $N = 0.47237$ and $M = 0.049787$. Equation (17-34) becomes

$$\theta_2(z) = \frac{28.070515z^2 - 1.4033898z}{z^3 + 0.148706z^2 - 0.630157z - 0.5185489} \quad (17\text{-}35)$$

It is appropriate to pause and ask whether the multiplication of the z-transforms in Eq. (17-34) violates the rule cited in Eq. (17-21). The multiplication is valid here, because the three variables at the summing point are all z-transforms. As shown in Fig. 17-24, the z-transform of the function entering the sampler must be used, and the z-transform $(1 - e^{-sT})Z\{G(s)H(s)\}$ develops a z-transform of θ_s, so it is appropriate that the change in reference temperature also be a z-transform.

The inversion of Eq. (17-35) may be performed by division, as explained in Sec. 17-8. If there should be a need for frequent application of the division process for inverting z-transforms, a short computer program can be written[2]

17-12. Proportional-Integral Control

to perform the division when given the coefficients and the highest power of z of the numerator and denominator. The results of the inversion are shown in Table 17-6 and closely match the computations from the numerical simulation. The deviations that do exist are due to the finite time steps in the simulation.

The final topic pertaining to this proportional controller is to suggest a program for the computer controller. Figure 17-34 shows a possible sequence that also provides time within the loop for the computer to attend to other tasks or to remain idle in order to develop the desired sampling interval.

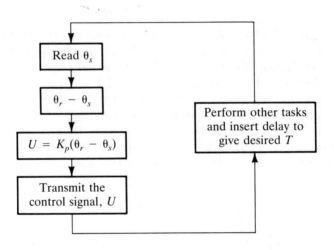

Fig. 17-34. Flow diagram of the control program for proportional control of the discharge air temperature.

17-12 Proportional-Integral Control

The integral mode in a continuous controller integrates the error between the sensed and the set temperature over time, multiplies the integral by a constant, and applies this magnitude as an adjustment to the control signal. In a computer controller of air temperature in the system of Fig. 17-29 the error E is

$$E = \theta_r - \theta_s$$

The control signal U to the valve is made up of the contributions of the P and I modes and is updated every time interval of T s. The P-mode contribution is

$$U = K_P E$$

or

$$U_{k+1} = U_k + K_P(E_{k+1} - E_k) \tag{17-36}$$

where the subscript k refers to the present time and $k+1$ to the time interval T later. When the contribution of the I mode is included,

$$U_{k+1} = U_k + K_P(E_{k+1} - E_k) + K_I(E_{k+1} + E_k)\frac{T}{2} \tag{17-37}$$

Example 17-7. The liquid level in the vessel having a cross-sectional area of $1.2\,\mathrm{m}^2$ shown in Fig. 17-35 is controlled by a PI computer controller with $K_P = 10\,\mathrm{m}^3/(\mathrm{s\text{-}m})$ and $K_I = 3\,\mathrm{m}^3/(\mathrm{s}^2\text{-}\mathrm{m})$. The time interval of sampling and control $T = 0.1\,\mathrm{s}$. The outlet flow rate has been $0.16\,\mathrm{m}^3/\mathrm{s}$ for a long time, and the controller has brought $x = x_{\text{set}}$. (a) What is the magnitude of the I-mode contribution at this condition? (b) At $t = 0$ just after a sample and control adjustment, w_o experiences a step increase in flow rate to $0.2\,\mathrm{m}^3/\mathrm{s}$. Compute the entering flow rate w_i at $0.2\,\mathrm{s}$.

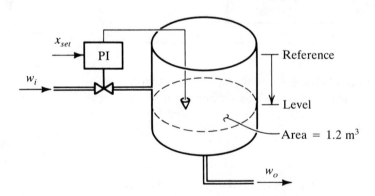

Fig. 17-35. Level controller in Example 17-7.

Solution. (a) $E = 0$ at $t = 0$, so the P contribution is zero and all of the control signal is attributable to the I mode, $U_{t=0} = U_0 = 0.16\,\mathrm{m}^3/\mathrm{s}$.

(b) $E_0 = x_{\text{set}} - x_0 = 0$

$$E_1 - E_0 = \left(\frac{0.20 - 0.16}{1.2}\right) 0.1 = 0.003333 \text{ m}$$

and
$$E_1 = 0.003333 \text{ m}$$

$$U_1 = 0.16 + 10(0.003333) + 3(0.003333)\frac{0.1}{2} = 0.1938 \text{ m}^3/\text{s}$$

$$E_2 - E_1 = \left(\frac{0.20 - 0.1938}{1.2}\right) 0.1 = 0.0005167 \text{ m}$$

and
$$E_2 = 0.003849 \text{ m}$$

$$\begin{aligned} U_2 &= 0.1938 + 10(0.0005167) + 3(0.003849 + 0.003333)\frac{0.1}{2} \\ &= 0.200044 \text{ m}^3/\text{s} \end{aligned}$$

so $w_i = 0.1992 \text{ m}^3/\text{s}$ at $t = 0.2 \text{ s}$. As action continues, w_i settles to $0.2 \text{ m}^3/\text{s}$ and $E \rightarrow 0$.

17-13 Forms of Actuator Signals

Several different ways of achieving modulating (in contrast to on/off) control are possible. One is to convert the digital (8-bit, 10-bit, etc.) signal from the computer through an ADC to an analog voltage that passes directly to a magnetic modulating operator (such as the magnetic valve in Sec. 6-8). For pneumatic operators the output of the ADC passes to an electric-to-pneumatic transducer (Sec. 6-10), which translates the voltage into a proportional air pressure. Such signals as these can utilize the P or PI control modes.

Many electric motorized operators can be controlled by two bits from the computer. When one bit is high the operator moves in one direction and when the other is high it moves in the opposite direction. When both bits are low, the actuator remains motionless. Two possible signals that interface to such actuators are the sense/delay/fixed change and pulse-width modulation (Fig. 17-36). In Fig. 17-36a if the sensed variable is higher than the set value a small adjustment is made in the actuator. At time interval T later another sense and adjustment are made. In principle the offset using this strategy approaches zero without the need of using PI control. The combination of T and the extent of the fixed change must, however, be tuned to the system characteristics.

The pulse-width modulation form of actuator signal in Fig. 17-36b can be adapted to the computer control signal represented by Eqs. (17-36) and (17-37) by setting the length of the pulse proportional to $U_{k+1} - U_k$.

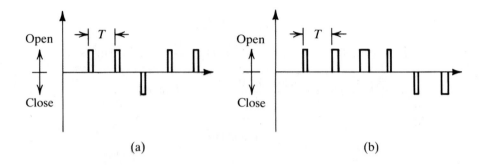

Fig. 17-36. (a) A sense/delay/fixed change signal and (b) a pulse-width modulation signal.

17-14 Non-linearities—Dead Time

The dynamic performance of many real components can be approximated by a combination of a dead time followed by a time-constant response, as shown in Fig. 17-37. The physical reason for this behavior of the coil in Fig. 17-29 is that there is a flush time required to expel the water from the tubes of the coil. Thereafter the thermal capacity of the metal of the coil responds as a time-constant block.

While this process is non-linear, which usually defies an analytical treatment, the z-transform is capable of accommodating such time delays.[3]

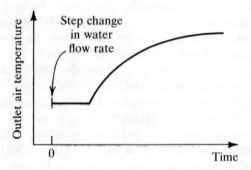

Fig. 17-37. Dead time followed by a time-constant response.

17-15 Non-linearities—Hysteresis

Because of friction the valve regulating the water flow in Fig. 17-29 is likely to exhibit hysteresis, as in Fig. 17-38. With only proportional control the precision of the control is likely to be diminished because the actuator may not be responsive to the control signal, particularly if a reversal of motion is called for. With the combination of P and I modes, the controlled variable should ultimately be pulled to the set value. Allowing the I mode to correct for hysteresis may require a long time, because K_I must be set small enough to maintain a stable loop. Another approach, readily adaptable to computer control, is hysteresis compensation whereby when the computer senses a change in direction of motion of the actuator, it applies a step change in the control signal (probably somewhat less than the Δ shown in Fig. 17-38) in order to shorten the time for the reversal of the actuator to begin.

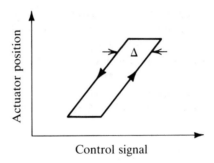

Fig. 17-38. Hysteresis of the actuator.

17-16 Summary

All the principles and knowledge of the characteristics of the microcomputer, the behavior of feedback control loops, and the response of thermal and mechanical systems all merge when computer control is finally applied to the system. Computer control exhibits some strengths not normally available in continuous controllers, such as the flexibility of actuation and the compensation for non-linearities. Also, certain features such as the integral mode can be incorporated at low cost in the software in comparison to a higher cost if done in hardware. There are certain precautions, however, that must be observed, particularly in the choice of the sampling interval.

References

1. B. C. Kuo, *Digital Control Systems*, Holt, Rinehart and Winston, Inc., New York, 1980.

2. C. Rohrer, *Digital Control of Discharge Air Temperature Including z-Transform Analysis*, M.S. Thesis, University of Illinois at Urbana-Champaign, 1985.

3. G. F. Franklin and J. D. Powell, *Digital Control of Dynamic Systems*, Addison-Wesley Publishing Company, Reading, MA, 1980.

Problems

17-1. The fluid temperature θ_f in Fig. 17-1 obeys a sinusoidal variation with an amplitude of 5°C varying about a mean temperature of 35°C. The period of the sinusoidal variation is 100 s, and the time constant of the process is 20 s. Use a computer program to perform a numerical simulation and determine for the steady-periodic condition: (a) the time lag in seconds of the temperature of the sensing bulb with respect to the fluid, and (b) the amplitude of the temperature variation of the bulb. **Ans.:** Check results against the exact equation:

$$\theta - \theta_{\text{mean}} = \frac{\text{amplitude of } \theta_f}{\sqrt{1 + 4\pi^2 \tau^2 f^2}} \sin\left[2\pi f t - \tan^{-1}(2\pi f \tau)\right]$$

where

$$t = \text{time, s}$$
$$\tau = \text{time constant, s}$$
$$f = \text{frequency, s}^{-1}$$

17-2. At time $t = 0$ the fan shown in Fig. 17-39 is at rest, but then a constant torque of 200 N-m is applied by the motor. The moment of inertia of the motor and fan is 10 kg-m^2, and the torque absorbed because of the delivery of air is proportional to the rotative speed, T_{load}, N-m $= 1.2\omega$, where ω is in rad/s. (a) Solve the differential equation to determine ω at $t = 1$ s (b) Determine ω from an Euler-type numerical simulation with a time step of 1 s (c) Determine

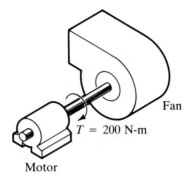

Fig. 17-39. Startup of fan in Prob. 17-2.

ω from a numerical simulation with a time step of 1 s but using a Taylor series that includes the second-degree term. **Ans.:** (c) 18.8 rad/s.

17-3. The torque delivered by a gasoline engine as a function of engine speed above a rotative speed of 1500 rpm can be represented by a second-degree equation. When this relation is translated to the tractive force at the wheels at full power for a specific automobile, the equation is

$$F = -110 + 170V - 2.5V^2$$

where F = tractive force, N, and V = speed, m/s. During an acceleration this force is distributed between the wind resistance force, $2.35V^{1.8}$ N and the force to accelerate the 1100 kg automobile.

Use a numerical simulation to determine for an acceleration from 10 m/s to 25 m/s: (a) the elapsed time, and (b) the distance traveled. **Ans.:** (b) 174.5 m.

17-4. If the time constant of the sensing bulb in Fig. 17-1 is 10 s, compute the temperature of the bulb at $t = 2.6$ s if the bulb is initially at a temperature of 30°C and then subjected to a variation in fluid temperature shown in Fig. 17-40. **Ans.:** 30.412°C

17-5. Substitute $F(t)$ into the definition of the Laplace transform and integrate to verify transform 5 in Table 17-1.

17-6. A water storage tank, Fig. 17-41a, has an area of 1.6 m² and experiences a continuous draw rate of 0.4 m³/s. From time $t = 0$ to $t = 4$ s the supply rate is 0.4 m³/s and the level in the tank, x, is 0.8 m. In the time interval of 4 to 6 s the supply rate steps to 0.5 m³/s, following which it reverts to 0.4 m³/s. (a) What is the transfer function of the process in Fig. 17-41b?

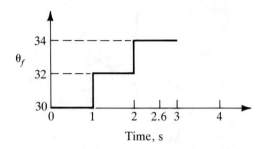

Fig. 17-40. Step changes in fluid temperature applied to sensing bulb in Prob. 17-4.

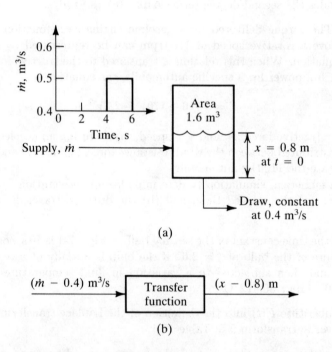

Fig. 17-41. (a) Water storage tank, and (b) transfer function of the process.

(b) What is the transform of $(x - 0.8)$ when the process is subjected to pulsed supply flow rate? (c) What is the equation for x in the time domain? **Ans.:** (c) A checkpoint is that when $t = 5\,\text{s}$, $x = 0.8625\,\text{m}$.

17-7. Verify the z-transform of e^{-at} that is shown in Table 17-3 by substituting the function into Eq. (17-12) and developing the closed form.

17-8. Invert $z/(z - e^{-0.5T})$ to find the magnitude of $F(t)$ at $0, T, 2T$, and $3T$. **Ans.:** at $t = 3T$, $F(t) = e^{-1.5T}$.

17-9. Invert $Tz^2/(z^2 - 0.16)$ to find the values in the time domain at 0, T, $2T$, $3T$, and $4T$. **Ans.:** at $t = 4T$, $F(t) = 0.0256T$.

17-10. A computer controller operating in the proportional mode regulates the concentration of an additive to a fluid stream in the arrangement shown in Fig. 17-42. The sensor detects the mass ratio of the additive to the mainstream fluid, thus $R_s = w_a^*/w_s$, where $w_a^* = $ flow rate of additive leaving the mixing tank, and w_s is the mainstream flow, which remains constant at $120\,\text{kg/s}$.

The combination of K_P and the valve characteristic is represented by the relation $w_a = 1000 \times \text{Error}$. The capacity of the tank is $2500\,\text{kg}$, and to facilitate modeling the system, use the approximation that a constant mass of $2450\,\text{kg}$ of mainstream fluid exists in the tank at all times. (a) What are the steady-state values of w_a^* and R_s leaving the tank when the setting of R_r is 0.02? (b) Construct the block diagram of the system with R_r as the reference value and R_s as the controlled variable, indicating s-domain transfer functions and the variables normalized to the initial steady state. (c) Determine the z-transform of the control loop. (d) Above what sampling interval T does the system go unstable? **Ans.** (c)

$$\frac{8.333(1 - e^{-0.04897T})}{z - 9.333e^{-0.04897T} + 8.333}$$

17-11. In the computer-controlled concentration regulator in Prob. 17-10: (a) What are the steady-state values of w_a^* and R_s leaving the tank when the setting of R_r is 0.02? (b) Construct the block diagram of the system with R_r as the reference value and R_s as the controlled variable, indicating s-domain transfer functions and the variables normalized to the initial steady state. (c) Perform a numerical simulation of the process that results from a step increase in R_r from 0.02 to 0.022 when the controller has a sampling interval of $4\,\text{s}$. Compute the values of R_s for 6 sampling intervals. **Ans.:** (c) R_s at $16\,\text{s}$ is 0.01930.

17-12. In the computer-controlled concentration regulator in Prob. 17-10: (a) What are the steady-state values of w_a^* and R_s leaving the tank when the

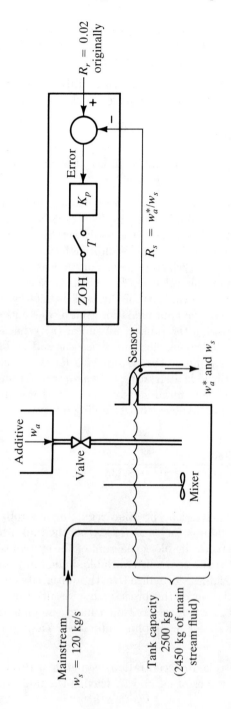

Fig. 17-42. Concentration controller in Probs. 17-10, 17-11, and 17-12.

setting of R_r is 0.02? (b) Using the z-transform of the loop shown in the answer of Prob. 17-10c, apply a step increase of R_r from 0.02 to 0.022 when the controller has a sampling interval of 4 s. Invert the transform to compute the values of R_s for 6 sampling intervals. **Ans.:** (b) R_s at 16 s is 0.01931.

17-13. Paper from a mill operation is wrapped on a spool whose rotative speed is regulated by a computer controller operating on the proportional mode that senses the position of the takeup pulley, as shown in Fig. 17-43. The windup spool is driven by a variable speed motor, and the motor-spool assembly has a moment of inertia of 16 kg-m². The analysis will be conducted at the time when the diameter of paper on the spool is 1.2 m. The torque provided by the motor is proportional to $x_r - x$, namely $\theta = 540(x_r - x)$, where θ is the torque in N-m and x_r and x are in meters. The torque must also overcome friction of 30ω, where ω is the angular velocity in rad/s (see Example 16-5). (a) When the setting $x_r = 2$ m and the paper velocity is 8 m/s, what are the steady-state values of ω and x? (b) Construct a block diagram of the control loop with x_r as the reference variable and ω as the controlled variable, indicating s-domain transfer functions and the variables normalized to the initial steady state. (c) Determine the transfer function in

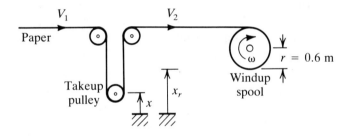

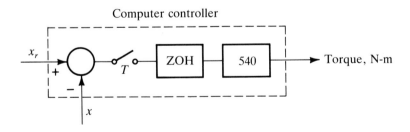

Fig. 17-43. Proportional mode speed control of a paper windup spool in Probs. 17-13, 17-14, and 17-15.

the z domain with x_r as the input and ω as the output of the loop. (d) Above what sampling interval T does the system go unstable? **Ans.** (c)

$$\frac{18a\left(1-e^{-aT}\right)(z-1)}{az^2+bz+c}$$

where

$$a = 1.875$$
$$b = 5.4\left(aT - 1 + e^{-aT}\right) - a\left(1 + e^{-aT}\right)$$
$$c = 5.4\left(1 - e^{-aT} - aTe^{-aT}\right) + ae^{-aT}$$

17-14. In the computer-controlled paper roller in Prob. 17-13, (a) when the setting $x_r = 2\,\text{m}$ and the paper velocity is $8\,\text{m/s}$, what are the steady-state values of ω and x? (b) Construct the block diagram of the control loop with x_r as the reference variable and ω as the controlled variable, indicating s-domain transfer functions and the variables normalized to the initial steady state. (c) Perform a numerical simulation of the process that results from a step increase of x_r from 2.0 to 2.4 m as V_1 remains constant at $8\,\text{m/s}$. The sampling interval is $0.2\,\text{s}$. Compute values of ω for 6 sampling intervals. **Ans.:** (c) $\omega = 15.31\,\text{rad/s}$ at $0.8\,\text{s}$.

17-15. In the computer-controlled paper roller in Prob. 17-13, (a) when the setting $x_r = 2\,\text{m}$ and the paper velocity is $8\,\text{m/s}$, what are the steady-state values of ω and x? (b) Using the z-transform of the loop given for the answer to Prob. 17-13c, apply a step increase of x_r from 2.0 to 2.4 as V_1 remains constant at $8\,\text{m/s}$ when the controller has a sampling interval of $0.2\,\text{s}$. Invert the transform to compute the values of ω for 6 sampling intervals. **Ans.:** (b) $\omega = 15.31\,\text{rad/s}$ at $0.8\,\text{s}$.

Chapter 18

Field Application of Microcomputer Controllers

18-1 Applying Microcomputer Controllers to Field Processes

Chapters 1 through 17 have provided a foundation for the understanding and design of microcomputer-based control systems. The final chapter in this book addresses considerations in the field application of microcomputer controllers to thermal and mechanical systems. The chapter opens with a review of the linear proportional-integral (PI) control laws previously discussed in Chapters 16 and 17 and a discussion of their limitations. PI control algorithms suitable for field application are then described in Secs. 18-3 and 18-4.

Microcomputer controller commissioning involves the field adjustment, or tuning, of software parameters to best meet the control objectives of the process. Standard performance criteria used in the tuning of microcomputer controllers are discussed in Sec. 18-5. The z-transform analysis techniques developed in Chapter 17 are rarely used for field tuning because of the complexity involved. Instead, a number of tuning procedures have been developed for determining tuning parameters at the field site. The three most popular tuning methods, the trial-and-error, closed-loop, and open-loop methods, are described in Secs. 18-7, 18-8, and 18-9.

An advantage microcomputer controllers have over their analog counterparts is the ability to compensate for process non-linearities in software. Section 18-10 describes a method for the compensation of actuator hysteresis.

Finally, the chapter concludes with a brief mention of some advanced microcomputer control applications.

18-2 Practical Control Algorithms

The proportional-integral (PI) control algorithm was first introduced in Secs. 16-7 and 16-20. Instead of representing the K_P and K_I terms independently as had been done there, they may be combined by factoring out the K_P term and then designating K_I/K_P as $1/T_I$,

$$U(t) = K_P \left(E(t) + \frac{1}{T_I} \int_0^t E(t)\,dt \right) \qquad (18\text{-}1)$$

where

$U(t)$ = controller output
$E(t)$ = controller error (setpoint − feedback)
K_I = integral gain
K_P = proportional gain
T_I = integral time, s

This convention is somewhat standard in the control industry and serves well for all but integral-only control, which is not very common. The $1/T_I$ term is often called the number of repeats per second or the number of repeats per minute.

Using backward-referenced rectangular integration to approximate the integral term, a position form digital proportional-integral algorithm may be derived. Backward-referenced integration increases stability at long sampling intervals.[1]

$$U_{k+1} = K_P \left(E_{k+1} + K_I^* \sum_{i=1}^{k} E_i \right) \qquad (18\text{-}2)$$

where

U_{k+1} = controller output at the $(k+1)$th sampling interval
E_{k+1} = controller error at the $(k+1)$th sampling interval
K_I^* = controller integral gain = $\Delta t/T_I = \Delta t K_I/K_P$
Δt = controller sampling interval

While the algorithm represented by Eq. (18-2) can be directly implemented in a digital controller, a number of potential problems may arise in practice. There are situations when the controller may not be able to meet the desired process conditions because of a process limitation. An example might be where the controller had opened a valve fully and was still not able to achieve the desired value of the controlled variable. Under those conditions the integral term will continue to build up beyond physically realizable limits, contributing to what is called integral windup error. A sustained, large error of the opposite sign is required to bring the process back into control after integral windup.

Another problem relates to the adjustment of controller tuning parameters while under control. Under normal operation, the integral term error summation is non-zero. When the proportional or integral gains are changed, there will be an undesirable step change in output through multiplication of the error summation.

18-3 Incremental PI Control Algorithm

An alternative to direct implementation of Eq. (18-2) involves computing an incremental control output instead of an absolute position. The incremental algorithm is derived by computing the difference between successive calculations of the position algorithm in a manner similar to that first shown in Eq. (17-37),

$$\begin{aligned}\Delta U_{k+1} &= U_{k+1} - U_k \\ &= K_P \left[E(n) - E(n-1) + K_I E(n-1) \right] \\ &= K_P \left[E(k+1) - E(k) + K_I^* E(k) \right]\end{aligned} \quad (18\text{-}3)$$

In many control applications, the incremental controller output is directly used to control devices such as motor-driven actuators (Sec. 6-7) and switching-type pneumatic interfaces. The incremental form of the algorithm may be applied when absolute indication of the controlled device position is not required. Incremental interfaces are often less expensive than position interfaces because they do not include feedback. It should be noted, however, that the incremental algorithm is not generally suitable for proportional-only control because the control action is independent of setpoint. The control output is a function of the change in error instead of the absolute error as in the position algorithm.

The incremental algorithm has a number of inherent advantages over the position algorithm. In the incremental algorithm, the controller integral term is stored by virtue of the current position of the control device and is not stored as a numerical value of the algorithm. Because the control device cannot travel beyond physical limits, the integral windup problem is eliminated. Parameter changes are bumpless with the incremental algorithm because an error summation is not stored in the controller.

Example 18-1. Consider the use of the incremental control algorithm and a solenoid switching-type pneumatic interface as shown in Fig. 18-1. The pneumatic incremental interface accepts a pulse-width-modulated control signal from the controller that is proportional to the incremental output. The controller can output a signal to increase or decrease control pressure of up to the 255 units in the 8-bit binary number. This number, called counts, corresponds in this example to a unit pulse width of 20 ms. The pneumatic interface changes output pressure at a rate of 10 kPa/s. The control valve modulates from fully open to fully closed over a pressure range of 50 kPa. If $K_P = 5$, $K_I^* = 0$, and $E_k = 0$, what change in error, ΔE, would cause a 10% change in valve position in one sampling interval?

Solution. The first step of the solution involves calculating a relation for the change in control valve position as a function of the change in control error. For proportional-only control, $K_I^* = 0$.

$$\begin{aligned}\text{Counts output} &= K_P\left(E_{k+1} - E_k + K_I^* E_k\right) \\ &= K_P \Delta E \quad \text{(if } E_k = 0\text{)}\end{aligned}$$

$$\begin{aligned}\Delta \text{Position} &= (K_P \Delta E \text{ counts}) \left(\frac{10\,\text{kPa}}{1\,\text{s}}\right)\left(\frac{5.1\,\text{s}}{255\,\text{counts}}\right)\left(\frac{100\%}{50\,\text{kPa}}\right) \\ &= (5)(10)(5.1/255)(100/50)\Delta E \\ &= (2\,\text{percent per unit error})\Delta E\end{aligned}$$

The change in error can now be computed.

$$\begin{aligned}\Delta E &= 0.50 \cdot \Delta \text{Position} \\ &= 5.0 \quad \text{for } \Delta \text{Position} = 10\%\end{aligned}$$

18-4 Position PI Control Algorithm

The position form of the proportional-integral algorithm is required in situations where positive indication of the controlled device position is critical for successful control application. The position form of the algorithm is also a common choice when existing analog-type controls are replaced with a microcomputer controller during a retrofit. In this way, the existing field devices may be used in the digital control application.

18-4. Position PI Control Algorithm

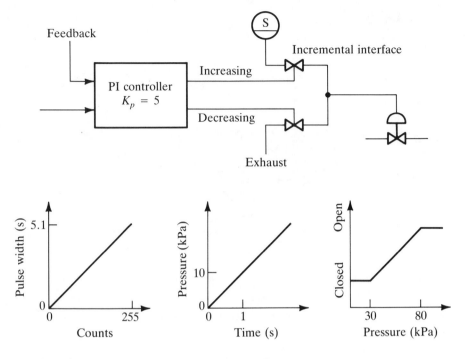

Fig. 18-1. Control system configuration for Example 18-1.

The incremental algorithm can be used as the basis for a practical position algorithm. In the position form implementation, the microcomputer integrates the output of the incremental algorithm and issues the resulting output to the control device.

$$U_{k+1} = U_k + \Delta U_{k+1} \tag{18-4}$$

This algorithm is initialized by storing the desired initial controller output in ΔU_{k+1}. This implementation eliminates controller output bumps during parameter changes, and by limiting the controller output value to one that is physically realizable, the problems of integral windup are avoided, as shown in Fig. 18-2.

$$\text{Minimum output} \leq U_k \leq \text{maximum output} \tag{18-5}$$

An additional mode of operation that should remain bumpless is the transfer of control from manual to automatic. This transfer can be accomplished by resetting the controller output term to the override value while in the manual

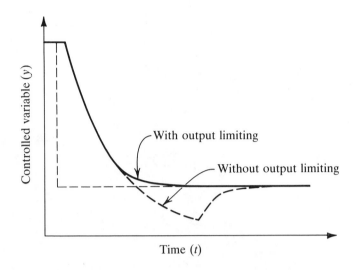

Fig. 18-2. Elimination of integral windup through output limiting.

mode. The incremental algorithm provides this functionality automatically as the controller commands a change in position from wherever the device is currently positioned.

The stability of all control algorithms can be improved with the addition of an error deadband. The deadband limits control actions when the control error is within an acceptable range of setpoint. This is desirable because process non-linearities and noise can produce small oscillations about setpoint even when the controller is well-tuned. These oscillations unnecessarily wear the physical components in the control system. The deadband can be implemented by computing an effective error, $E'(k)$, which is then used in the control algorithm calculations. This implementation provides for smooth control transitions in and out of the deadband as shown in Fig. 18-3.

$$
\begin{aligned}
E'_k &= 0: & -\text{deadband} \leq E_k \leq \text{deadband} \\
E'_k &= E_k - \text{deadband}: & E_k > \text{deadband} \\
E'_k &= E_k + \text{deadband}: & E_k < -\text{deadband}
\end{aligned}
\tag{18-6}
$$

18-5 Criteria for Tuning

The objective of tuning is to select the appropriate control parameters such that the closed-loop system (1) is stable over a wide range of operating condi-

18-5. Criteria for Tuning

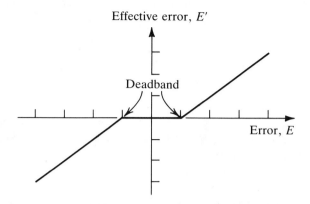

Fig. 18-3. Controller deadband showing smooth transitions.

tions, (2) responds quickly to reduce the effect of disturbances on the control loop, and (3) does not cause excessive wear of mechanical components through continuous cycling. These are often mutually exclusive criteria, and a compromise must generally be made.

The microcomputer controller parameters that need to be specified during the commissioning process include the proportional gain, integral gain, controller sampling interval, deadband, and hysteresis compensation constant. The initial controller output and the minimum and maximum output limits must also be specified for position-type controllers.

The first step in commissioning control systems is deciding on the criteria that will be used to determine the quality of control. A standard tuning criterion[2] used for over 40 years in the process control industry is the Ziegler-Nichols or "quarter amplitude decay" criterion. In this criterion, the controlled response decays in amplitude by a factor of one-fourth each period of oscillation. A system tuned to satisfy this criterion will respond quickly to disturbances and will eventually damp out to steady state. A problem with this tuning criterion is that the gain may be so high that a small change in process characteristics can cause the control loop to go unstable. Another popular criterion results in critical damping of the control loop. A system tuned with a critical gain, as discussed in Sec. 16-18, will respond quickly to a setpoint change without overshooting the setpoint. In practice, critical damping is a difficult criterion on which to base tuning because of the difficulty in visually differentiating a critically damped system from an overdamped system.

A number of statistically based tuning criteria are commonly used as a basis in evaluating controller performance. Two particularly useful measures include the integrated absolute error (IAE) and integrated squared error (ISE) criteria:

$$\text{IAE} = \int_0^t |E(t)|\, dt \qquad (18\text{-}7)$$

$$\text{ISE} = \int_0^t E^2(t)\, dt \qquad (18\text{-}8)$$

Controllers tuned to minimize these error functions generally have damping characteristics somewhere between the critically damped and quarter-amplitude decay responses, as shown in Fig. 18-4.

Another factor to be considered in selecting a tuning criterion is the nature of the disturbances to be rejected. The dynamic response and stability of a control system depend to a great extent on where disturbances enter the control loop. A change in controller setpoint tends to generate more oscillations than a disturbance at an input of the process block such as a load disturbance. Controller parameters that satisfy a tuning criterion for one type of disturbance will be sub-optimal for other types of disturbances, as shown in Fig. 18-5.

The various tuning criteria described assume that the controlled processes are linear with constant static and dynamic characteristics. This is typically not the case for thermal and mechanical systems where non-linearities can be quite pronounced. In heat exchanger control applications, for example, the

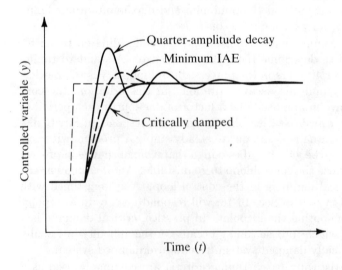

Fig. 18-4. Common PI controller tuning criteria.

18-5. Criteria for Tuning

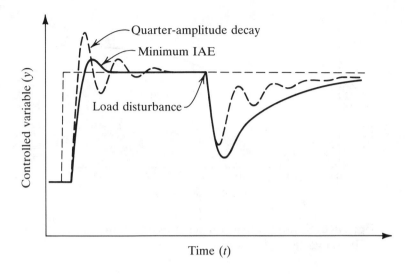

Fig. 18-5. Tuning constants are often optimal for only one disturbance type.

steady-state process characteristics change with changes in operating conditions (Sec. 17-11).

Example 18-2. Consider a coil that heats air with hot water, having the steady-state characteristics shown in Fig. 18-6 and given by the relation:

$$\text{Temperature rise} = (0.6°\text{C}/\%)U_{ss} - (0.003°\text{C}/\%^2)U_{ss}^2$$

where U_{ss} is the steady-state output of the position form algorithm and ranges from 0 to 100%. Using the process gain, which is the change in process output with respect to the process input as a measure of the process sensitivity: (a) What is the process gain at a controller output of 60%? (b) What is the process gain at a controller output of 20%? (c) Nonlinear processes should be tuned to guarantee stability under worst case operating conditions. Where should this process be tuned?

Solution. The process gain is equal to the derivative of the steady-state characteristic equation with respect to the control output. Let y_{ss} equal the temperature rise in °C.

$$\frac{dy}{dU_{ss}} = 0.6 - 0.006\, U_{ss}\, (°\text{C}/\%)$$

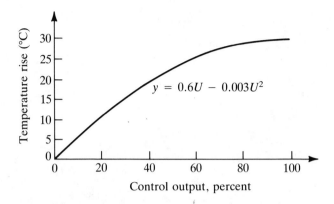

Fig. 18-6. Steady-state characteristics of the air-heating coil in Example 18-2.

(a)
$$\text{Process gain} = \left.\frac{dy}{dU_{ss}}\right|_{U_{ss}=60\%} = 0.24\,°\text{C}/\%$$

(b)
$$\text{Process gain} = \left.\frac{dy}{dU_{ss}}\right|_{U_{ss}=20\%} = 0.48\,°\text{C}/\%$$

(c) In order to ensure stability, this process should be tuned under low part-load conditions (i.e., low controller output).

Regardless of the tuning criterion used, it is important to tune controllers under conditions that result in the highest process gain. When tuned under these circumstances, the control system will satisfy the desired tuning criterion under worst-case conditions and will be less responsive, but stable, under all other operating conditions. A controller tuned to produce a small overshoot, on the order of 5 to 25%, during setpoint changes under worst-case conditions probably represents a reasonable compromise between absolute stability and an ability to reject load disturbances.

18-6 Manual Control Test

The first step in the commissioning of process controllers should always be to control the process manually. Manual control is best accomplished through overriding the output of the controller and issuing direct commands to the control devices. The manual control test involves moving the process between

18-7. Trial-and-Error Tuning

various setpoints and evaluating a number of questions. After completing the manual control test, one of the three controller tuning methods, trial-and-error, closed-loop, or open-loop, can be used to determine the controller parameters.

1. *How difficult is it to keep the process on setpoint?* Difficulty could indicate that disturbances are frequently entering the control loop. If there is great difficulty in maintaining control, some control or mechanical system problem may be indicated. At this point, the design and installation of the control and mechanical systems should be reviewed for possible faults.

2. *How quickly does the process respond to changes in the manipulated variable?* A significant delay between control action and the start of process response indicates that deadtime is appreciable. Processes with significant deadtime are difficult to control.

3. *How much hysteresis is apparent when the control device changes direction of travel?* Hysteresis, or backlash, is common to some extent in all control processes. Processes with significant hysteresis are difficult to control, and software compensation may be indicated.

4. *In what operating region does the process have the highest sensitivity?* High sensitivity is indicated by a large change in the controlled variable for a given controller output. This is the operating region where tuning should be performed as it represents the worst case from a stability standpoint.

5. *How much process noise is observed during manual control?* Rapid fluctuations in the controlled variable will tend to overexercise control devices and should be compensated for through filtering or with the use of the controller deadband.

18-7 Trial-and-Error Tuning

In the trial-and-error tuning method, the operator's experience is combined with a step-by-step procedure to experimentally determine the appropriate controller parameters. Without an established procedure, controller tuning can be a frustrating experience due to the interaction of the tuning parameters, all of which can force control loop instability. The following procedure is applicable for all controllers that use the controller error in calculating the proportional term. Some controllers use the change in control variable instead of the change in control error in the calculation of the proportional term. This modification has the effect of limiting oscillations during setpoint changes. For controllers of this design, step load disturbances may be introduced into the control loop instead of setpoint disturbances.

Step 1. *Configure the controller as a proportional-only controller.* For an incremental controller, the integral gain should initially be very small. The sampling interval should be set as short as practical, and other controller parameters such as hysteresis compensation and deadband should be disabled.

Step 2. *Introduce setpoint disturbances into the closed control loop.* The setpoint changes should be within the high process gain operating range as observed during the manual control test. The setpoint changes also should avoid saturation of the control devices.

Step 3. *Adjust the proportional gain for a slight overshoot of steady-state.* The control loop should rapidly rise to setpoint and dampen to steady-state without significant oscillation, as shown in Fig. 18-7. Changes in the tuning constants should, at least initially, be on the order of 50 percent, since smaller changes do not significantly change the control response. The steady-state offset from setpoint is not important at this point.

Step 4. *Configure the controller as a proportional-integral controller.*

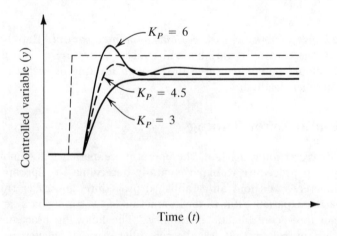

Fig. 18-7. Proportional-only control where $K_P = 4.5$ is the desired response.

18-7. Trial-and-Error Tuning

Step 5. *Adjust the integral gain until the desired transient response is centered at the setpoint.* With small integral gains, the transient response of the controller is primarily determined by the proportional term. The integral gain should be increased just to the point where the proportional-only response is now centered at the setpoint, as illustrated by Fig. 18-8. If the integral gain is increased further, the integral term will tend to dominate the transient response, which is undesirable.

Step 6. *Set the sampling interval to between one-fifth and one-tenth the control loop rise time.* The selection of the controller sampling interval is a compromise between obtaining an ideal control system response, wear of mechanical components, and the utilization of limited computational resources. The compromise can be evaluated by considering the transient characteristics of the controlled process. The rise time for an underdamped loop is the time it takes the controlled process to reach the new setpoint after a setpoint change, and is a useful measure of the time scale of the control loop. A good compromise for sampling intervals generally falls within the range of one-fifth to one-tenth of rise time, with shorter sampling intervals being appropriate for processes subject to frequent disturbances.

Step 7. *Verify that the control response is acceptable after changing the sampling interval.* The integral gain parameter will require an adjustment

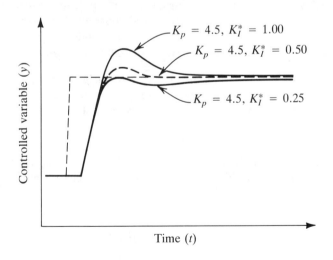

Fig. 18-8. Proportional-integral control where $K_P = 4.5$, $K_I = 0.50$ is the desired response.

after the sampling interval change.

$$K_{I,\text{new}} = K_{I,\text{old}} \left(\frac{\Delta t_{\text{old}}}{\Delta t_{\text{new}}} \right) \tag{18-9}$$

18-8 Closed-Loop Tuning

The closed-loop tuning procedure uses the results of a closed-loop control test and a set of standard tuning rules to reduce the time required in tuning the controller. The closed-loop tuning procedure is derived from the classical "ultimate sensitivity" tuning method developed by Ziegler and Nichols[2] for analog controllers. The procedure involves forcing the control loop under proportional-only control into a marginally stable condition. The ultimate period, P_u, is the period when the oscillations become steady periodic. The proportional gain that brought about marginal stability is called the ultimate gain, K_u. Frequency analysis methods can then be used to determine the controller tuning constants required to achieve a desired system phase lag. Rules are available for mapping the ultimate period and ultimate gain into proportional-integral tuning constants that satisfy a variety of tuning criteria.[3]

Step 1. *Configure the controller as a proportional-only controller.*

Step 2. *Introduce disturbances into the closed loop.* Setpoint changes or other process disturbances should exercise the process within the high process gain operating range as observed during the manual control test. Load disturbances should be used for incremental controllers and controllers with proportional terms based only on the controlled variable. The disturbances should not cause saturation of the control devices.

Step 3. *Adjust the proportional gain for marginal stability.* A marginally stable system will oscillate with a constant amplitude and period. Changes in the proportional gain should be on the order of 50 percent.

Step 4. *Determine the period of oscillation.*

Step 5. *Compute the proportional and integral tuning constants.* The tuning rules for computing control parameters that satisfy the quarter amplitude decay criterion are:

$$K_P = 0.45 K_u \tag{18-10}$$
$$K_I^* = 1.2(\Delta t / P_u) \tag{18-11}$$

Step 6. *Set the sampling interval to between one-fifth and one-tenth of the phase-crossover period.*

18-9. Open-Loop Tuning

Step 7. *Update the control parameters, and confirm system stability.* A series of setpoint changes within the high process gain operating range should be used to verify proper tuning. Fine-tuning adjustments can be made with the proportional gain constant, as it should have the greatest effect on system stability.

Example 18-3. In performing the closed-loop tuning procedure on a process, the data shown in Fig. 18-9 are collected. Using the tuning rules of Eqs. (18-10) and (18-11), determine (a) K_P, (b) the sampling interval, and (c) K_I^*.

Solution. (a) $K_P = 0.45 K_u = 5.6$

(b) The sampling interval should be set to between $P_u/10$ and $P_u/5$, depending on the expected frequency of disturbances. In this case, choose $\Delta t = P_u/5 = 4$ s.

(c) $K_I^* = 1.2(\Delta t / P_u) = 0.24$

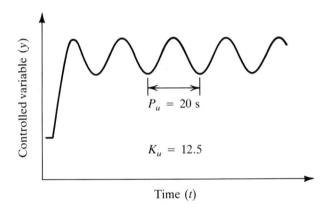

Fig. 18-9. Closed-loop control data for Example 18-3.

18-9 Open-Loop Tuning

The open-loop tuning method is a practical compromise between using a detailed dynamic model of the controlled process to determine the tuning parameters and using the procedural tuning methods described previously. In this method, the control loop is opened at the output of the controller, and experimental data are collected during a step test. The step test data are then plotted and used to estimate an approximate model of the process. Tuning rules based on the model parameters are then used to determine the controller turning parameters. A major advantage of this method is that the control loop is not required to oscillate during tuning; oscillation is generally objectionable.

The model typically chosen to represent thermal and mechanical processes is the first-order plus deadtime model first shown in Fig. 17-37, which has the form

$$y(t) = y_0 + \underbrace{U(t)\,K\,e^{-\tau(t-D)}}_{\text{for } t \geq D} \qquad (18\text{-}12)$$

where

$$\begin{aligned}
y_0 &= \text{initial value of controlled variable} \\
y(t) &= \text{controlled variable} \\
U(t) &= \text{controller output} \\
K &= \text{process gain} \\
\tau &= \text{process time constant (lag)} \\
D &= \text{process deadtime (delay)}
\end{aligned}$$

The model predicts that the process can be modeled as having one dominant time constant and an effective deadtime. All high-order effects contribute to, and are lumped into, the effective deadtime parameter. This simple model approximates the dynamics of a wide range of control processes.

The following step test procedure is used in determining the controller tuning parameters using the open-loop method.

Step 1. *Open the control loop at the output of the controller.*

Step 2. *Override the process to a control point in the operating range of highest process gain.*

Step 3. *Correct for any hysteresis in the control devices.* When performing the step test, it is important that the step change in controller output does not cross the hysteresis band of any control devices. A sequence to correct for hysteresis involves driving the control device toward one extreme in position and then bringing the controlled device back to an appropriate starting position on one side of the hysteresis band.

Step 4. *Wait for the controlled variable to stabilize.*

Step 5. *Issue the step controller output command, and record the controlled variable over time.* The step change must not involve a crossing of the hysteresis band and should thus be in the same direction as the last output command during the hysteresis correction sequence. The change should be large enough to cause a significant change in the controlled

18-9. Open-Loop Tuning

variable but not so large as to extend beyond the region of high process gain. The controlled variable data should be recorded at intervals that will allow accurate plotting of the response curve.

Step 6. *After the controlled variable stabilizes, estimate the process model parameters.* The procedure for estimating the model parameters is summarized in Fig. 18-10 and the description below.

(a) Plot the data versus elapsed time, and draw a smooth curve through the data.

(b) Identify y_i, y_f, U_i, and U_f, the initial and final controlled and output variables.

(c) Determine t_{slope}, the elapsed time between the start of the test and the time at which a line of maximum slope drawn through the response curve intersects the initial controlled variable.

(d) Determine $t_{63\%}$, the elapsed time between the start of the test and the time at which the controlled variable reaches 63.2% of its final steady-state value.

(e) The process gain, K, is equal to $(y_f - y_i)/(U_f - U_i)$.

(f) The process deadtime, D, is equal to t_{slope}.

(g) The process time constant, τ, is equal to $t_{63\%} - t_{\text{slope}}$.

Step 7. *Compute the control parameters from the tuning rules.* The following tuning rules estimate control parameters that result in a 5% overshoot on setpoint changes.

$$K_P = \frac{0.45\tau}{K(D + \Delta t/2)} \qquad (18\text{-}13)$$

$$K_I = \Delta t/\tau \qquad (18\text{-}14)$$

$$\Delta t : \tau/10 < \Delta t < \tau/3 \qquad (18\text{-}15)$$

The controller sampling interval should range between one-third and one-tenth the process time constant with processes subject to frequent disturbances using the shorter sampling intervals.

Step 8. *Update the control parameters, switch to automatic control, and confirm system stability.* A series of setpoint changes within the critical operating range should be used to verify proper tuning. Fine-tuning adjustments can be made with the proportional gain constant, as it will have the greatest effect on system stability.

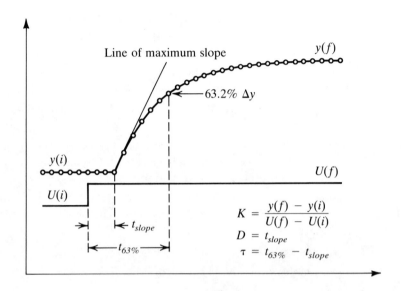

Fig. 18-10. Graphical procedure for estimating the open-loop model parameters.

18-10 Hysteresis Compensation

The performance of a proportional-integral controller can often be improved through the addition of software hysteresis compensation. As discussed in Sec. 17-15, hysteresis is a non-linearity common to most thermal and mechanical control systems. Without hysteresis compensation, the integral action of the controller slowly drives the control device across the hysteresis band. This process can take a long time if the magnitude of the error is small or the integral gain is small. Hysteresis can be compensated for in software by making an additional step change in control output each time the incremental controller output changes sign. This gives the control device an initial push across the hysteresis band whenever a change in direction is needed.

The addition of hysteresis compensation to the proportional-integral control algorithm involves the calculation of an effective output change, $\Delta U'_{k+1}$:

$$\Delta U'_{k+1} = \Delta U_{k+1} + \Delta \text{Hys} \tag{18-16}$$

where

$$\Delta\text{Hys} = 0 \quad \text{if } \Delta U_{k+1} \text{ does not change sign or is zero}$$
$$\Delta\text{Hys} = +(\text{Hysteresis constant})$$
$$\quad \text{if } \Delta U_{k+1} \text{ changes from negative to positive}$$
$$\Delta\text{Hys} = -(\text{Hysteresis constant})$$
$$\quad \text{if } \Delta U_{k+1} \text{ changes from positive to negative}$$

A number of warnings related to hysteresis compensation are appropriate. Hysteresis compensation puts a lower limit on the change in controller output whenever the control device changes direction of travel. If the hysteresis compensation applied is greater than that required to cross the hysteresis band, continuous cycling of the control loop may result. It is also important to note that hysteresis often changes with respect to the operating range of the control device. It is important to conservatively apply hysteresis compensation in field applications.

A procedure for evaluating the extent of hysteresis in field applications is outlined below.

Step 1. *Manually override the control device position toward one end of travel.* The change in position should be large enough to cross the hysteresis band and should remain within the normal operating range of the controlled process.

Step 2. *Issue a sequence of small output changes in the opposite direction of travel until some indication of control device movement is observed.* For modulating control devices such as valves and dampers, positive indication of movement can be made by having an assistant hold the end control element (e.g., valve stem, damper linkage) during the sequence and responding when motion is detected. If this is not practical, waiting for a change in the controlled variable is another method.

Step 3. *Sum up the total output change and use about half of that value as the hysteresis constant.* Using half of the measured hysteresis builds in a safety factor against overcompensation.

18-11 Summary

In this chapter, the field application of microcomputer controllers has been covered, including the development of practical PI control algorithms, the selection of appropriate criteria for evaluating controller performance, and three methods for determining tuning parameters in the field. These methods included the trial-and-error, closed-loop, and open-loop tuning techniques.

The compensation of process non-linearities in software was briefly described using hysteresis compensation as an example. Indeed, the real advantage of microcomputer control involves the ability to routinely perform complex computations in support of advanced control applications.

The development of automatic tuning and self-tuning controllers are demonstrations of the capabilities of modern microcomputer-based controls. The open-loop and closed-loop tuning methods discussed in this chapter can be automated and directly implemented in microcomputer controllers.[4,5] Self-tuning controllers have also been developed that estimate process models and update control parameters in real time using closed-loop control data.[6] A self-tuning microcomputer controller reduces the time required for field commissioning and continues to tune the control parameters as operating conditions change. It is capabilities such as this that place the microcomputer controller at the very center of developing control technology.

References

1. E. H. Bristol "Designing and Programming Control Algorithms for DDC Systems," *Control Engineering*, January, pp. 24–26, 1977.

2. J. G. Ziegler and N. B. Nichols, "Optimum Settings for Automatic Controllers," *Transactions of ASME*, vol. 64, pp. 759–765, 1942.

3. D. W. Clarke "PID Algorithms and their Computer Implementation," Oxford University Engineering Laboratory Report No. 1482/83, Oxford University, 1983.

4. K. J. Astrom and T. Hagglund, "Automatic Tuning of Simple Regulators with Specifications on Phase and Amplitude Margins," *Automatica*, vol. 20, pp. 645–651, 1984.

5. C. G. Nesler "Automated Controller Tuning for HVAC Applications," *ASHRAE Transactions*, vol. 92, pt. 2, pp. 189–201, 1986.

6. C. Park and A. J. David, "Adaptive Algorithm for the Control of a Building Air Handling Unit," National Bureau of Standards, NBSIR 82-2591, 1982.

Problems

18-1. Consider the incremental controller and pneumatic interface described in Example 18-1 but with $K_P = 5$ and $K_I = 0.5$. Assume that the software is written such that all computed outputs less than one count are truncated to zero. (a) What is the smallest error that will result in a change in controller output? (b) Suggest a change in the software that would reduce this output quantization error.

18-2. Consider the incremental controller and pneumatic interface described in Example 18-1. Assume that the actuator has 20% hysteresis. What value of the hysteresis constant will exactly compensate for the non-linear characteristics of the actuator? **Ans.:** $\Delta\text{Hys} = 25$.

18-3. In a mixed-air control process, dampers mixing outdoor and room return air are modulated to maintain a mixed-air temperature setpoint. In performing the open-loop tuning procedure on this process, the following data were collected.

Time, s	Output, %	Mixed Air Temp., °F	Time, s	Output, %	Mixed Air Temp., °F
0	20	64.5	120	60	56.0
10	20	64.5	130	60	55.5
20	20	64.5	140	60	54.9
30	60	64.5	150	60	54.5
40	60	64.8	160	60	54.2
50	60	64.5	170	60	53.9
60	60	63.3	180	60	53.6
70	60	61.5	190	60	53.5
80	60	59.9	200	60	53.3
90	60	58.7	210	60	53.1
100	60	57.7	220	60	52.9
110	60	56.7	230	60	52.8

Using the graphical technique described in Sec. 18-9, estimate the (a) process gain, (b) process deadtime, and (c) process time constant. (d) Using the open-loop tuning rules, compute the controller tuning parameters using $\Delta t = \tau/4$. **Ans.** (d): $K_P = 3.4$ percent per °F and $K_I^* = 0.25$.

18-4. A common problem in commissioning environmental control systems is that process characteristics change with weather conditions. Consider a mixed-air damper process first shown in Fig. 3-10 and given by the relation:

$$T_m = \frac{\text{Output}}{100\%}T_r + \frac{100\% - \text{Output}}{100\%}T_o$$

where
T_m = mixed-air temperature
T_o = outdoor-air temperature
T_r = return-air temperature

Assume the process is controlled with a position form algorithm whose output ranges from 0 to 100%. (a) What is the process gain when the outdoor air temperature is 13°C and the room return-air temperature is 24°C? (b) What is the process gain when the outdoor air temperature is 4°C and the room return-air temperature is 27°C? (c) Process changes of this magnitude can have a negative effect on controller tuning. Assume that the process deadtime and time constant are dependent only on the physical configuration and do not change with temperature. Develop a relationship that can be used to adapt the controller proportional gain to changes in the outdoor and return-air temperature.

Index

741 op amp   39, 165
8080, see Intel 8080/8085
8155 RAM   268
8155/8156 RAM and I/O   286
8205 decoder   266
8212 I/O chip   286
8251A serial I/O   302
8279 keyboard/display controller   266
8355 ROM   266
8755 EPROM   266

A accumulator   217, 223, 232
absolute assembler symbol   257
ac (alternating current)   11
AC flag   230
accumulator   210, 217, 223, 232
accumulator addressing   238
ACIA   297–299
acoustic coupler   311
actuator   105–122, 411
   hydraulic   114
   magnetic   113
   motor   112
   pneumatic   114
   signals   399
   voltage   159–173
Adams predictor-corrector   361
ADC, see analog to digital converter
add with carry   202
add without carry   202
addition, binary   126, 137
address bus, see bus
addressing modes   218, 228, 238
agriculture   8
air compressor   78
air conditioning   4, 7, 35
air flow rate   79
air mixture   35
air supply system   336
alternating current   11
ALU   210
ammonia sensor   92
amplification ratio   31
amplifier, see also operational amplifier
amplifier
   buffer   33
   current   47, 106
   follower   33

   inverting   30
   multiplying   34
   summing   34
   voltage   49
analog electronics   123
analog sensor   318, 320
analog switch   150
analog to digital converter (ADC)   173–181
   ADC 0800   177, 178
   DAC based   159, 174, 180
   dual-slope integration   176
   noise immunity   177
   pinout   177
   resolution   181
   selecting   180
   staircase   176
   start conversion   178
   successive approximation   176
AND   126, 127
anemometer   79, 85
anode   107
answer mode   311
application of controllers   409–430
arithmetic logic unit (ALU)   210
arithmetic, see binary arithmetic
ASCII, American Standard Code for Information Interchange   313
assembly language   249–260
   directive   251, 253
   label   252
   separator   252
   statement   251
asynchronous communication   291
Asynchronous Communications Interface Adapter (ACIA)   297–299
automobiles   6
auxiliary carry flag   230

B register   217, 223, 232
backlash   419
backward-referenced integration   410
base 10   125
base 2   125
base of transistor   46
base-emitter voltage drop   49
basic oxygen plant (BOP)   4
baud rate   290

431

BCD (binary-coded decimal)   151
BCD to seven-segment   151
Bell System 103F modem   311
bellows   90
Bernoulli equation   79, 83, 85
beta of transistor   47, 49
binary   125
binary arithmetic   195–208
  addition   126, 137, 198–205
  multiplication   196
  subtraction   195–205
binary numbers   123–138
  signed   198
  unsigned   198
binary to BCD   151
binary to decimal   125
binary-coded decimal (BCD)   151
bipolar-junction transistor (BJT)   49
bit   125, 195
BJT   49
Boltzmann's constant, k 21
BOP (basic oxygen plant)   4
branching   238, 240
breakdown voltage   21, 53, 56, 108
bridge circuit
  amplification   69
  rectifying   22, 23
  temperature sensing   67
buffer   33
bumpless changes   411
bus   263, 313
  address   209, 261, 262, 266, 270
  control   266, 270
  data   209, 261, 262, 266, 270
  multiplexed   186
  structure   262, 268, 270

C, capacitance   16
C (carry) flag   206, 237
C register   223
CALL subroutine   224, 231
capacitance, C 16
capacitor   16, 37
carbon monoxide sensor   93
carbon resistor   18
carry flag   200, 206, 230, 237
cascading gates   129
cascading z-transforms   382
cathode   107
CC (common collector)   52
CE (common emitter)   52
central computer   5
chemical industry   2

chemical sensor   92
clear to send (CTS)   297, 307
clock   142
closed-loop tuning   422
CMOS   124
cold air mixture   35
collector of transistor   46
color code   18, 21
combining gates   129
comment, assembly language   252
commissioning of controllers   409–430
common collector (CC)   52
common emitter (CE)   52
comparator   30, 150
complementary metal oxide semiconductor
  (CMOS)   124
compressed air   4
compressibility   75
compressible fluid   83
compressor   3, 7, 78
computer controller   123
computer, high level   261
condition codes   237
constant current   54–56, 70
constantan   62
control bus, see bus
controller, programmable   123
cooling, buildings   4
copper-constantan   62
counter   144
CR (carriage return)   313
critical damping   340, 415
cross assembler   250
crystal oscillator   142
CTS (clear to send)   297, 301, 307
cube root   346
current amplifier   47
current source   54–56, 70

D register   223
D5BUG monitor ROM   270
DAC, see digital-to-analog converter
data bus, see bus
data carrier detect (DCD)   297, 301, 307
data communications equipment (DCE)
  307
data direction register (DDR)   277
data set   311
data set ready (DSR)   307
data terminal equipment (DTE)   307
data terminal ready (DTR)   307
dc (direct current)   11–27
DCD (data carrier detect)   297, 301, 307

Index

DCE (data communications equipment) 307
DDRA (data direction register A) 277
DDRB (data direction register B) 277
De Morgan's laws 130
dead time 400
deadband 321, 414
debounced switch 141
decimal 125
decimal to binary 125
decimal to hexadecimal 199
decode instruction 211
decomposition 381
decrement accumulator 217
decrement index register 220
demultiplexer 167
derivative, transform of 330
dial-up modem 313
diaphragm 90
differential equation 330
differential transformer sensor 94
digital electronics 123–158
digital signal conversion 159–173
digital-to-analog converter (DAC) 160
   1408 163, 165
   R-$2R$ ladder 162
diode 20
   germanium 20
   silicon 20
   zener 53
DIP (dual in-line package) 39, 131
direct addressing 238
direct current 11–27
directive, assembler 253
displacement sensor 94
distributed control 5
divide-by counter 144
downloading 250
drain, FET 49
DRAM 188, 192
DSR (data set ready) 307
DTE (data terminal equipment) 307
DTR (data terminal ready) 307
dual addressing 238
dual in-line package (DIP) 39, 131
dynamic analysis 327
dynamic memory 188, 192

E register 223
EEPROM 192
electric current sensor 91
electric motor actuator 112
electric power generation 7

electrically erasable programmable read-only memory (EEPROM) 192
electric-pneumatic transducer 115
elementary microcomputer 261–272
emitter of transistor 46
emitter-follower 52
energy conservation 4, 6, 78, 90, 91, 318
environmental control 4
EPROM 187
equivalent resistance, R_{th} 13, 14
erasable programmable read-only memory (EPROM) 187
Euler method 360
evaluation kit 261
even parity 292
execute instruction 211
expansion factor 84
extended addressing 238

feedback loop 333, 382
ferromagnetic 95
FET 49, 150
fetch instruction 211
field application 409–430
field-effect transistor 49, 150
first-order hold 364
flag register, Intel 8080/8085 229
flip-flop 143
flow rate sensing 78–90
follower 33
food processing 3, 4
force sensor 91
forward current 21
forward reference 258
four-wire sensing 70
framing error 301
full duplex 311
full-wave rectification 22

gate, FET 49
gate, logic 126–141
gate, SCR 107
gate turn-off (GTO) 107
generic microprocessor 209
germanium diode 20
GTO (gate turn-off) 107

H (half carry) flag 206, 237
H register 223
half carry flag 206, 237
half duplex 311
Hall effect 94, 96
handshake 307

heat pump   7
heating, buildings   4
Herschel venturi tube   79
hexadecimal number system   197
hexadecimal to decimal   199
hierarchical control   5
high level language (HLL)   250
high-order byte (HOB)   202, 214
HLL (high level language)   250
HOB (high-order byte)   202, 214
holding current   108
home appliance   6
hot-wire anemometer   85
humidity sensor   92
hydraulic actuator   114
hysteresis   114, 401, 419, 424, 426

I^2R heating   64
IAE (integrated absolute error)   416
IC (integrated circuit)   131
ice bath   63
IEEE-488 protocol   313
immediate addressing   218, 228, 238
impulse   370
impulse response   376
I_N (Norton current)   14
increment accumulator   217
incremental PI control   411
index register   218, 219, 240
indexed addressing   238, 240
indirect flow sensing   89
indoor air   35
inductor   16
infrared sensor   93
inherent addressing   229, 238
input/output (I/O)   1, 210, 261, 273
instruction decode   211
instruction execution sequence   211
instruction fetch   211
instruction sequence   211
instruction set   212, 224, 234
instrumentation   61
integral control   344
integral windup   411, 413
integrated absolute error (IAE)   416
integrated circuit   131
 temperature sensing   71
integrated squared error (ISE)   416
integrator   37, 326
Intel 8080/8085   222–232, 261, 264
intercooler   90
interrupt   273, 275, 279
interrupt mask   237

interrupt request   301
interrupt service routine (ISR)   275, 280
inverter   128
inverting a transform   328
inverting a z-transform   380
inverting amplifier   30
I/O, see input/output
ISE (integrated squared error)   416
ISR, see interrupt service routine

J-K flip-flop   143
JTS (jump to subroutine)   221
jump instruction   214

Kirchhoff's law   11, 47

L register   223
label, assembly language   252, 255
ladder diagram   123, 131–135
LAN (local area network)   7
Laplace transform   327, 367, 370
 table   329
latch   149
LDX (load index register)   219
least-significant bit (LSB)   125
LED (light-emitting diode)   151
level conversion   308
LF (line feed)   313
light-emitting diode (LED)   151
linear variable displacement transformer
 (LVDT)   96
linearity, temperature-to-voltage   65
liquid flow rate   79
liquid level
 capacitance sensor   93
 pressure sensor   93
 ultrasonic sensor   93
liquid temperature sensing   71
listing   249
load shedding   7
loader, relocating   257
LOB (low-order byte)   202, 214
local area network (LAN)   7
location counter   254
logic   123–153
 AND   127
 combining   129
 NAND   129
 NOR   128
 NOT   128
 OR   127
 XOR   129
logic voltage range   124

Index

logical one   124
logical zero   124
long division   381
loops   240
low-frequency pulses   148
low-order byte (LOB)   202, 214
LSB (least-significant bit)   125
LVDT   96

machine language   249
Maclaurin series   373
magnetic actuator   113
make/break sensor   318, 319
manual control test   418
manufacturing   7
mark, serial communication   290, 291
MC1488 line driver   308
MC1489 line receiver   308
MC6821 PIA   276
MC6850 ACIA   297
MEK6802D5 Evaluation Kit   261, 269
memory   185–193
memory addressing   228
memory map
  6802D5   270
  SDK-85   266
methane sensor   93
microcomputer   1, 195, 261
microprocessor   1, 261
microprocessor, programming   209–247
minuend   196
modem   305, 310
modulating control   321
monostable multivibrator   148
most-significant bit (MSB)   125
motion sensor   94
motor control   3, 325
Motorola 6800   232–244, 261
  programming guide   243
Motorola 6802   268
MSB (most-significant bit)   125
multichannel control   172
multiplexed, see bus
multiplexer (MUX)   159, 167, 172
  4051   167
  fidelity   169
multiplication   196
multiplier   34
MUX, see multiplexer

N (negative) flag   237
NAND   126, 129
negative flag   237

non-linearities   400–401
NOR   126, 128
Norton current, I_N 14
Norton equivalent   14, 15, 25
Norton resistance, R_N 14
NOT   126, 128
npn 45
numerical integration   410
numerical simulation   358
nutating-disc   79

object module   249
odd parity   292
offset null   39
Ohm's law   11
one-pass assembler   258
one's complement   196
one-shot   148
on/off control   7, 106, 318, 319
op amp, see operational amplifier
op-code   252
open collector output   139
open-loop tuning   423
operand, assembly language   252
operational amplifier   29–44
  741   39
  3140   172
  application   29
  buffer   33
  characteristics   30
  dynamic response   40
  follower   33
  generalized circuit   36
  input resistance   40
  input voltage   40
  integrator   37
  inverting   30
  limitations   40
  multiplying   34
  non-inverting   32
  operating limits   40
  output current   40
  output power   40
  pin diagram   39
  ratings   40
  summing   34
  supply voltage   40
optically-isolated switch   110
OR   126, 127
orifice, liquid flow   79, 81, 89
originate mode   311
oscillator   142
outdoor air   35

436 Index

overdamped 340, 415
overflow flag 237
overflow, two's complement 204
overrun 301

P (parity) flag 230
package
  DIP 39
  op amp 39
  transistor 57
paper making 4
parallel I/O 266, 273–288
  generic chip 274
parallel to serial 292
parity 292
parity error 301
parity flag 230
partial fraction 328, 381
PC (program counter) 214, 234
peripheral interface adapter (PIA) 276–286
petrochemical plants 3
petroleum refiners 3
phase-crossover period 422
PI control 410, 412
PIA 276–286
piezoelectric transducer 90
pipeline 4
pitot tube 79, 85, 89
platinum 64
pneumatic control 114, 322, 323, 411
pnp 45
position feedback 411
position form PI control 410, 412
position sensor 94
potentiometric sensor 94
power factor 92
power generation 7
power resistor 18
power supply 11, 24
practical control algorithms 410
pressure transducer 90–97
process gain 417
program counter 214, 234
program listing 249
programmable controller 123
programmable read-only memory (PROM) 187, 262, 266
programmable unijunction transistor 107
programming guide
  Intel 8080/8085 232
  Motorola 6800 243

PROM 187
proportional band 323
proportional control 321, 322, 390, 411
  air supply system 336–344
proportional-integral (PI) control 321, 324, 350, 397, 411
proportional-integral-derivative (PID) control 321, 327
protective ground 307
proximity sensor 94, 95
pseudo-op 253
pull-up resistor 138
pulse-width modulation 112, 399, 412
PUT (programmable unijunction transistor) 107

quarter amplitude decay 415

radiation 75
random-access memory (RAM) 187, 209, 261, 263, 266
  connections 188
  dynamic 188, 192
  MCM2114 190
  MCM6810 189
  static 188
RC circuits 16
RDRF (receive-data-register full) 296, 300
read-only memory (ROM) 186, 210, 261, 262, 263, 266
receive-data register 294
receive-data-register full (RDRF) 296, 300
receive-shift register 294
received data 307
receiver/controller 323
receiver overrun 301
rectifying circuit 22, 27
reference junction 63
register addressing 218, 219, 228
registers 223, 232
relative addressing 238
relay, electromechanical 123, 132
relocatable assembler symbol 257
relocating assembler 257
relocating loader 257
reluctance sensor 94
request to send (RTS) 307
resistance temperature device (RTD) 61, 63, 64, 70
resistor 18, 32
  pull-up 138

Index

Reynolds number 82
ripple, voltage 23
R_N (Norton resistance) 14
ROM (read-only memory) 186
rotating-vane anemometer 79
rotative-speed sensor 97
RS-232-C 304–316
RTD (resistance temperature device) 63, 64, 70
R_{th} (Thévenin resistance) 13, 14
RTS (request to send) 297, 307
RTS (return from subroutine) 243
Runge-Kutta method 361
Rx, receive serial data pin 293

S (sign) flag 206
sample-and-hold 160, 170
sampled control 357–408
  response 365
sampled data 322, 362
sampled value response 365
sampling interval 370, 421
satellite controllers 5
saturation, transistor 51
Schmitt trigger 145
SCR 107
SDK-85 System Design Kit 261, 266
sense terminal 24
sense/delay/fixed change actuator 399
sensor heating 65
sensor, see transducer
sequence controller 123
sequential control 131
sequential logic 137
serial I/O 273, 289–316
series circuit
  RTD sensing 64
seven-segment LED 151
shift register 292
sign flag 206, 229
signal conditioning 33
signed binary numbers 198–205
silicon-controlled rectifier (SCR) 107
silicon diode 20
simulation, numerical 358
sing-around 86
slewing 119
solid state relay 110
source, FET 49
source program 249
SP (stack pointer) 214, 242
space, serial communication 290, 291
SSR (solid state relay) 110

stability 334, 338, 386, 414
stack pointer (SP) 214, 223, 242
stack, program 221, 242
stagnation temperature 75
start bit 291
static memory 188
status register 206, 214
steelmaking 4
Stefan-Boltzmann equation 75
step response 339
step test 424
stepping motor 115–121
  performance 118
  permanent magnet 115
  running torque 119
  slewing 119
  start-without-error 119
  torque 119
  variable reluctance 115
sterilization 3
stop bit 291, 303
strain gauge 94
stratification 76
Strouhal number 87
subroutine 221, 231, 243
subtract with carry 203
subtract without carry 203
subtraction 195
subtrahend 196
summing amplifier 34
switch 106
  analog 150
  debouncing 141
  optical 110
  transistor 51
synchronous communication 291

table lookup 381
Taylor's series 361
TCO (two's complement overflow) 204
TDRE 296, 300
telephone transmission of data 310
temperature control 6
temperature sensing 61–78
  air and gas 75
  errors 73
  liquid 71
temperature-sensitive IC 71
temperature transmitter 323
thermistor 63
thermocouple 61, 62
  chromel-alumel 62
  compensating junction 63

copper-constantan 62
iron-constantan 62
reference junction 63
thermometer 61, 73
Thévenin equivalent 12, 13, 25
Thévenin voltage, V_{th} 13
three-state output 139, 140, 263
three-wire sensing 70
throttling range 323
thyristor 107
time constant 342
timer 318
torque sensor 91
totem pole output 139
transducer 61
  chemical 92
  current 91
  electric-pneumatic 115
  force 91
  Hall-effect 96
  humidity 92
  level 93
  motion 94
  position 94
  pressure 90
  rotative 97
  strain 94
  torque 91
transfer function 332
transform of a derivative 330
transistor 45–60
  base 46
  BJT 49
  characteristics 47
  collector 46
  common collector 52
  common emitter 52
  drain 49
  emitter 46
  FET 49, 150
  gate 49
  impact 45
  npn 45
  operating limits 56
  package 57
  pnp 45
  power dissipation 56
  PUT 107
  saturation 51
  source 49
  switch 51
  symbol 45
  terminology 45
  voltage 49
transistor-transistor logic (TTL) 124
transmit-data register 294
transmit-data register empty (TDRE) 296, 300
transmit-shift register 294
transmitted data 307
triac 107, 108
trial and error tuning 419
tristate output 140
truth table 127
TTL output 124, 139
tungsten 64
tuning 353
  closed-loop 422
  criteria 414
  of controllers 409–430
  open-loop 423
  trial and error 419
turbine 8
turbine flow meter 86
turn-down ratio 87
two-pass assembler 258
two-wire sensing 70
two's complement 195, 196
two's complement overflow (TCO) 204
Tx, transmit serial data pin 293

UART 293, 302
ultimate gain 422
ultrasonic flow meter 86
ultrasonic sensor 93
underdamped 340
universal asynchronous receiver/transmitter (UART) 293, 302
unsigned binary numbers 198, 200
USART 302

V (overflow) flag 237
valve 323
  chatter 320
variable-frequency inverter 3, 325
velocity sensing 78–90
vena contracta 81
venturi tube 79, 83, 89
voice grade telephone lines 310
voltage amplifier 49
voltage follower 52
voltage regulator 24
voltage ripple 23
vortex-shedding flow meter 87
V_{th} (Thévenin voltage) 13

Index

warm air mixture 35
water flow rate 79
water heater 417
windup 411, 413
wobble plate 79

XOR 126, 129

Z (zero) flag 206, 229, 237
zener diode 53
zero flag 206, 229, 237

zero-order hold (ZOH) 364, 378
Ziegler-Nichols 415, 422
ZOH (zero-order hold) 364
z-transform 370–408
    cascading 382
    definition 370
    inverting 380
    stability 386
    table 374